Residential Wiring

to the 2008 NEC®

by Jeff Markell

Free CD-ROM inside the back cover:

- ***The Interactive Study Center,*** with all the end-of-chapter test questions in interactive self-test software. Use Study Mode or Exam Mode. The ISC will load automatically when you put the CD in your computer's drive.

- ***A PDF of the entire book.*** Read the book on your computer, and use the *Adobe Reader* search feature to instantly find any topic. To load this file:
  1. Go to *My Computer*, or *Computer*, and navigate to your CD drive.
  2. Doubleclick on the folder called *PDF*.
  3. Doubleclick on the *Residential Wiring* file.

If you don't have *Adobe Reader*, go to www.adobe.com and download it free.

Craftsman Book Company
6058 Corte del Cedro / P.O. Box 6500 / Carlsbad, CA 92018

Production Manager, Christine Bruneau; design, layout and photographs, Joan Hamilton; image conversion, Lori Boon; cover photos, Ed Kessler Studios.

Library of Congress Cataloging-in-Publication Data

Markell, Jeff.
Residential wiring to the 2008 NEC / by Jeff Markell. -- 8th ed.
p. cm.
Rev. ed. of: Residential wiring to the 2005 NEC.
Includes index.
ISBN-13 978-1-57218-204-2
ISBN-10 1-57218-204-0
1. Electric wiring, Interior--Handbooks, manuals, etc. 2. Electric wiring, Interior--Standards--Handbooks, manuals, etc. 3. Dwellings--Electric equipment--Handbooks, manuals, etc. 4. Electric wiring--Insurance requirements--Handbooks, manuals, etc. 5. National Fire Protection Association. National electrical code (2008) I. Markell, Jeff. Residential wiring to the 2005 NEC. II. Title.

TK3285.M179 2008

621.319'24--dc22

2008020254

CONTENTS

INTRODUCTION

This book was written for anyone who intends to make a living wiring residential buildings. If you can understand and follow the instructions in this manual, you should have no trouble installing safe, modern, efficient electrical systems in homes and apartments.

As an electrician, you need to know how to use a wide variety of tools and materials. This manual describes the tools that should be in every electrician's tool box, and suggests how they can be used to best advantage. I'll also explain what you should know about electrical materials: wire, cable, conduit, fixtures, boxes, switches, breakers and panels. There's a correct tool and a right material for every purpose. Sometimes selecting the right tools and materials isn't easy. After reading this book, you should have little trouble choosing both tools and materials appropriate for the work you do.

This manual isn't a book of electrical theory. But every professional electrician needs some background on how electricity is generated and distributed. And, of course, you should know how Ohm's Law and Watt's Law are used to design electrical systems. The first two chapters cover these important subjects.

If you've worked as an electrician for some time, you know that nearly everything an electrician does is governed by the *National Electrical Code*®. For our purposes, the only right way is the code way. Until you're comfortable with the code, doing everything the code way can be a nuisance. Once you understand the code and the reasons for code requirements, you may have a different perspective. Most experienced electricians would agree that the *NEC*® protects everyone (including electricians and electrical contractors) and is a good guide to professional practice — even if the building inspector didn't enforce it.

This book will help you follow the code. But it isn't a substitute for the *NEC*. Every professional electrician needs a copy of the current code. Many bookstores sell the *NEC*, or you can order a copy from this publisher, using the order form bound into the back of this book. But just having the current *NEC* isn't enough. Many cities and counties don't adopt the model code exactly as published by the National Fire Protection Association. Instead, they supplement the code with amendments or changes that will be enforced on jobs in that city or county. Once you have the current *NEC*, ask at your local building department about amendments or changes that apply in that jurisdiction. Keep those changes with your copy of the code.

I'll explain floor plans, cable plans and wiring diagrams in detail. This is important information for every electrician. The code has a lot to say about types of outlets, spacing of outlets, what must be switch-controlled and what need not be switch-controlled. The work you do will have to follow the plans and comply with the code. The information in this book should help you understand and follow plans prepared for your jobs.

Finally, I'll explain how to diagram the circuits you're likely to find in a home or apartment. As a teacher of electrical wiring for many years, I've found that a student who can diagram a circuit correctly has a reasonably good chance of wiring it correctly as well. And a student who can't diagram a circuit probably can't install it either!

Now let's get down to business — what you need to know to wire homes and apartments.

Jeff Markell

1 ELECTRICAL ENERGY

In order for an electrician to understand his work sufficiently and keep current in his field, he needs a good background in electricity and its capabilities. But he also needs to know some basics about the nature of matter, as the creation and transfer of electrical energy is primarily a function of the properties of matter at the molecular level. Electricity has always been, and typically still is, mysterious. It's an invisible form of energy, but as you've probably discovered, it can make its presence extremely evident. This is a practical book on how electrical wiring in a small building should be done to meet accepted standards of good workmanship, and to comply with the provisions of the *National Electrical Code*® (*NEC*®). It doesn't focus on theory, so the discussion of theoretical matters will be minimized.

Historical Introduction

It might be surprising that, as far back as 600 BC, the Greeks amazed themselves with elementary uses of static electricity. For example, they discovered that a piece of amber rubbed with cloth attracted bits of straw, hair, etc. Their word for amber was "elektron," which is the root of "electron," "electricity," "electronics," and other words containing "electro."

"The ancients" also discovered that certain heavy black stones they occasionally found mysteriously attracted iron. Since they were often found in a part of Asia Minor called Magnesia, they were called magnets. Naturally, all manner of hocus pocus was created to explain these curious phenomena — none of them with much semblance to the facts as we now know them.

Over many centuries, observations indicated that various other materials had characteristics similar to amber. These materials could also be rubbed to attract light objects. Scientists developed a theory that the rubbed materials would leak a "fluid-like substance" that caused the attraction. This fluid was called *electricity*. Theorists of the early 18th century, discontent with just one "fluid," hypothesized that there were two fluids. One was called "vitreous" and the other was "resinous." The difference was based on the nature of the substance being rubbed. By the middle of the 18th century, Ben Franklin went back to the "one fluid" theory. He decided that the two fluids were simply different aspects of the same thing. When an object had too much of this electric fluid it was "positive," if it had too little it was "negative," and if it was neither, it was "neutral." While those in scientific fields were dissatisfied with this theory, it was the only one available until the early 20th century. At that time, investigating the structure of matter produced a more satisfactory alternative.

The Composition of Matter

Matter is anything that has mass and occupies space. It exists in the following three states:

1. Solid — such as rock
2. Liquid — such as water
3. Gaseous — such as the air around us

With variations in temperature and pressure, matter can be changed from one state to another. Remove enough heat from a quantity of water, by reducing the temperature, and at 32 degrees F it'll change from a liquid to a solid — ice. Add enough heat to the same quantity of water, increasing the temperature, and at 212 degrees F it'll start to vaporize, changing from a liquid to a gas.

Although a particular type of matter may change state from solid to liquid to gaseous, the component building blocks it's made of remain the same, so "What's matter made of?" To find out, we must divide, subdivide and subdivide again to reach the smallest particle that maintains the characteristics of that type of matter, like water, steel, or foam plastic. The smallest particle that maintains the characteristic of the material is a "molecule." Each kind of matter has a corresponding different molecule. But the molecule definitely isn't the smallest part.

Molecules are composed of even smaller parts called "atoms." Water, for example, is composed of molecules made of two hydrogen atoms plus one oxygen atom — H_2O. All matter, then, consists of the atoms of some 120 elements combined in different compounds to form the molecules that distinguish different substances from each other. We will discuss how atoms accomplish combining into molecules after we look more closely at the atom itself.

The atoms that compose molecules are quite complicated structures. Each one seems to be a miniature solar system, consisting of a nucleus surrounded by varying numbers of revolving electrons. The nucleus contains various particles, such as protons, neutrons, positrons, neutrinos, mesons, and even a few odd bits called "quarks" and "charms." We're primarily concerned with the bulk of the nucleus consisting of the protons and neutrons. The number of protons in the nucleus differentiates the atoms of the 120 elements from each other. The number of protons in the atom's nucleus is its "atomic number," for example hydrogen is #1, helium is #2, and so on.

Protons are positively charged, neutrons have no electrical charge, and the orbiting electrons are negatively charged. Since, under normal conditions, atoms are electrically neutral, an atom of any element will contain equal numbers of electrons and protons. The number of neutrons, along with the various other nuclear components (neutrinos, mesons, etc.), has nothing to do with the electron-proton balance. Hydrogen has no neutrons, while the 92 protons of uranium are outnumbered by 146 neutrons.

While the magnitude of the opposite electrical charges in electrons and protons is equal to each other, the difference in mass between the two is staggering. The mass of a proton is 1,840 times that of an electron. There's a similarity between what's observed on a huge scale in the solar system to the minute scale in the atom. All but a tiny part of the solar system's mass is contained in the sun. Similarly, all but a tiny part of an atom's mass is contained in the nucleus.

The solar system's planets are held in their orbits around the sun by complex factors involving the mutual attraction of their gravitational fields with the sun's; and their masses and velocities. Electrons are held in their orbit around the nucleus by the electrostatic attraction between their negative charges and the proton's positive charges in the nucleus, including a relationship between mass and velocity.

At this point, the parallel between the atom and the solar system breaks down. Each planet of the solar system differs greatly from the other in mass, composition, orbital velocity, and other characteristics. However, the electrons orbiting the nucleus of an atom do not differ.

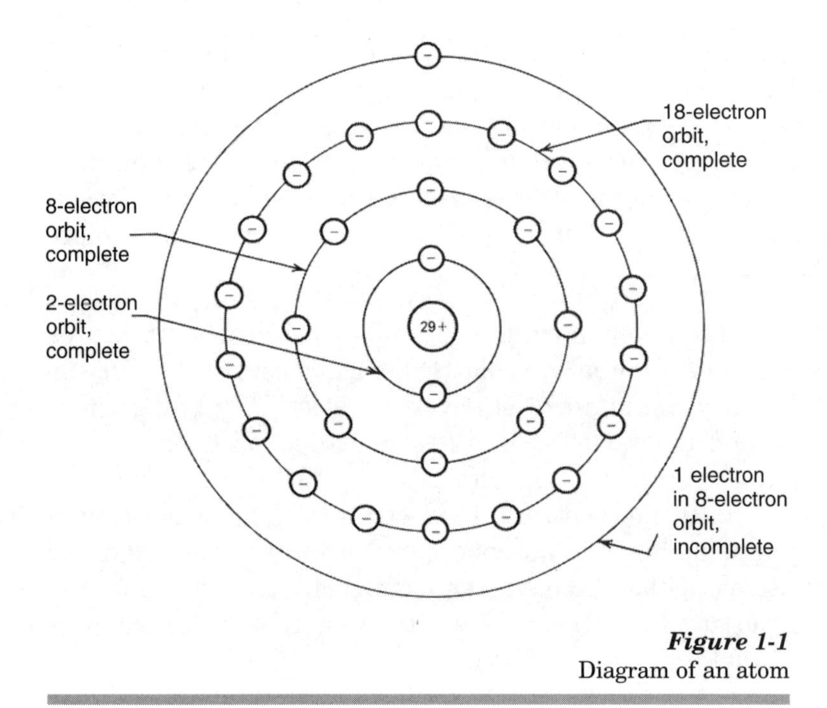

Figure 1-1
Diagram of an atom

The electron must maintain a constant speed to sustain the centrifugal force that keeps it from falling into its nucleus or spinning away from its nucleus. Because of its mass, it must also have a level of energy resulting from a combination of its mass and velocity. Only a very limited number of specific energy levels are possible for electrons; there are seven altogether. As an electron can only occupy an orbital path suitable to its energy level, there are seven possible orbits.

The more complex atoms might have as many as 100 electrons, but since only seven possible energy levels exist, the electrons must group at various appropriate orbit distances from the nucleus, forming "shells" in layers around it. See Figure 1-1. A consistently repeated pattern is found in the formation of these shells. The innermost shell (#1) can hold no more than two electrons. Any number above two starts the second shell, which holds up to eight. When it's filled, the third shell is started.

At this point, the picture becomes a little more complicated. The third shell (#3) holds up to 18 electrons; however, the *outermost* shell of any atom, regardless of which one, cannot hold more than eight. So, when shell #3 is on the outside, with eight electrons, the next electron must orbit in shell #4. Only after shell #4 has one or two occupants can the rest of the 18 possible spaces in #3 be filled. When shell #4 has eight electrons, shell #3 will already have its allotted 18. When #4 is the outermost shell, holding eight electrons, then shell #5 starts. Shell #4 can hold 32 electrons. When it has 32, and shell #5 is up to eight, shell #6 is started. Shell #6 is completed and shell #7, the last possible electron shell, is started in a similar way. With all elements, it's the spare electrons of the outer shell — whatever shell number that is — that take part in any of the various chemical and electrical phenomena. These are called "valence electrons."

Formation of Molecules

Regardless of its shell number, the outermost shell of any atom cannot contain more than eight valence electrons. Any atom that has all eight is stable, and doesn't normally combine with other atoms. The atoms with valence electrons anywhere between one and seven, trying to attain stability, are available to combine with other atoms to form molecules. The process of molecule formation is called atomic bonding. This process occurs in any one of the following three ways:

1. Ionic bonding
2. Covalent bonding
3. Metallic bonding

Ionic Bonding

An atom alone will contain matching numbers of electrons and protons, which, since they have opposite electrical charges, results in a neutral charge for the atom as a whole. However, this matter of valence electrons gets in the way. An atom with more than four but fewer than eight valence electrons is unstable. It tries to obtain whatever number of valence electrons is missing to fill its outer shell to eight. In contrast, an atom with fewer than four valence electrons is also unstable, but willing to unload its excess.

Where an atom with one valence electron meets another one with seven, there's a tendency for the one to join the seven, stabilizing both atoms. However, in the process, something else happens. The atom that lost an electron now has a net *positive* electrical charge of one. The atom that picked up an electron has a net *negative* electrical charge of one. An atom that's no longer electrically neutral but has a net positive or negative charge has become an "ion." Ions with opposite electrical charges are attracted to each other, tending to combine via "ionic bonding" to form molecules.

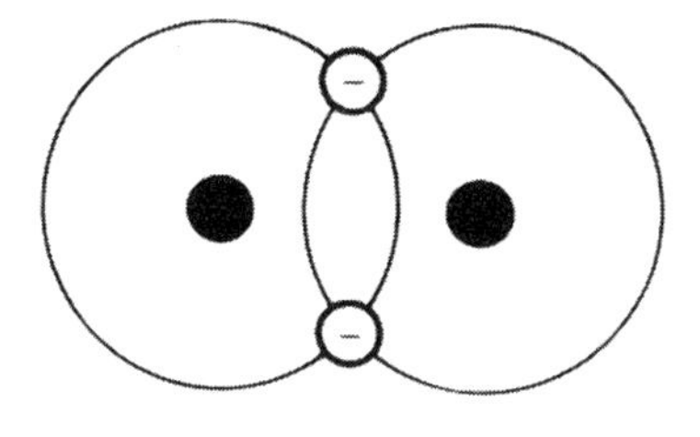

Figure 1-2
Covalent bond of two hydrogen atoms

Covalent Bonding

Hydrogen with atomic number "1" has only a single electron in the #1 shell. It's unstable because that shell is incomplete without two electrons. One way it stabilizes is to join with another hydrogen atom to form a hydrogen molecule in which the two component atoms share their two electrons. See Figure 1-2. This is an example of "covalent bonding."

Metallic Bonding

Copper is a good example of "metallic bonding" because it's the most commonly-used material for electrical wires. The atom in this case has 29 electrons. Shell #1 is complete with two, shell #2 is complete with eight, and shell #3 is filled with the next 18. That totals 28. The 29th is a lone valence electron in shell #4, which is loosely held and has a tendency to wander off, becoming a "free electron." The copper atom has become a positive ion, and so have a lot of other atoms that have also lost their single valence electrons. Although, like-charges repel, the copper ions don't simply fly apart as one might expect. They're immersed in a sort of soup of free electrons. The mutual attraction between the positive copper ions and the negatively charged electron mass around them holds the whole substance together by "metallic bonding."

That same soup of free electrons, unattached to specific atoms, flows as an electrical current through a metal when it's connected to a source of electrical pressure. We'll measure that pressure in volts and measure the current it creates in amperes.

The more free electrons available in a given material, the more readily they'll move in response to a given electrical pressure; the more free electrons in a material, the less "resistance" it'll have to the flow of those electrons as electrical current.

Materials containing large numbers of free electrons, and therefore offering little resistance to the flow of electron current, are called "conductors." Those with very few free electrons will inevitably have a high resistance to the flow of electron current, since there are only a few available electrons to participate in the process. These materials, due to their high resistance, are good insulators.

Metals in general, because of metallic bonding, have many free electrons and are good conductors. Glass, rubber, wood, cloth, and plastics, having few free electrons, are good insulators. A few materials exist that don't fall into either conductors or insulators. They have some of the characteristics of both, so they're called semiconductors. Silicon and germanium are two. These types of materials are used in various electronic devices, but not directly in building wiring; so they won't concern us.

Static Electricity

Under normal conditions, the atoms of a substance are neutrally charged, since the negative charges of the orbiting electrons are exactly balanced by the positive charge of the protons in the nucleus. When two electrically unbalanced atoms bond ionically to form a molecule, that molecule also becomes neutral, since the net positive charge of one atom has been offset by the net negative charge of the other.

However, when an outside influence forces many atoms of a material either to gain or lose an electron, that material either becomes negatively or positively charged. This charge collects on the object's surface and

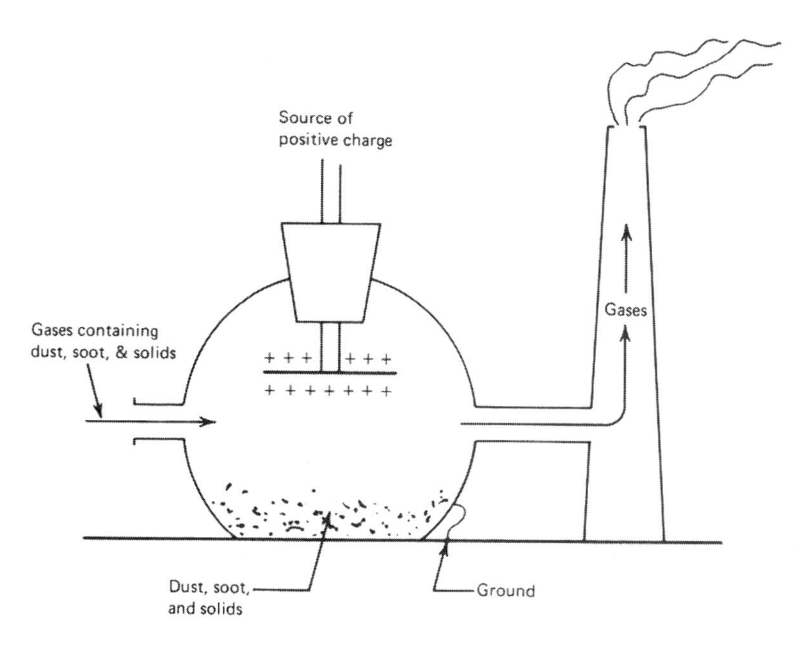

Figure 1-3
Smoke precipitator

tends to stay there until it's conducted away. The pieces of amber that were rubbed with cloth by the Greeks in 600 BC were charged in this way. When you walk across a thick carpet and touch a door knob, the small spark you receive is the same kind of charge. This type of surface charge is called a *static charge*.

One important use of static electricity is in cleaning the solid pollutants, such as soot and dust, from the exhausts of industrial plants. We rarely see black smoke belching from factory smoke stacks the way we used to. Now, those gases are vented into a *precipitation chamber*, where a positively-charged plate attracts the solids suspended in the gas. The moment they touch it, they become positively charged and are strongly repelled. They then drop to the bottom of the chamber, where they are collected and disposed of safely. See Figure 1-3.

While this and a few other constructive uses of static electricity exist, the static form is generally useless because it's essentially an instantaneous rather than a steady, dependable force.

Current Electricity

When a neutral atom loses an electron it becomes a *positive* ion. A neutral atom that gains an electron becomes a *negative* ion. Between any two charged particles, a force field exists in which like charges are repelled and unlike charges are attracted. This force field is called an "electrostatic field." In response to the force being exerted by the field, charged particles move. This movement constitutes an electrical current. In a solid conductor, the only mobile particles are free electrons that have escaped from the outer shell of an atom, leaving it as a positive ion. In liquids and gases, the positive ions are also free to move. This effect is encountered with certain types of lighting equipment.

When an excess of electrons causing a negative charge is built up at one end of a conductor, and a deficiency of electrons causing a positive charge is built up at the other end, the pressure caused by the field existing between the two ends will cause the loose electrons in the conductor to flow from the area of excess to the area of deficiency, if permitted to do so. As the electron differential between the area of excess and deficiency increases or decreases, the pressure differential between them varies as well.

The difference in electrical pressure between two points is measured in units called volts. The volt is named after an 18th century Italian experimenter named Alessandro Volta, the inventor of the battery. One volt is defined as the pressure necessary to force one ampere of electrical current through a resistance of one ohm. This definition isn't too helpful until we understand what is meant by *ampere* and *ohm*.

The ampere, the unit used to measure current flow, is named in honor of Andre Marie Ampere, also a late-18th century electrical experimenter. His experiments dealt in part with the flow of current in a conductor. Since an electrical current consists of a flow of electrons through a conductor, then the measurement of that flow is a count of the electrons passing a designated metering point in a specific length of time. As a comparison, amperage measures the flow of electricity per second the same way gallons per minute measures the flow of water. An electrical flow of 6,250,000,000,000,000,000 electrons per second equals one ampere, and that's what a pressure of one volt will push through a resistance of one ohm.

For an electrical pressure (voltage) to push a current (amperage) through any substance, that voltage must be sufficient to overcome the resistance of the substance. All substances have some kind of resistance to the flow of electrical current. Conductors such as metals have low resistance. Various insulators, such as plastics, paper, glass, or rubber, have high resistance, but no material exists that has no resistance. Since the resistances of different materials vary so widely, it's necessary to have a means for measuring these differences.

It was internationally agreed on long ago to accept a unit called the ohm as the measure of resistance. The ohm is named after another late-18th and early-19th century investigator of electrical phenomena, Georg Simon Ohm. Ohm recognized resistance as an inherent property of all materials. He also worked out *Ohm's Law* that explains the relationship among voltage, amperage, and resistance.

Ohm's Law

Ohm's Law states that an absolute fixed relationship exists between current, voltage, and resistance such that the current flowing in a circuit is directly proportional to the applied voltage, and inversely proportional to the resistance. Expressed in words, this sounds rather complicated, but it can be reduced to a very simple and easy to understand mathematical formula. This formula can be stated three ways. For mathematical purposes, the following symbols are used:

- Electrical current in amperes = I
- Electrical pressure in volts = E
- Electrical resistance in ohms (Ω) = R

1. The current in amperes is equal to the pressure in volts divided by the resistance in ohms.

$$I = \frac{E}{R}$$

2. The resistance in ohms is equal to the pressure in volts divided by the current in amperes.

$$R = \frac{E}{I}$$

3. The pressure in volts is equal to the current in amperes multiplied by the resistance in ohms.

$$E = I \times R$$

With any two factors known, the third can easily be calculated by either division or multiplication. All that is necessary is to keep track of when to multiply and when to divide. Use the diagram shown in Figure 1-4 as a guide.

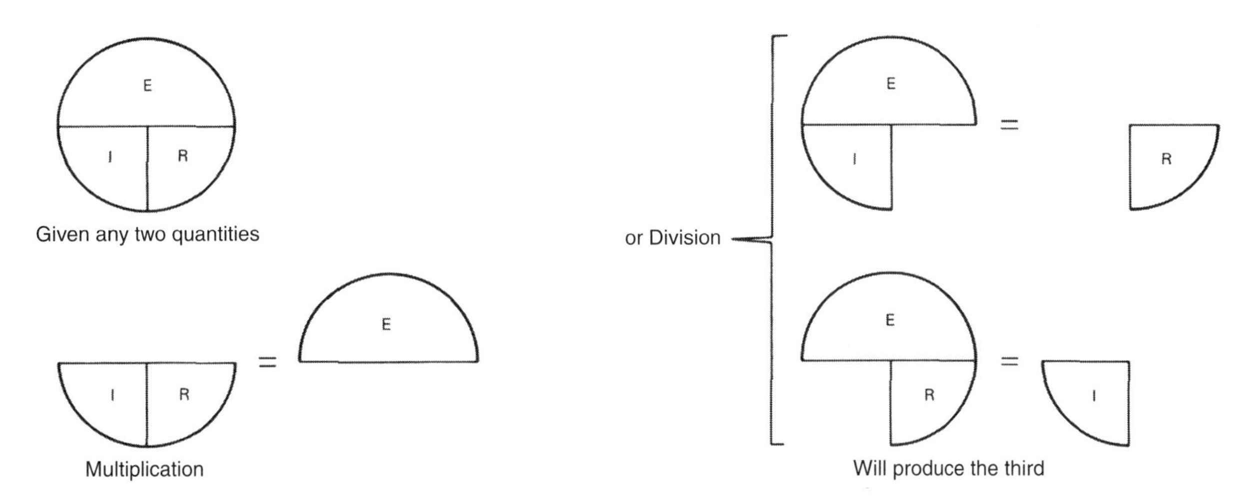

Figure 1-4
Ohm's Law

EXAMPLE Using Ohm's Law, a device with a resistance of 18 ohms will be connected to a 120-volt circuit. What amperage will it draw?

$$I = \frac{E}{R} = \frac{120}{18} = 6.6667 \text{ amperes}$$

EXAMPLE A countertop appliance draws 4 amperes at 120 volts. What is its internal resistance?

$$R = \frac{E}{I} = \frac{120}{4} = 30$$

EXAMPLE An electric dryer has a resistance of 10.67 and draws 22.5 amperes. What voltage should be supplied?

$$E = I \times R = 10.67 \times 22.5 = 240 \text{ volts}$$

An ohmmeter can be used to read resistances directly, but only when the circuit is off. However, many electrical devices show very little resistance when turned off and cold, but will increase in resistance dramatically when they're turned on and hot. A broiler, toaster, or tungsten light bulb are examples of this. A 60-watt tungsten light bulb has a cold resistance of only 5 ohms. A resistance of 5 ohms with a pressure of 120 volts would mean that a current of 24 amperes would be drawn by that bulb.

$$I = \frac{120}{5} = 24 \text{ amperes}$$

What actually happens is completely different. The filament heats instantaneously, which increases its resistance instantaneously from 5 ohms to 240 ohms. This resistance, however, can be found only by computation rather than direct measurement. This particular computation requires using another formula in addition to Ohm's Law. This one is known as Watt's Law and it was formulated by James Watt.

Watt's Law

This law is named after the same James Watt who invented the reciprocating steam engine. After his invention, he found it difficult to sell it to a skeptical public until he could work out a way to compare its performance with that of a horse. The horsepower ratings for not only steam engines, but also gasoline, diesel, and electric motors are derived from his basic formulations.

Just as a fixed relationship exists in Ohm's Law between voltage, amperage, and resistance, so in Watt's Law a fixed relationship exists between power expressed in watts, amperage, and voltage. Watt's Law

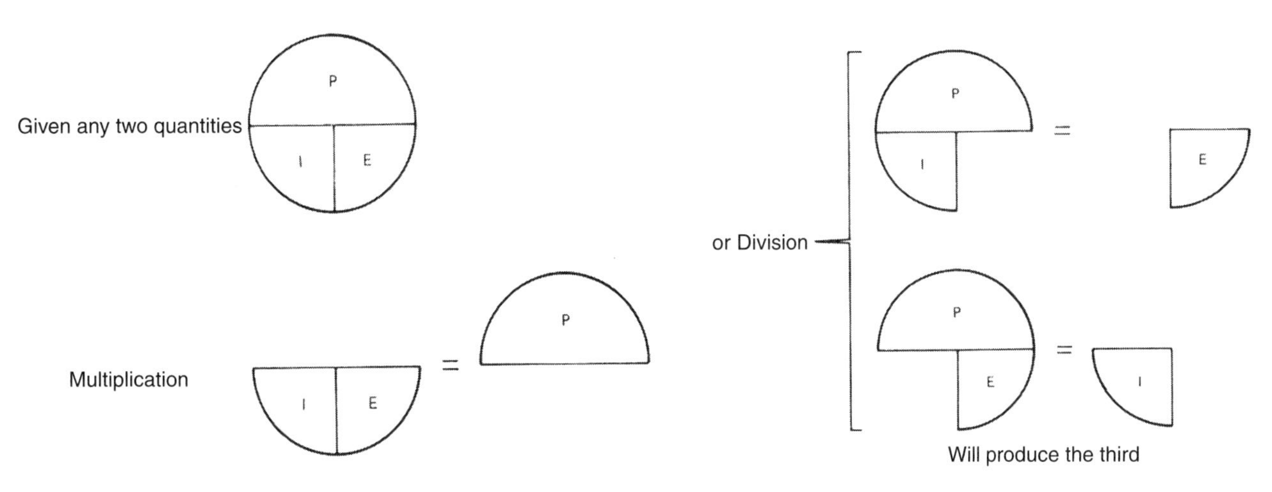

Figure 1-5
Watt's Law

states that the power available in watts is equal to the amperage multiplied by the voltage.

$$P = I \times E$$

There are two other common versions of this formula. One is that the current in amperes is equal to the power in watts divided by the voltage.

$$I = \frac{P}{E}$$

The other version is that the voltage is equal to the power in watts divided by the amperage.

$$E = \frac{P}{I}$$

As with Ohm's Law, when any two quantities are known, the third is obtained by simple multiplication or division. This method also applies to Watt's Law.

Watt's Law provides a simple method for converting watts to equivalent amperage, and vice versa. See the diagram in Figure 1-5. This type of computation is needed to determine connected loads on circuits, as we'll discuss in Chapter 8. Loads must be known accurately in order to insure that proper wire and breaker sizes are specified. Load computations are also used in troubleshooting to determine when a circuit is overloaded, and help determine proper action when an overload is found.

EXAMPLE A 1 horsepower electric motor draws 746 watts of power at 120 volts. Will it operate satisfactorily on a 15-ampere circuit?

$$I = \frac{P}{E} = \frac{746}{120} = 6.22 \text{ amperes} \qquad \text{No problem!}$$

EXAMPLE A coffee maker drawing 1000 watts, a toaster at 1200 watts, and an electric skillet at 600 watts are plugged into a 20-ampere, 120-volt counter-top appliance circuit. Any two will operate satisfactorily, but as soon as the third one (no matter which one it is) is turned on the breaker trips. What is the matter?

$P = I \times E = 20 \times 120 = 2{,}400$ watts is the maximum the circuit can handle. Above that wattage the breaker should trip.

Coffee maker	1,000	
Toaster	1,200	
Total	*2,200*	No problem
Coffee maker	1,000	
Electric skillet	600	
Total	*1,600*	No problem
Toaster	1,200	
Electric skillet	600	
Total	*1,800*	No problem
Coffee maker	1,000	
Toaster	1,200	
Electric skillet	600	
Total	*2,800*	Overloaded by 400 watts

Electrical Measurements

The electrician wiring residences or other small buildings actually takes very few electrical measurements. However, when he needs a measurement, he must know what instrument to use and how to use it. His workhorse and most commonly used instrument is the multimeter, as shown in Figure 1-6. This instrument gives readings in Alternating Current (AC) volts, Direct Current (DC) volts, ohms (resistance), or milliamperes. Prices can vary from as little as $30.00 for a basic model to as much as $1,000.00 for a top-of-the-line, high-precision one.

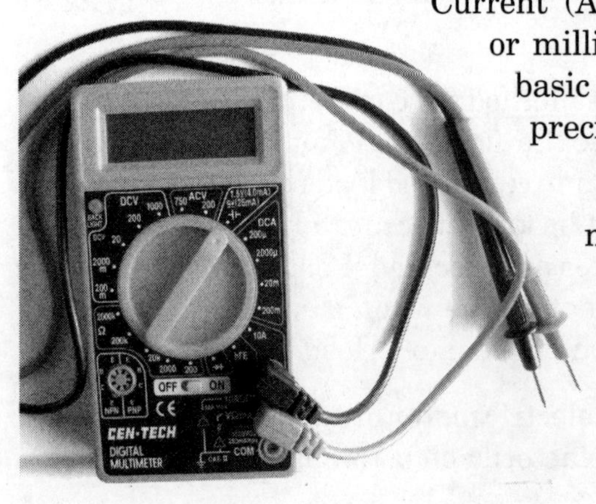

Figure 1-6
Digital multimeter

The precision of the expensive instruments isn't necessary for building wiring. A small, inexpensive instrument is ideal. In fact it's preferable, since it's compact enough that an electrician can carry it in his pocket while he squirms in and out of nooks and corners during his normal work day. If it's accidentally smashed in the process, he hasn't lost much.

The multimeter will have a central function selector switch to shift among the various AC, DC, ohms, and milliampere scales. In normal building wiring, only AC

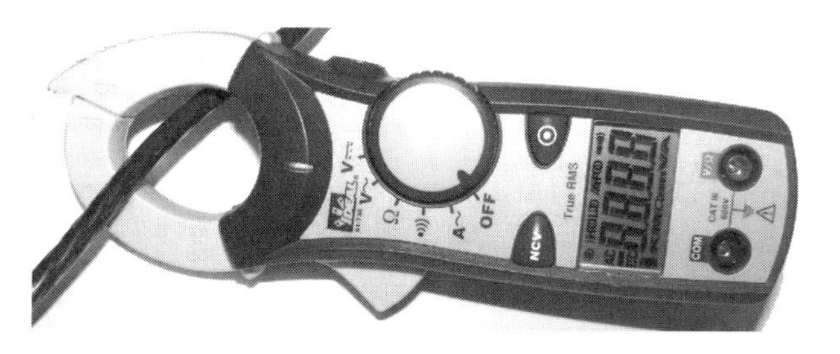

Figure 1-7
Clamp-on field-sensing ammeter

voltage and ohms scales are used. AC voltages in a building will be either 120 nominal, or 240 nominal. *Nominal* means that the actual voltage at any given time might vary anywhere between 110 volts and 120 volts, or between 220 volts and 240 volts. When reading building voltages, make sure the selector switch is set to AC. If the multimeter selector is accidentally set to DC, and you're reading a receptacle that has 240 volts AC, then you'll get a faulty reading of "0" volts because there's no DC *voltage* at that point. Your instrument will give you a correct reading if you enter accurate information. If you try to get a DC voltage reading from an AC outlet, the multimeter will correctly read "0." If you try to get an AC voltage reading from a car battery, the correct reading will also be "0."

The ohms scales on the multimeter will primarily be used to check circuit continuity, and to test for shorts. When using the ohms scales *be sure the power is off*. Resistances cannot be read on a live circuit, only on a dead one. If power is on, it'll burn out the meter. After setting the selector to ohms, and before taking any readings, short the test probes to each other, and use the "ohms adjust" control to set the meter accurately on "0." If this isn't done, you might receive misleading readings.

The measurement of amperage isn't usually necessary when wiring small buildings. However, it's a very helpful measurement to have for troubleshooting, such as when tracking an overload that keeps tripping a breaker. The correct instrument to use is a clamp-on field-sensing ammeter, like the one in Figure 1-7. This meter, when clamped around the power lead from a breaker, will detect the electrical field around it and translate the intensity of that field into a measurement of the amperage flowing into that wire. In order to read amperage, this meter must be clipped around the hot wire only. If it's clipped around the complete cable feeding an appliance, it'll read "0" instead of the amperage being drawn by the appliance. The reasons for this will be discussed in the next chapter. To read the exact amperage drawn by a plug-in appliance, an adapter is needed to separate the hot wire from the common in the appliance feed.

Besides voltage, resistance and amperage, wattage, the fourth factor in electrical computations, can be directly measured with — you guessed it, a wattmeter. The wiring of buildings discussed here doesn't require this measurement since wire sizing, breaker sizing, and circuit loading are specified and limited in the code by amperage rather than wattage. Probably the only contact the average electrician will have with wattage measurement will be the installation of the meter box in the service entrance. See Chapter 9. It's where the utility company mounts their watthour meter to record electrical power usage for billing purposes. See Figure 1-8.

Basic Electrical Circuits

Figure 1-8
Watthour meter

Since an electrical current will flow readily through a conductor, it's a simple matter to direct electrical energy from a remote source to a desired point by connecting conductors to form a low resistance path from one point to the other. Conductors connected this way become an electrical circuit. The simplest electrical circuit consists of a minimum of four parts, as shown in the diagram in Figure 1-9. They're a source of electrical pressure, or voltage; conductors to connect the source to the use point; an electrical load or using device; and a switch or other mechanism to control that load. Since a current will only flow when the path is complete, from the high pressure or hot side of the source back to the low pressure or grounded side, a return conductor is necessary to complete the circuit.

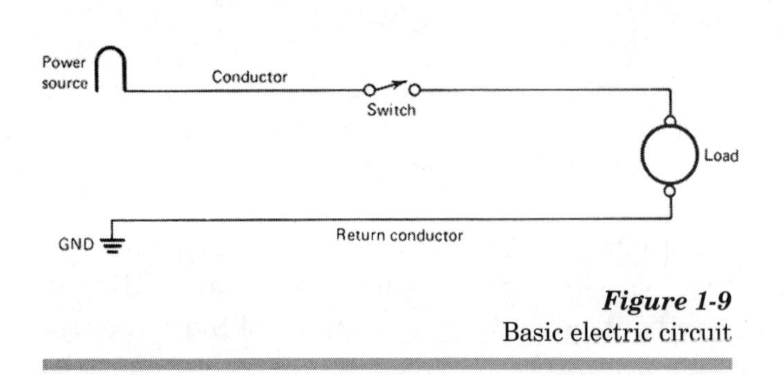

Figure 1-9
Basic electric circuit

Electrical loads can be connected to a power source in either of two ways. See the diagram in Figure 1-10. They can be connected in *series* or in *parallel*. With a series circuit, there's only one path through which current can flow, therefore the same amperage flows through all parts of the circuit. And with the parallel connection, there's a separate electrical path through each load with part of the amperage coming from the source passing through each path. The part of the total amperage drawn that passes through each load is proportional to its wattage and inversely proportional to its resistance.

Three loads connected in parallel on one circuit are shown in Figure 1-11. The total wattage being drawn is the sum of the three loads:

Television	300 watts
Lamp	100 watts
Fan	75 watts
Total	*475 watts*

Using Watt's Law to determine the total amperage being drawn:

$$I = \frac{P}{E} = \frac{475}{120} = 3.958 \text{ amperes}$$

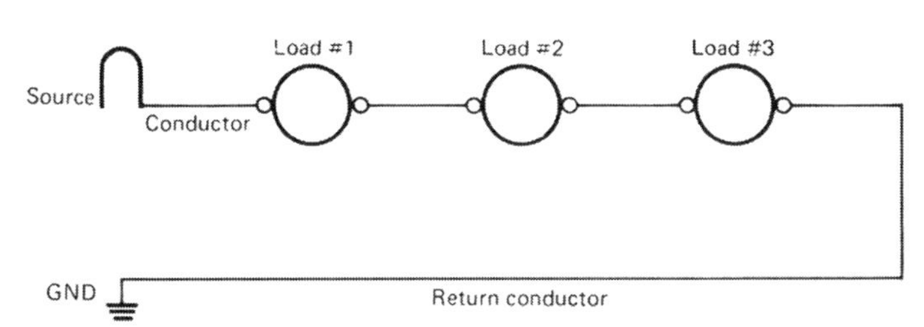

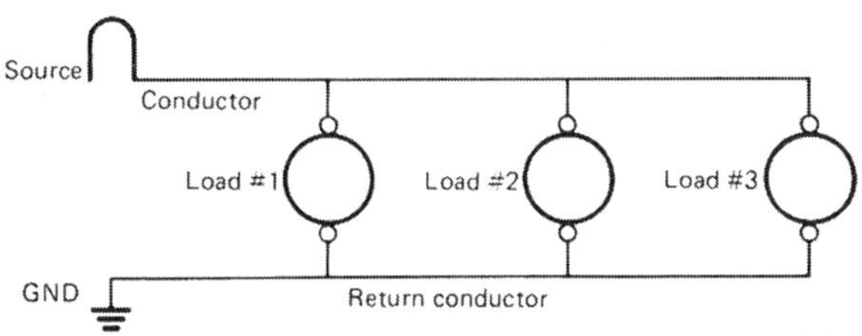

Figure 1-10
Series and parallel circuits

The amperage being drawn by the television alone:

$$I = \frac{P}{E} = \frac{300}{120} = 2.5 \text{ amperes}$$

Its resistance by Ohm's Law:

$$R = \frac{E}{I} = \frac{120}{2.5} = 48 \text{ ohms}$$

In contrast, the amperage being drawn by just the fan:

$$I = \frac{P}{E} = \frac{75}{120} = .625 \text{ amperes}$$

but its resistance is much greater:

$$R = \frac{E}{I} = \frac{120}{.625} = 192 \text{ ohms}$$

Something else is happening in this parallel circuit that'll help you understand overloads. The total resistance on this circuit (look again at Figure 1-11) using Ohm's Law, is:

$$R = \frac{E}{I} = \frac{120}{3.958} = 30 \text{ ohms}$$

Now, we already have a resistance of 48 ohms, and another of 192 ohms. We haven't calculated the third one, but we know it'll be somewhere between the two figures. How can the total resistance of the circuit be only 30 ohms? The truth is, in parallel circuitry the current may pass through multiple paths. Regardless of how high the resistance of an individual path is, once that path exists, *some* current can pass through it, current that couldn't and wasn't passing through other existing paths. Therefore, the more paths available, regardless of how high their resistances are, the more current the circuit allows to pass through, because at the opening of each new path the total resistance of the circuit is reduced.

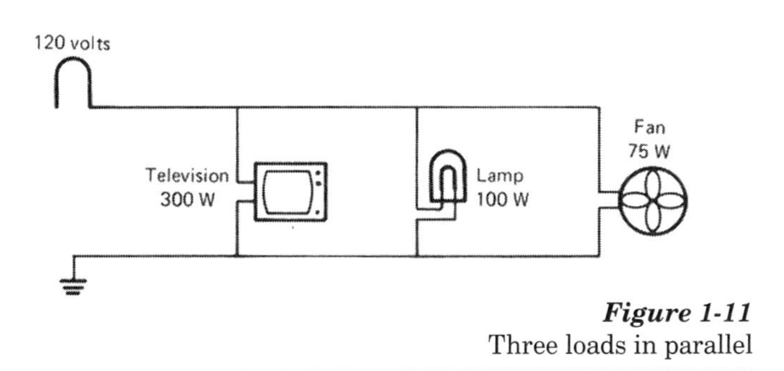

Figure 1-11
Three loads in parallel

In building wiring, all power-using devices are wired in parallel to keep each one independent of the others. Refer to the series circuit in Figure 1-10; if Load #2 were to break down, both #1 and #3 will stop because the only existing electrical path has been broken. If Load #2 in the parallel circuit failed, it wouldn't affect either #1 or #3, because each has independent access to the power source.

In building wiring the basic rule is, "All loads are wired in parallel; all switches are wired in series." A switch completes or breaks an electrical path, but it doesn't use any power. It offers either no resistance, or infinite resistance to the passage of an electrical current. Its purpose is merely to open or close the path to some electrical equipment.

Effects of Electrical Energy

Electrical energy can easily be channeled to produce heat, magnetism, chemical reactions, and even physiological effects as well. All of these effects involve the conversion of energy from one form to another. Such conversions always involve some loss.

Mechanical energy applied to an apparatus encounters resistance in the form of friction within the mechanism. When transmitting power from the engine to the wheels of an automobile, some of that power, despite the best lubrication, is lost in friction. Actually, it isn't lost; it's still present as heat that develops at friction points. A transformation of mechanical energy into heat energy has taken place.

Similarly, electrical energy is transformed into heat in the process of overcoming the resistance in a conductor. Conductors specifically designed to maximize this transformation are used in the heating elements of certain electrical appliances, such as toasters, broilers, electric ranges, water heaters, clothes dryers, and other electrical heating equipment. These are all common uses of the heating effect that can be produced with electrical energy.

The incandescent light bulb is another, but less common, example of the heating effect of electricity. Inside the bulb, an electrical current passes through a filament of tungsten wire. The resistance of the wire causes it to heat white hot, producing light. This process produces considerable unwanted heat, or "waste heat," as you'll notice if you touch a burning bulb.

The fluorescent light, as shown in Figure 1-12, is another example of the heating effect of electricity. In this case, filaments don't produce light, but act as heaters and ionizing electrodes. Air is pumped out of a fluorescent tube, then a bit of argon gas and a few drops of mercury are introduced. The heater current passed through the filaments vaporizes the mercury. The higher ionizing voltage then ionizes first the argon, then the mercury vapor,

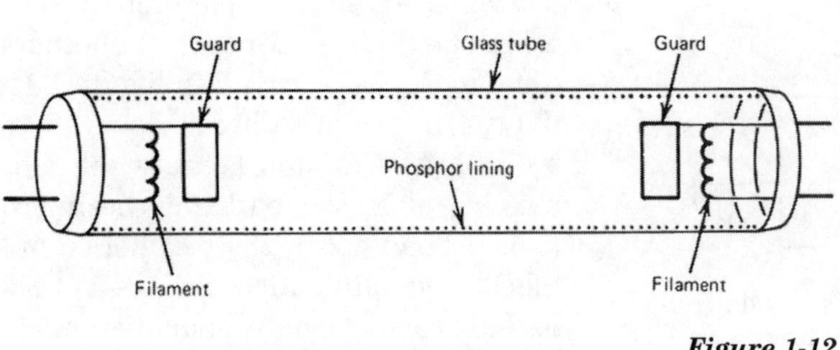

Figure 1-12
Fluorescent tube

producing ultraviolet light. The ultraviolet hits the phosphor coating of the tube, producing visible light. The fluorescent tube is a far more efficient light than the incandescent; a far higher percentage of the electrical energy used appears as visible light and far less is wasted in heat. Touch an operating fluorescent tube and you'll feel the difference.

Other electrical lighting systems such as neon, metal halide, sodium vapor, LED, and mercury vapor lamps are all examples of electrical heating effects.

In addition to producing heat, electricity can be used to produce many other useful results through magnetic effects. As we'll discuss in Chapter 2, a magnetic field can produce an electrical current. The reverse is also true. An electrical current can produce magnetic effects. The electric motor in its many forms is probably the most important use of the electromagnetic effect, but there are many others.

Some examples of other everyday items whose operation is based on magnetism are doorbells, buzzers, telephone transmitters and receivers, solenoid controls, electromagnets, dynamic stereo loudspeakers, and all material recorded on magnetic tape.

Chemical effects of electricity aren't as commonly encountered as heating or magnetic effects; an example is electroplated silverware. An electro-chemical effect produces the power to turn the car engine over every time it's started, and the dry cell batteries used in flashlights are chemical effect items.

The physiological effects of electricity usually aren't the ones we're most eager to encounter. However, while we tend to think of these effects as generally unpleasant, some are extremely useful. For example the pacemaker, that many heart patients depend on, regulates the heartbeat electrically. The lifesaving defibrillator in coronary care units is also very important in regulating the heart. And the medical arsenal contains a lot of other important electrical equipment for saving lives and improving patient comfort.

In Summary

We now know some of the basic facts about electricity. We know it flows easily through metallic substances that contain an abundance of free electrons. We know its pressure is measured in volts, its current flow in amperes, and the resistance to its flow is measured in ohms. We have seen that there are two types of electrical current: direct and alternating. We know of two types of electrical circuits — a series circuit provides only one path for the electrical current to take, while a parallel circuit provides alternate paths for the current. In addition, we now know that electricity can produce chemical and magnetic effects, as well as the physiological effects. In the next chapter we'll see how electric power is produced.

STUDY QUESTIONS

1. How many possible electron shells can an atom have?

A) 4
B) 5
C) 7
D) 9

2. What type of resistance to the flow of electrical current will a material containing an abundance of free electrons have?

A) Negative
B) Positive
C) Very high
D) Very low

3. Which is a practical use of static electricity?

A) Drycell battery
B) Exhaust cleaners on industrial plants
C) Carpet-cleaning attachment on a vacuum cleaner
D) Defibrillator in coronary care units

4. What unit is used to measure electrical current flow?

A) Ampere
B) Ohm
C) Volt
D) Watt

5. How many amperes will be drawn by a device with a resistance of 30 ohms connected to a 120-volt circuit?

A) 4
B) 30
C) 40
D) 360

6. Which electrical formula was devised to show the relationship between power, amperage and voltage?

A) Ohm's Law
B) Watt's Law
C) Power Conversion Formula
D) Fulton's Law

7. Which of the following won't a multimeter measure?

A) AC volts
B) DC volts
C) Watts
D) Ohms

8. Which formula will give the total resistance of a parallel circuit?

A) The voltage divided by the amperage being drawn
B) The voltage multiplied by the amperage being drawn
C) The sum of the resistances of the devices connected
D) The sum of the resistances divided by the voltage

9. Which of the following is true regarding how all power-using devices in a building are wired?

A) Depending on the use, they may be wired either in parallel or in series
B) They are wired in series to keep each independent
C) They are wired in parallel
D) If wired in parallel, each device must be independently grounded

10. Which of the following groups contain *only* examples of devices using the magnetic effects of electricity?

A) Doorbell, drycell battery, electric motor
B) Doorbell, electric motor, fluorescent light
C) Drycell battery, incandescent light, telephone receiver
D) Electric motor, stereo loudspeaker, telephone transmitter

2

DISTRIBUTION OF ALTERNATING CURRENT

As discussed in the previous chapter, electrical current exists in two forms: *Direct Current* (DC) and *Alternating Current* (AC). Direct current is produced in several ways. Alternating current is produced only by magnetic mechanical generators. Direct current flows steadily in one direction and often at constant voltage. Alternating current rhythmically reverses its direction, constantly increasing and decreasing in voltage.

DC Sources

One form of DC is thermoelectricity. When two dissimilar metals are connected and heated, a small electrical current is generated at one end. See Figure 2-1. This is the principle behind the safety of the thermocouple used in gas water heaters and ovens. One end of the thermocouple is mounted in the flame of the pilot light, generating a small current that tells the control unit the pilot light is on, and it's safe to turn on the main burner.

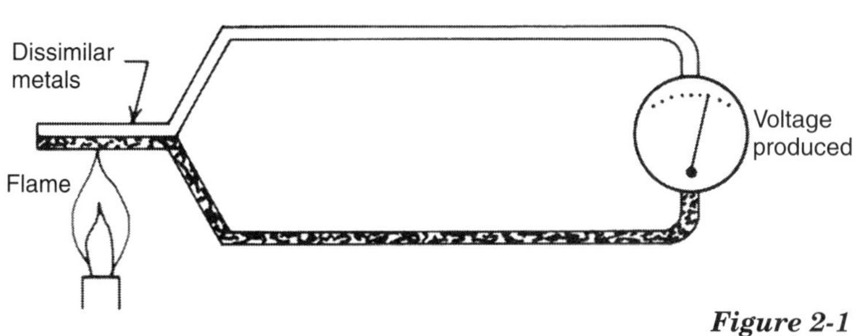

Figure 2-1
Thermoelectric diagram

Another form of DC is pizeoelectricity. Pressure applied to certain crystalline substances such as Rochelle salts or barium titanate produces a weak DC current. The pressure of sound waves on the diaphragm of a crystal microphone causes the crystal to produce an electrical signal. The same principle applies in the case of the crystal phonograph pick-up.

Photovoltaic cells are another source of DC that uses silicon to generate an electrical current in the presence of light. Photovoltaic cells have been a primary source of power for the electrical equipment used in many satellites. Extensive development is ongoing to reduce the cost and increase the output of photovoltaics.

The DC source that many are most familiar with is electrochemical, like drycells and storage batteries (wet cell). The familiar drycell powers flashlights, portable radios, pocket calculators, cell phones, MP3 Players, BlackBerries, and many other items. The storage battery is probably known to most as the battery in cars, boats or recreational vehicles.

Both the drycell and the storage battery produce electricity through chemical interactions. Many drycells are rechargeable. When the chemicals in the cell are used up, the cell dies and must be replaced. All storage batteries *are* rechargeable. Its chemical reaction is reversible, so the battery can be recharged and used again. It's recharged with the use of electrical power produced by magnetic-mechanical means.

Magnetic-Mechanical Generation and Alternating Current

The electrical power produced and delivered by the utility companies is alternating current, or AC; therefore, the building power and lighting circuits installed by the electrician will use AC. Alternating current refers to the voltage in this type of power that constantly alternates over time from 0 volts to a controlled maximum voltage, back to 0, then down to the same maximum in the opposite direction, and finally back to 0 to restart the cycle. We'll look at this cycle in detail later in the chapter. The current used in the United States cycles at the rate of 60 times a second; one complete cycle takes 1/60th of a second. Hertz is the name used when referring to cycles per second in electrical and radio frequencies. AC alternates at the rate of 60 Hertz, abbreviated as 60 Hz.

In order to understand why AC voltage alternates, you need to see how it's generated. But an understanding of AC generation depends on being acquainted with magnetism and electromagnetic induction.

Magnetism and Electrical Induction

As mentioned in Chapter 1, the ancient Greeks discovered a peculiar black stone that could attract iron. They also found that if a bar of iron was stroked with such a stone, that bar would also attract other pieces

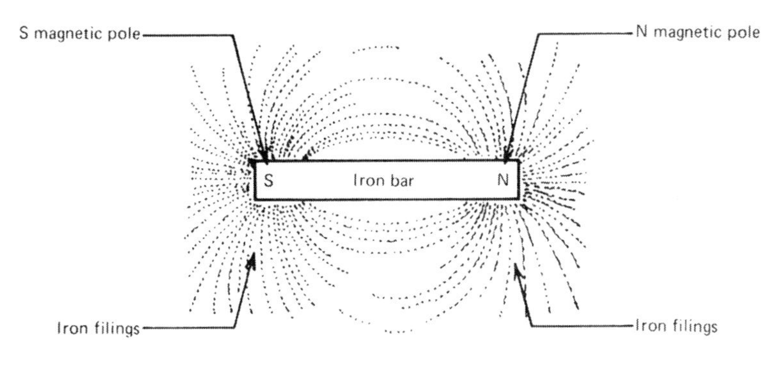

Figure 2-2
Bar magnet force field

of iron. The Chinese discovered long ago that if a magnet could be suspended, allowing it to move freely, one end would always point fairly close to the direction we call north. This was the origin of the magnetic compass, which is still standard equipment on all ships and aircraft, regardless of how elaborate their electronic guidance systems are.

Near the turn of the 16th century, William Gilbert started studying magnets and found that the attracting property of magnets is concentrated at the two ends, called "poles," with little force visible in between. This observation, along with the earlier Chinese discovery, pointed to the idea that the entire earth is a colossal magnet with opposing magnetic poles, located close to the geographic poles.

It was also noted that if the north-seeking pole of one magnet were brought close to the north-seeking pole of another, they repelled each other. Similarly, two south-seeking poles repelled each other. However, any two opposite poles, a north-seeking and a south-seeking pole, when brought together, are definitely attracted. These observations indicate clearly that a force field exists in the area of a magnet. This field can readily be demonstrated by placing a thin piece of cardboard on a bar magnet, and then sprinkling the cardboard with iron filings. The filings will form a pattern resembling the one in Figure 2-2.

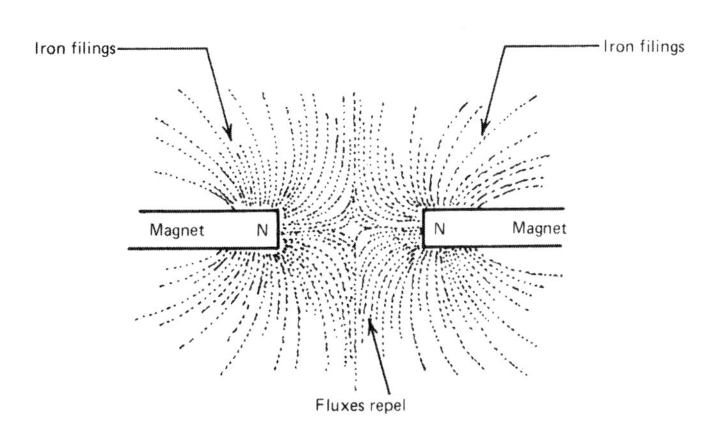

Figure 2-3
N-N or S-S repulsion pattern

The mutual repulsion of like poles, north to north or south to south, is also demonstrated by placing similar poles of two magnets under a thin piece of cardboard and sprinkling iron filings. This time the pattern will look like the one in Figure 2-3. In contrast, the mutual attraction of north and south poles, using the same demonstration, will produce a pattern like the one in Figure 2-4.

As I mentioned in the previous chapter, positive and negative electrical charges behave oddly similar to polar magnetism. Positive repels positive, negative repels

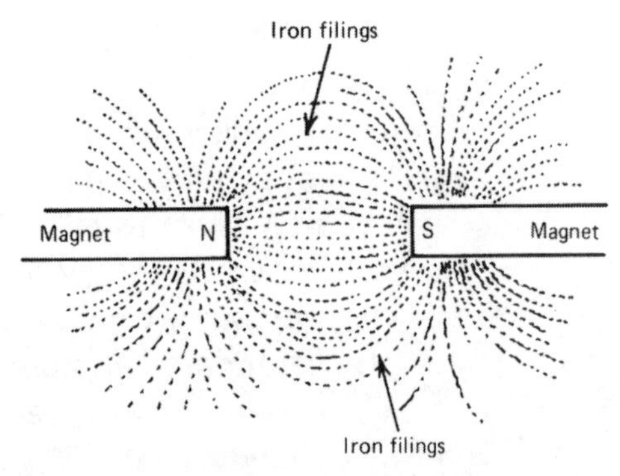

Lines of force between opposite poles

Figure 2-4
N-S attraction pattern

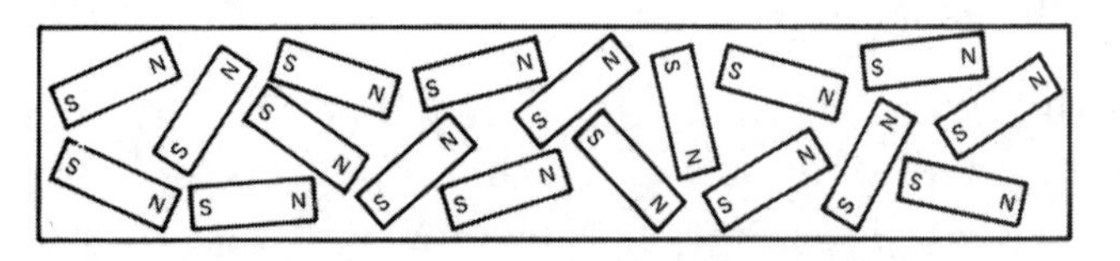

Molecular pattern in non-magnetized material

Figure 2-5
Molecular theory random pattern

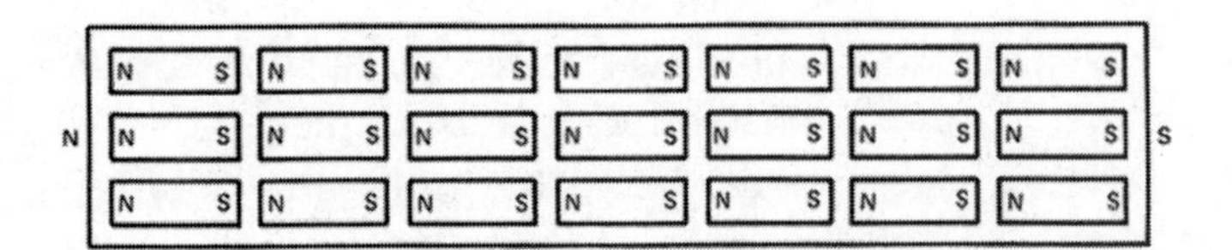

Molecular pattern in magnetized material

Figure 2-6
Molecular theory alignment pattern

negative, but positive and negative attract each other. We'll see, by looking closely at magnetism, that these similarities are indeed related.

In the 19th century, investigations by German physicist Wilhelm Weber resulted in the molecular theory of magnetism. This theory proposes that the individual molecules of a material that can be magnetized, such as iron or steel, are each miniature magnets with their own north and south poles and their own magnetic fields. When a piece of such material is unmagnetized, the molecules are arranged in random fashion, as shown in Figure 2-5, resulting in the magnetic fields of the molecules canceling each other out. This neutralizes the material so no external magnetic field results. However, when that material is magnetized, the molecules align themselves in an orderly manner like the pattern in Figure 2-6. The north pole of one molecule is facing the south pole of the next one, resulting in the various individual molecular magnetic fields reinforcing each other. This produces a magnet with an external field.

Weber's model explained much of what had been observed regarding magnets and was only modified in the light of atomic research. Magnetism and electricity had long been recognized as integrally interrelated, but it hadn't been proposed that their interrelationship was consistent with subatomic particles.

Refer back to the model of the atom in Figure 1-1, Chapter 1 and note that the electrons orbit in concentric *shells* around the nucleus. In addition to rotating around the nucleus, it was discovered that the electrons each spin on their own axis just as the planets of the solar system spin on theirs. The discovery of electron spin led to the discovery that magnetism appears to be related to the spin of the electrons, particularly in the third shell of atoms of magnetic materials.

As mentioned earlier, the electron has an electrically negative charge. When it spins, its electrical field spins with it, creating a magnetic field whose polarity depends on the electron's direction of spin. With most materials, half the electrons spin clockwise and the other half counter-clockwise. So, the opposing polarities cancel each other out, leaving no external magnetic field.

A magnetic material, for example, has an imbalance. More electrons rotate in one direction than in the other, so the magnetic fields don't cancel out completely. The magnetic material iron is a good example. Its atomic number is 26, meaning it has 26 protons and 26 electrons. Those 26 electrons are arranged in shells like any other atom.

The first shell has the normal two electrons. Since they spin in opposite directions, they neutralize each other magnetically. The second shell is complete with eight electrons, four spin one way, and four the opposite way, so the second shell internally cancels itself out the same as the first. However, the third shell is incomplete. It can hold 18 electrons but with an iron atom there's only 14. Of those 14, half of the number that would form a complete third shell spin one way, and the rest spin the opposite. Half of 18 equals nine that spin one way, which leaves only five to oppose them. This results in an external magnetic field caused by the four uncanceled fields. The fourth shell has only two valence electrons spinning in opposite directions, thus they cancel each other out. The outcome is that the atom, as a unit, has an external magnetic field due to the imbalance of the four fields in the third shell.

The effect of the force fields of individual atoms on each other causes groups of nearby atoms to align themselves so that their various fields reinforce each other. Such a group is called a *domain*. For example, in a piece of unmagnetized iron, these domains are randomly placed in relation to each other, similar to Weber's molecules shown back in Figure 2-5. When the material is magnetized, the domains are forced into alignment like the molecules in Figure 2-6. While Weber spoke of *molecules,* modern physicists speak of *domains*; but the two explanations of magnetic behavior are essentially the same regarding observed phenomena. The advantage of the modern explanation is that it clearly shows *why* things happen as they do.

Electrical Induction It's been known that the reactions of electrical charges and magnetic poles follow the same rules, for example, like charges and like poles repel each other; while unlike charges and unlike poles attract each other. This similarity between electricity and magnetism raised the question whether there's any connection between the two. The first connection discovered was that an electrical current flowing in a conductor creates a magnetic field around that conductor. See Figure 2-7. The field created is a different shape from that of a bar magnet. It's circular, with a clockwise direction related to the current's

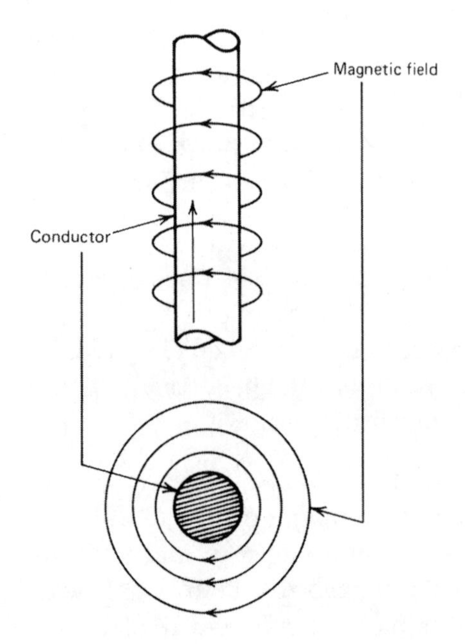

Figure 2-7
Magnetic field around a conductor

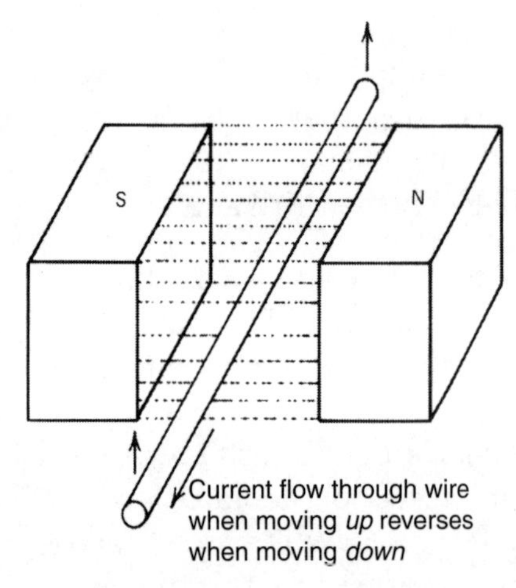

Figure 2-8
Conductor moving in a magnetic field

direction of flow. However, it's clearly a magnetic field regardless of its shape. This discovery questioned whether the reverse would work. "Could a magnetic field create an electrical current in a conductor?" Michael Faraday found that this could be done. The following are two classic demonstrations of the ways this can be accomplished.

First, if a conductor is moved up or down in the attraction field between the north pole of one magnet and the south pole of another, or between the poles of a horseshoe magnet, a current will flow in that conductor as long as it's in motion. See Figure 2-8. When it stops, even though it remains within the field, the current will stop as well. Therefore, current can be generated by moving a conductor in a magnetic field.

The second way of generating electricity by the use of magnetism is to coil the conductor, leave it stationary, and move the magnet relative to the coiled conductor. Again, when the magnet stops, whether inside or outside the coil, the current also stops. Thus, given the two basic components — a conductor and a magnetic field — an electrical current will be produced as long as the two are moving relative to each other.

After finding that moving a conductor and a magnetic field relative to each other produces voltage, it was discovered that any increase in the speed of that movement increases the magnitude of the voltage produced. It was also found that the stronger the magnetic field becomes, the greater the voltage induced. Additionally, the greater the number of turns in a coiled conductor, the greater the induced voltage becomes. Thus, by various means, the physical energy of movement in conjunction with a magnetic field and a conductor can be used to produce electrical energy.

Induction and AC Generation

Since the movement of a conductor in a magnetic field will induce an electrical voltage, and the voltage varies with the movement, then if constant and steady movement could be maintained, it would be possible to make a steady and constant flow of electrical energy. Let's see how this is done in Figure 2-9.

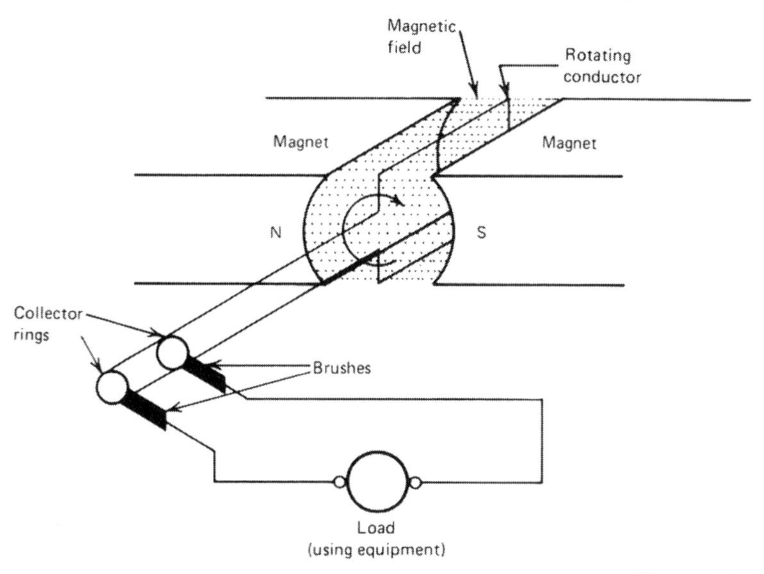

Figure 2-9
Simplified AC generator diagram

A simplified AC generator consists of a magnet providing the field, a conductor loop arranged so it rotates within that field, and a pair of collector rings to collect the electrical energy produced by the rotating conductor. As the loop rotates, it cuts across magnetic lines of force, inducing voltage. The voltage induced at any given time is a function of the number of lines of magnetic force being cut at that time. Look at Figure 2-10 to see that the number of lines of force changes continually as the conductor rotates in the field of the magnet.

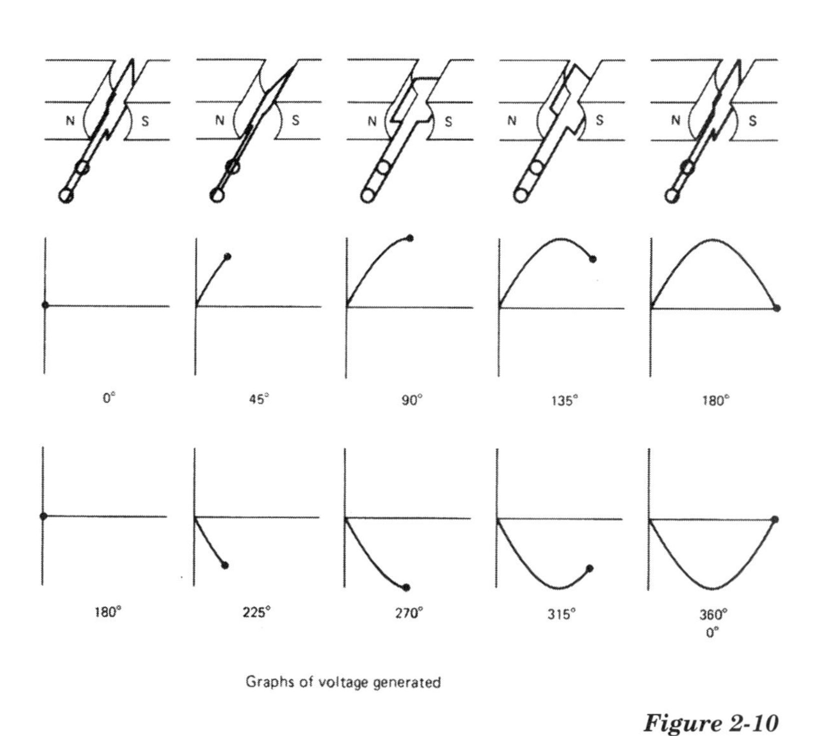

Figure 2-10
AC generator cycle

At 0 degrees, the conductor loop is moving in line with the flux of the magnetic field and cutting no lines of force, so no voltage is generated. As it turns through 45 degrees, it begins cutting lines of force and continues to cut more and more as it turns. Continuing through 90 degrees, it cuts the maximum number of force lines for this magnetic field, and reaches the maximum possible voltage. Beyond 90 degrees it begins to cut fewer and fewer force lines as it continues to turn until it reaches 180 degrees. This is where it again moves parallel to the lines of force of the magnetic field, cutting no lines nor producing any voltage. Everything is as it was at 0 degrees *except* that the side of the rotating loop that just moved past the south pole of the magnet is now starting past the north pole, and vice versa.

This reversal of the magnetic field relating to the moving conductor also reverses the flow of voltage induced in that conductor. During the second half of the rotation, from 180 degrees back to 0 degrees, the curve representing the magnitude of the voltage being generated will be the same as the curve for the first half of the rotation from 0 to 180 degrees, except the *direction* is reversed. It's this reversal of direction in the generating process that gives it the name *alternating current.*

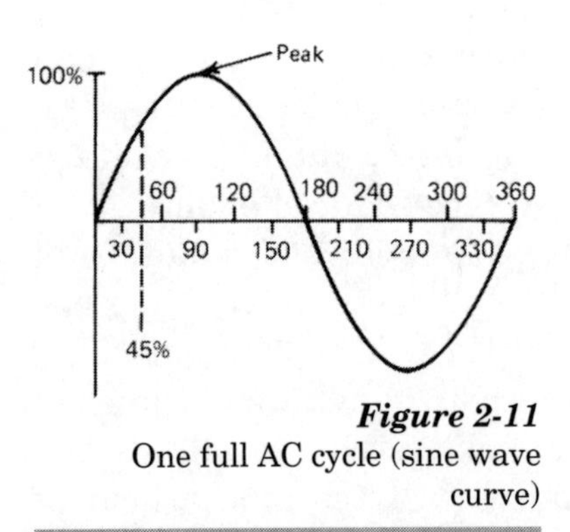

Figure 2-11
One full AC cycle (sine wave curve)

The graphic curve representing the voltage changes that occur during one full rotation of the conductor in the magnetic field of a generator is called a *sine curve.* Simply stated, AC voltage follows a *sinusoidal* curve. See Figure 2-11. One rotation of the armature (rotating part of an AC generator) takes that voltage curve through one full cycle from 0 volts up to maximum on one side, back to 0, down through the maximum in reverse and back to 0 at a rate or *frequency* equal to the speed of rotation of that armature. The standard frequency of AC voltage in the United States is 60 cycles per second or 60 Hz. This frequency is held steady in power-generating plants by carefully controlling the speed at which the armatures of the large generators are turned by either steam turbines or other mechanical drive systems.

Single-Phase and Three-Phase Power Generation

Look at Figure 2-11 to see the voltage produced by a generator with a single armature coil. This type of generator is called a *single-phase* generator. To produce the voltage illustrated, the armature coil had to be mounted on a shaft and rotated in a magnetic field. Since mechanical energy must be used to rotate that shaft — why not add more coils, each producing a similar voltage? This results in vastly greater electrical energy output in return for relatively little increase in the mechanical energy input.

This is exactly what's normally done. Currently, three coils generating three phases is the optimum system. See Figure 2-12. Virtually all utilities generate their power this way. Coil A produces phase A, coil B produces phase B, and coil C produces phase C. Each phase alternates independently of the others at a frequency of 60 Hz, but they follow each other by 180th of a second since they necessarily pass through the force line variations of the magnetic field in sequence.

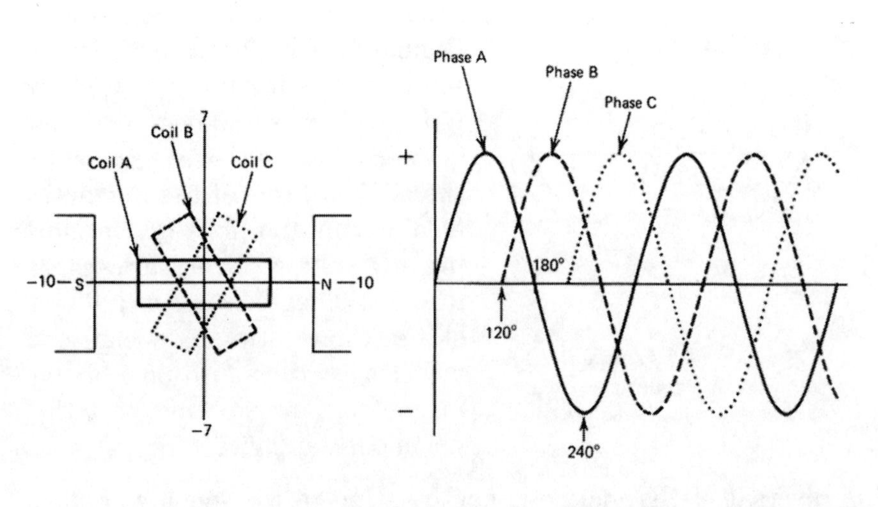

Figure 2-12
Three-phase generator and voltage wave form diagram

Since each of the three armature coils has two ends, you might expect that six wires would be required to transmit three-phase current. However, using one of two connection methods, shown in Figure 2-13, it's possible to join the three coils at the generator so that the three phases can be transmitted by one wire for each phase. Using the *delta* connection each armature coil is connected to one of the others at each end. With the *wye* (Y) connection, all three coils meet at a single point. In either configuration, only three wires are needed leading out.

While a lot of electrical power is generated as three phase, the majority of it is finally used as a single phase. The portion used as three phase primarily powers heavy industrial machinery. In this application, three phase is far superior to single phase. An electric motor gets 120 pushes per second from single phase (60 cycles per second with two peaks per cycle). On the surface, it might seem as though that would be fast enough for any normal purpose, but an average electric motor turns at about 1,800 rpm, which means the armature turns at 30 revolutions per second. With 120 electrical pushes turning the armature at 30 revolutions per second, the armature is getting four pushes per turn. In a little electric drill motor with an armature diameter of about 2 inches, the circumference is something over 6 inches, in which case it's getting a push every 1½ inches. However, when the armature diameter of a larger motor is in the vicinity of 12 inches, the circumference is now around 38 inches. At the same four pushes per revolution, it's now more than 9 inches between them. By changing over to three-phase current, the larger motor is now going to get 12 pushes per revolution, or one almost every 3 inches of its circumference. Obviously, this will provide a much steadier drive and smoother operation.

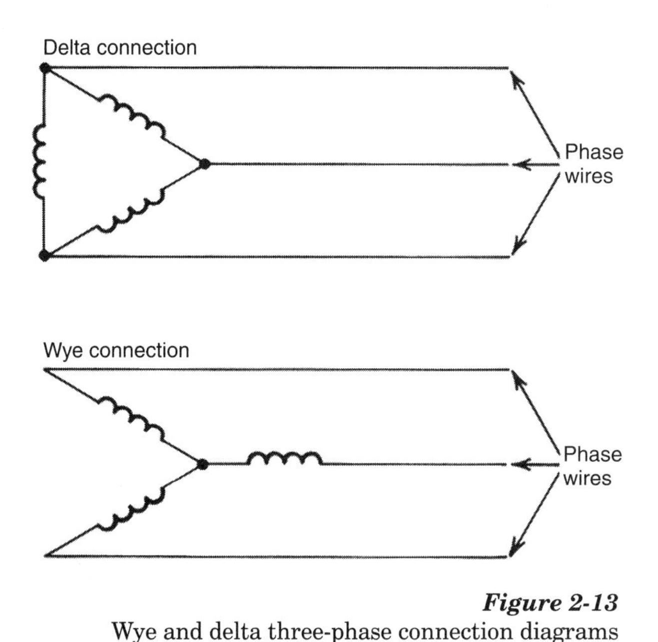

Figure 2-13
Wye and delta three-phase connection diagrams

While the larger electric motor benefits both in smoothness of operation and in efficiency through three-phase power, smaller equipment is much simpler to construct and operate using single-phase power. For this reason, although the majority of electrical power is generated and transmitted over long distances as three phase, most of it's ultimately broken down into single phase before it's used. When three-phase power is split into single phases, voltage is reduced. The device used to increase or decrease voltage is called a transformer.

Transformers The operation of a transformer takes us back to magnetic fields. An electrical current creates a magnetic field around its conductor, as diagrammed back

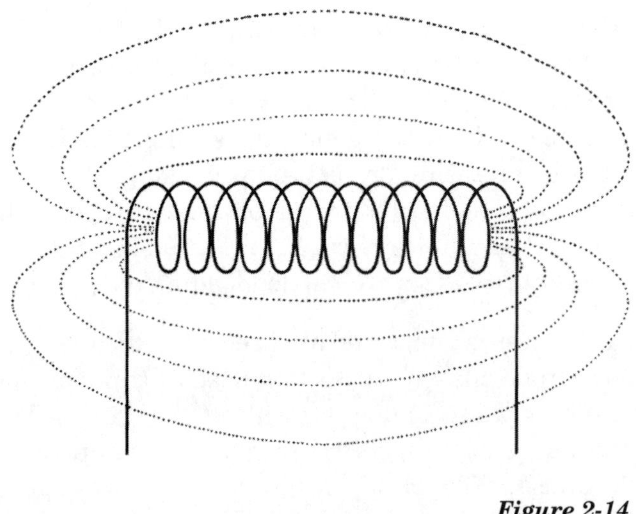

Figure 2-14
Magnetic field around a coiled conductor

in Figure 2-7. When that conductor is coiled, shown in Figure 2-14, the resulting magnetic field closely resembles a simple bar magnet. Look back to the example in Figure 2-2. However, when that field is created by an alternating current in the coil, there's a significant difference. The field created by the bar magnet is constant in both polarity and intensity. The field produced by the alternating current is constant in neither respect. Since the voltage of alternating current is constantly changing, and its direction regularly reverses, the magnetic field it produces when passed through a coil is constantly changing in intensity as well as periodically reversing in polarity.

This constantly changing and reversing field around a coil has a very interesting and useful capability. As we've already seen, a conductor and a magnetic field moving relative to each other will induce a voltage in the conductor. It's the conductor cutting magnetic lines of force, due to the movement, that produces this result.

In the case of a coil through which alternating current is passing, the magnetic field around that coil builds up in intensity as the voltage goes up, and then collapses as the voltage goes back down. Since the voltage is constantly changing, the field is in constant change as well. As the voltage goes up, the magnetic lines of force around the coil move outward. As the voltage decreases, they contract. Thus, a conductor placed within the area of influence of such a field is actually being crossed and recrossed by moving lines of magnetic force, even though no physical movement is involved. Since magnetic lines of force are being cut, voltage is induced in the second conductor. When the second conductor is also coiled, the result is a device in which the magnetic field of the current passing through the first or *primary* coil produces a voltage in the second or *secondary* coil.

By varying the number of turns in the two coils relative to each other, the voltage induced in the secondary coil can be made either greater or smaller than that passing through the *primary* coil. As an example, if the primary coil has 100 turns and the secondary has only 10 turns, the voltage in the secondary will be 1/10 of the primary. If the primary is plugged into a 120-volt line, the secondary will produce 12 volts. The initial 120 volts has been transformed to 12 volts, and this pair of coils working together is called a *transformer*. See Figure 2-15.

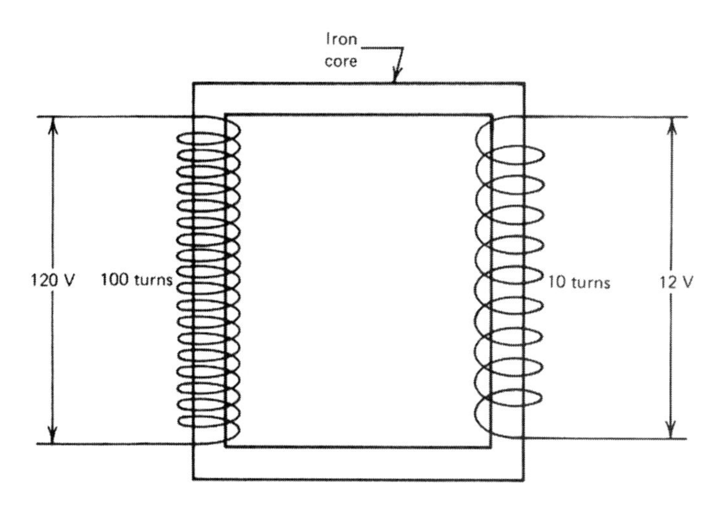

Figure 2-15
120 volt to 12 volt step-down transformer

A transformer that changes 120-volt current to 12 volts is called a *step-down* transformer because the voltage is being reduced. A transformer that does the reverse is called a *step-up* transformer. A step-up transformer has more turns in the secondary coil than in the primary, and transforms the voltage up in proportion to the ratio of turns in the two coils. Take a primary coil with the same 100 turns, but this time put 200 turns in the secondary coil. For example, when the primary is connected to the same 120-line voltage, the secondary will yield 240 volts. The transformer's efficiency improves considerably when both primary and secondary coils are wound around an iron core; look again at Figure 2-15.

The fact that AC can be easily transformed up or down in voltage is one of its major advantages. It can be generated, then transmitted over long distances, and finally put to various uses. The voltage changes at each step to one that's optimum for the immediate purpose. We'll examine more closely the what and why of some of these voltage changes in connection with commercial power generation and distribution. First, it'll be useful to investigate a few other aspects of AC behavior.

Inductive Reactance

As we've seen, when a coil is cut by a magnetic field, however that field is produced, voltage is induced in the coil. It could also be produced by a permanent magnet (look back to Figure 2-8), or it could be produced by an electrical current originating at some external source, and flowing through the coil. When the current flowing through a coil starts to build up a magnetic field around that coil, the field, as it builds up, cuts across the turns of its own coil and induces a second voltage.

Such an induced voltage, and the current that flows because of it, are always due to a polarity opposing any change in the existing magnetic field. As the source current increases, causing the magnetic field around the coil to expand, the induced voltage builds an increasing field in opposition to that of the source current. When the source current starts to decrease, causing its field to collapse, the induced current reverses to support that field. Because it acts in opposition to the source, the induced voltage is termed *Counter Voltage* or *Counter Electromotive Force* (CEMF).

Obviously, with steady DC this induced counter voltage is purely a momentary phenomenon occurring only when the current is turned on and again when it's shut off. As I mentioned earlier, an induced voltage is produced in a conductor that's cutting magnetic lines of force. The conductor may be moving in a stationary magnetic field (look again at Figure 2-8), or the conductor may be stationary in a moving magnetic field, or with a transformer the magnetic field itself may be expanding and collapsing relative to a stationary conductor. The critical factor for induction is that magnetic lines of force must move relative to the conductor in which a voltage is induced. In the case of the counter voltage produced by an alternating current passing through a coil, it's the alternately expanding and collapsing field of the coil itself that induces its own opposing voltage.

The property of opposition to change in current flow, and consequent change in magnetic field intensity, is called *inductance*. A component that produces inductance, such as a coil, is called an *inductor*. The amount of inductance produced depends on the strength of the magnetic field of the coil, the number of turns in the coil, and the frequency with which the field changes and cuts across these turns.

In a DC circuit the only opposition to the flow of current is resistance. That resistance is made up of the individual resistances of the various components of the circuit. In an AC circuit, the presence of an inductor in the circuit causes the appearance of a counter voltage that further opposes the flow of current. When a force, in addition to resistance, opposes the flow of a current, the total combined opposition is called *impedance*. In electrical formulas, the symbol for impedance is *Z*. Since Z, like R, opposes the flow of electrical current, it too is measured in ohms and may be substituted for R in the Ohm's Law formulas as follows:

$$E = I \times Z \qquad I = \frac{E}{Z} \qquad Z = \frac{E}{I}$$

Impedance exceeds resistance because of a factor called *reactance*. Reactance is the counter-electromotive force, or counter voltage, that only appears in AC circuits because of the continuously fluctuating source voltage. In the case of an inductor, such as the coils we've been discussing, the reactance (symbol X) is identified as *inductive reactance* (symbol X_L).

Inductance and Phase Relationships

In an AC circuit containing only resistive loads, voltage and current remain constantly in phase, as shown in Figure 2-16. Both voltage and current rise together, fall back to 0, reverse to maximum, and return

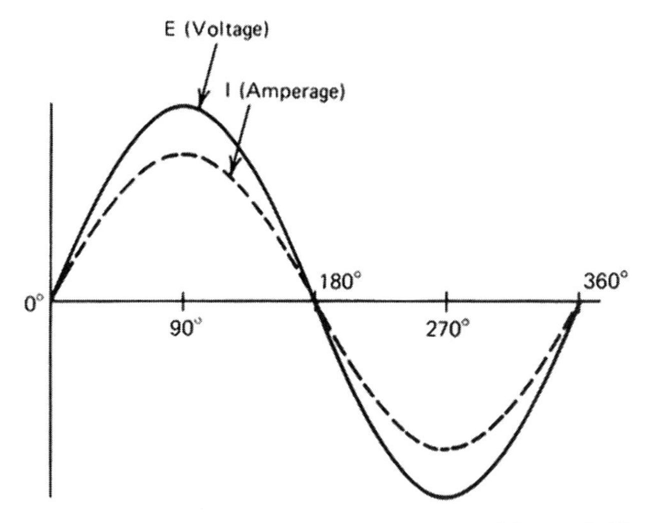

Figure 2-16
Phase relations of voltage and current in a resistive loaded circuit

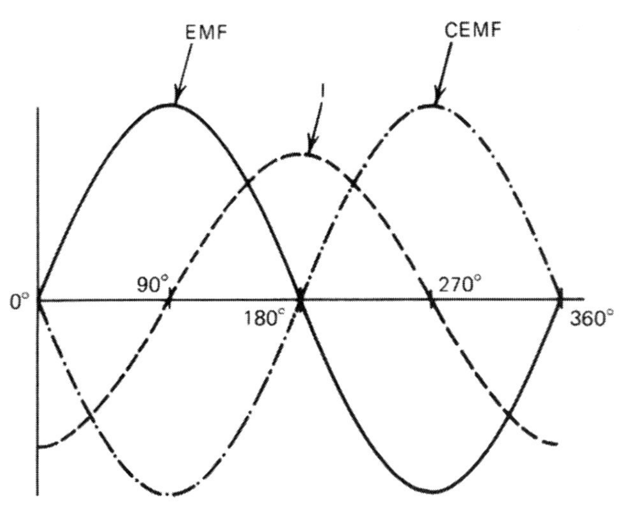

Figure 2-17
Voltage and current phase relationships — purely inductive circuit

to 0 again — still together. The relative amplitudes of the voltage and amperage curves as illustrated are meaningless. They differ only for clarity of illustration.

With a circuit containing induction, the induced counter voltage always opposes change in the source current, so that as the current is rising, the CEMF tends to hold its value below what would be attained by the source voltage alone. Then, as the current decreases, the CEMF again opposes the decrease. The result is that both the rise and fall of the current lag behind the rise and fall of the source voltage through the entire cycle. If the circuit contained only pure inductance, the phase relationships between voltage, counter voltage, and current would appear as in Figure 2-17.

The induced voltage is greatest when the current is *changing* most rapidly. This is when the current passes through 0 at 90 degrees and 270 degrees. Since the source voltage is opposite to the induced voltage, it too peaks at the same time, but in the opposite direction. The induced voltage drops to 0 when the current is changing most slowly, which is when it peaks at 0 degrees and 180 degrees. Since the source voltage is opposite to the induced counter voltage, it too drops to 0 at the same times. Thus, in a hypothetical circuit that contains only inductance, the current would lag behind the voltage by 90 degrees. In practice, it's impossible to build a circuit containing only inductance and no resistance whatsoever. So, in practice, due to the inevitable presence of resistance to a greater or lesser degree, the 90 degrees phase lag between current and voltage never does occur.

Capacitive Reactance

The capacitor is another component that acts much differently in an AC circuit than it does in the presence of DC. A capacitor is a device that can store an electrical charge. The size of the charge it can store is the measure of its *capacitance*. It's an extremely simple construc-

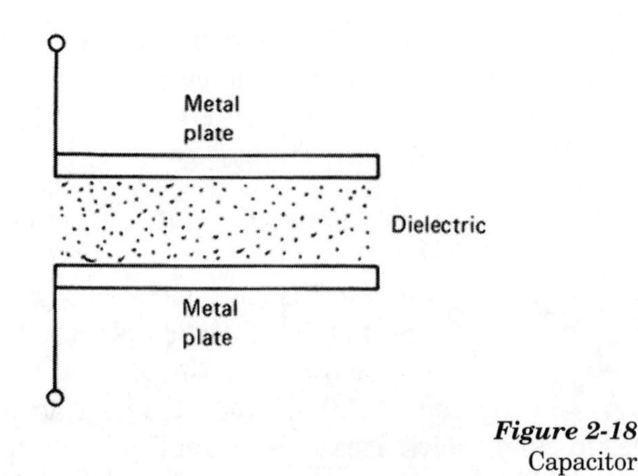

Figure 2-18
Capacitor

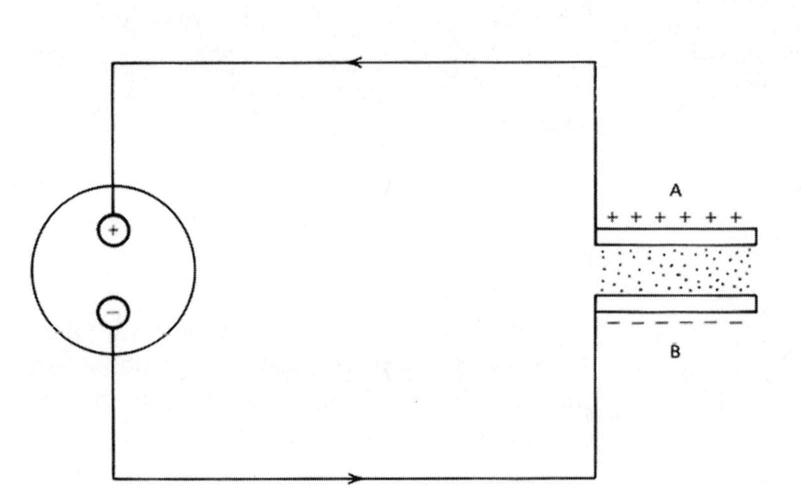

Figure 2-19
Charging of capacitor

tion consisting of two plates of conducting material separated by a layer of dielectric or nonconducting material. See Figure 2-18. The dielectric can be paper, mica, ceramic, even air. Simply placing a dielectric between two plates creates the capability for storing an electrical charge. The only other requirement is a source of voltage. A battery will do quite well. When connected to DC, the capacitor is charged by the movement of electrons from plate A to the positive terminal of the battery, and the movement of other electrons from the negative terminal of the battery to plate B. See the diagram in Figure 2-19. When the potential (voltage) across plates A and B is both equal and opposite to that of the battery, the capacitor is fully charged, or as fully charged as it can be from that power source. A charged capacitor will hold its charge for varying lengths of time, but it can be discharged at will for various purposes, such as in an automotive ignition system.

When the DC power source is replaced by AC, the picture changes considerably. As the AC voltage increases from 0 to its maximum value, current flows from the source to one side of the capacitor, building up a counter voltage in the capacitor just as the battery did. But as the source voltage moves on to its decreasing curve, the counter voltage drives a current back from the capacitor to the source. The source voltage now reverses, rising to its opposite maximum. At this point, the current driven by the counter voltage in the capacitor and the current driven by the source are both moving in the same direction — which is toward the other side of the capacitor. A new counter voltage now builds up on the opposite side of the capacitor. As the source voltage fades from its second maximum, the counter voltage in the capacitor drives a current back from the capacitor to the source to start the cycle all over again.

During this entire cycle, current has been flowing throughout the circuit, except through the dielectric of the capacitor. If a device, such as a light bulb, had been connected to this circuit, it would light, indicating a steady flow of current in the circuit. For this reason, although nothing actually passes through the dielectric layer of a capacitor, it's said that AC can flow through a series circuit containing a capacitor.

The plates of a capacitor will readily store a charge; thus, current can flow onto the plate of an uncharged capacitor essentially unopposed. With no resistance initially, there's no voltage drop, which means current will lead voltage in a capacitive circuit. In the inductive circuit, the inductor opposed changes in current. In a capacitive circuit, the capacitor opposes changes in voltage. This opposition is called *capacitive reactance* (symbol X_C). Like inductive reactance, it's measured in ohms.

Capacitance and Phase Relationships

In a circuit containing pure capacitance, as in Figure 2-20, current will flow from the source toward the capacitor only when the source voltage is rising. The maximum current will flow when the voltage is rising most rapidly. This will be at 0 degrees and 180 degrees. At 0 degrees (with the voltage crossing 0 volts) the current flow is at its maximum positive value. From 0 degrees to 90 degrees, the voltage continues to rise, but at a rate that steadily slows. As the voltage has been rising, the current has been dropping. At 90 degrees, the voltage has peaked, the capacitor is fully charged, the counter voltage has also peaked, and the current has reached 0 amperes.

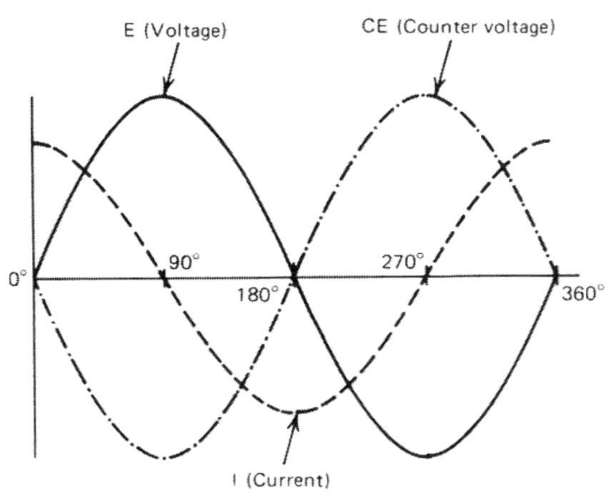

Figure 2-20
Voltage and current phase relationships — purely capacitive circuit

Source voltage now starts to fall; as it does, current starts to flow from capacitor to source, and counter voltage decreases as well. At 180 degrees, the source voltage has reached 0 volts, current has reached maximum negative value, and with the capacitor now discharged, the counter voltage has reached 0 volts.

Source voltage now reverses and steadily increases, but at a diminishing rate. Meanwhile, current begins to decrease from its maximum negative value toward 0 amperes while the capacitor is becoming recharged in the opposite direction from its first charge. At 270 degrees, the source voltage is again at maximum, and current again at 0 amperes. As the source voltage falls, the counter voltage drives the current back up to start the cycle again at the 360-degree mark.

The effect of capacitance is to cause the current to lead the voltage, an effect opposite to that of reactance. In this hypothetical circuit containing only capacitance, the current leads the voltage by 90 degrees. For a circuit to contain only pure capacitance is just as impossible as for one to contain only inductance. Either one is necessarily combined with resistance, thus the phase difference won't ever be a full 90 degrees.

Power in an AC Circuit

As we saw earlier in Watt's Law, the power consumed by a circuit is the product of the voltage times the amperage, as in this equation:

$$P = E \times I$$

This law holds true as long as voltage and amperage are in phase. The relationship between voltage, current, and power are graphed in Figure 2-21 for such a situation. Where both voltage and current are positive, power is positive as well. During the second half of the cycle, when both voltage and current are negative, power is still positive — since the product of two negative numbers is a positive one.

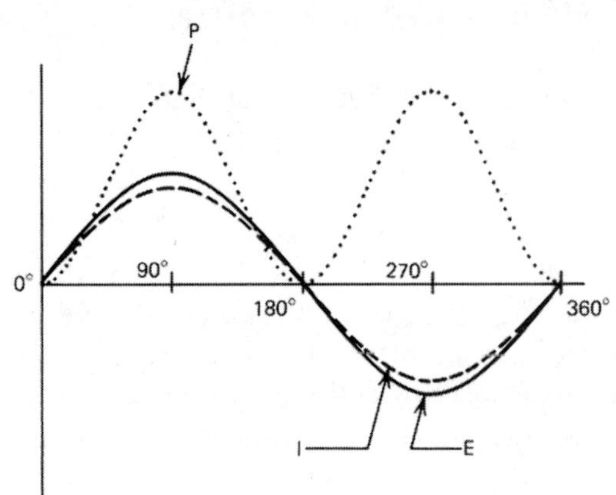

Figure 2-21
Power, current and voltage graph with current and voltage in phase

In a purely resistive circuit, voltage and current will be exactly in phase as shown. However, as we've just seen in a circuit containing inductance, current will lag behind voltage, and in a circuit containing capacitance the reverse occurs, with current leading voltage. The phase differences in such circuits have a significant effect on the power those circuits consume.

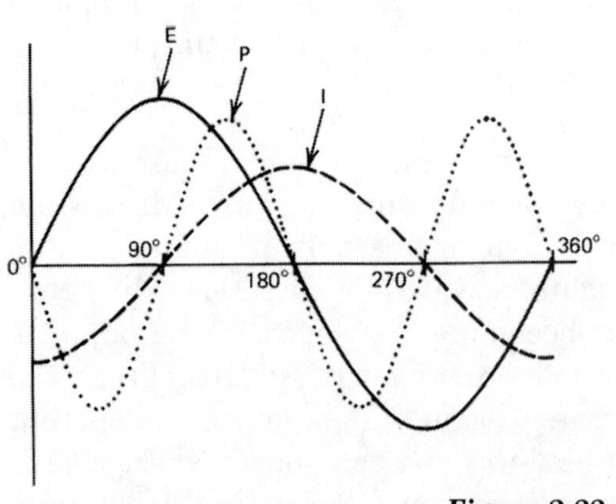

Figure 2-22
Power, current and voltage graph with E and I 90 degrees out of phase

To demonstrate, let's look at the purely inductive circuit with current lagging behind voltage by 90 degrees and examine its power graph in Figure 2-22. For the first 90 degrees of the cycle, although the voltage is positive, the current is negative. The multiplication of a positive and a negative gives a negative result, showing that power is flowing from the circuit back to the source. During the second 90 degrees, both voltage and current are positive; thus, the multiplication of the two produces a positive result. During the third 90 degrees, current is positive, but voltage is negative. Again, positive times negative results in a negative product. During the final 90 degrees of the cycle, both current and voltage are negative, giving a positive result. For two quarters of

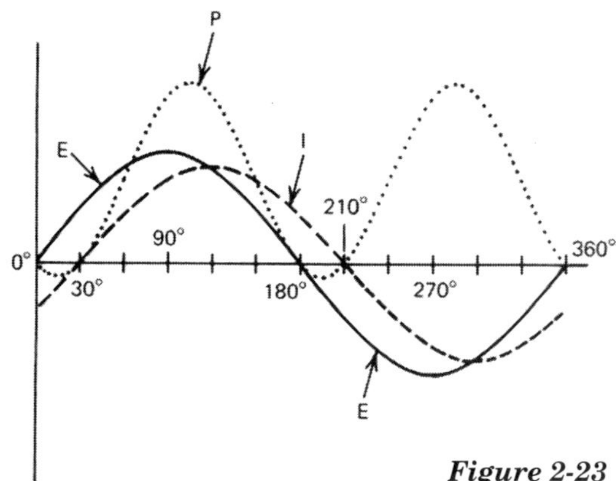

Figure 2-23
Power, current and voltage graph with I lagging E by 30 degrees

the cycle, power flowed from the circuit back to the source, and for the other two quarters, power flowed out from the source to the circuit. At the end, they cancel each other out, with the result that the circuit finally consumes no power. Power merely flows from the source out into the circuit, and back.

Of course, this pattern will only hold when current and voltage are 90 degrees out of phase with each other, that is, only in purely inductive or purely capacitive circuits. We've also noted that the hardware just doesn't exist that will allow construction of such completely pure circuits. It's impossible to eliminate resistance. Therefore, in the real world, current and voltage get out of phase in circuits containing inductance or capacitance, or even combinations of both, by much less than 90 degrees. As an example, let's examine what happens when forces combine to cause current to lag behind voltage by 30 degrees instead of 90 degrees. See the graph in Figure 2-23. Between 0 degrees and 30 degrees and between 180 degrees and 210 degrees, power is negative, flowing back to the source. However, between 30 degrees and 180 degrees, as well as between 210 degrees and 360 degrees, power is positive; thus, the net power consumed is by no means 0, but rather a considerable positive quantity. This positive quantity, however, is considerably lower than a comparable circuit in which current and voltage are perfectly in phase, due to the periods of negative flow. Refer back to the graph in Figure 2-21.

Power Factor

When current and voltage are out of phase, power is flowing back to the source rather than from it, during parts of the cycle. This means that the *true power* actually consumed is necessarily less than the *apparent power* mathematically derived by multiplying voltage by amperage. The ratio of true power to apparent power is called the *power factor*. The formula for finding the power factor is:

$$\text{Power factor} = \frac{\text{True Power}}{\text{Apparent Power}}$$

The power factor may be expressed as a fraction, a decimal, or a percentage. A power factor could be given as three quarters, 0.75 or 75 percent. At 1.0 or 100 percent, power factor, wattage, and volt-amperage are equal — the ideal condition.

The practical importance of the power factor is that it directly measures the efficiency with which the energy fed into a circuit is utilized. A system with a 75 percent power factor is getting 75 watts of useful power out of every 100 volt-amperes being delivered (and billed) by the utility company. The other 25 watts are used fighting counter voltage generated in the circuit by the power-using devices connected to it.

Apparent power is easily calculated by measuring voltage with the voltmeter, amperage with any of several types of ammeters, and then multiplying the values obtained. True power, however, can be determined only by measuring with a wattmeter. Since true power can only be found by wattmeter measurement, it's referred to in watts to distinguish it from apparent power, which is given in volt-amperes.

Commercial Generation and Distribution of AC Power

The AC power supplied by utility companies is generated continuously in very large three-phase alternators at a voltage of 13,800 volts alternating at 60 Hz. This 13,800 volts, 60 Hz power passes through a bank of step-up transformers that greatly increase the voltage for transmission purposes. The magnitude of the step-up in any particular instance depends on how much power is required and how far it must be transmitted. The step-up could be to as low a value as 23,000 volts, or could go as high as 765,000 volts. As the power requirement at the destination increases, or the distance to the destination increases, or both, the voltage increases.

The reason for stepping the voltage up to very high values for transmission purposes is that current causes power loss during transmission, not voltage. Since power (symbol *P*) is the product of voltage and amperage, then any given power value can be transmitted at very, very low amperage, as long as the voltage is made sufficiently high. Extremely high voltage levels can easily be obtained with step-up transformers.

Wherever power is to be drawn from the main transmission line for use in a local area, it passes through a substation containing banks of step-down transformers. Here, it's reduced in voltage to either 2,300 or 4,100 volts for distribution through that area. At this point, most of it is split into single phase. At the final delivery point, it passes through another step-down transformer. In residential and in much commercial use the power is cut to 120/240 volts and enters the building. This will be discussed in Chapter 9. Industrial and commercial uses requiring 440-volt supplies are beyond the scope of this book.

STUDY QUESTIONS

1. In which of the following forms can direct current (DC), produced by chemical interactions, be found?

A) Coiled conductor
B) Magnetic field
C) Storage battery
D) Generating plant

2. Which of the following describes what happens when the north poles of two bar magnets are brought together?

A) Iron molecules repel
B) They repel each other
C) They attract each other
D) Random patterns form

3. Moving a conductor in which of the following will generate current?

A) Clockwise circular motion
B) Counter-clockwise circular motion
C) Coil
D) Magnetic field

4. What is produced by rotating a conductor loop in a magnetic field?

A) Alternating current
B) Direct current
C) Static electricity
D) A frequency of 60 cycles per second, or 60 HZ

5. What is produced by rotating three coils on the same generator shaft?

A) Low-voltage transmission
B) Alternating current
C) Three-phase power
D) Three-wye output

6. What device is used to split three-phase power into single phases?

A) Coil
B) Wye device
C) Armature
D) Transformer

7. Which of the following describes the magnetic field produced by alternating current (AC) in a coiled conductor?

A) Varies in intensity, but is constant in polarity
B) Periodically reverses in polarity, but is constant in intensity
C) Constantly varies in intensity and periodically reverses in polarity
D) Maintains constant polarity and intensity

8. In a step-up transformer, what would the primary coil voltage have to be if the secondary coil voltage were 240 volts?

A) 60 volts
B) 120 volts
C) 480 volts
D) 600 volts

9. When a force, in addition to resistance, opposes the flow of a current, what is the total combined opposition called?

A) Inductive reactance
B) Inductance
C) Reactance
D) Impedance

10. Reactance is the counter-electromotive force that only appears in what kind of circuits?

A) AC
B) DC
C) AC or DC
D) Delta wye

3

TOOLS AND SAFETY

Many tools used by electricians are everyday and commonplace, and used by carpenters and plumbers as well. Because the tools are common, they're often taken for granted, and as a result frequently misused and improperly cared for. A craftsman in any field can't do first-class work with second- or third-class tools. You've undoubtedly heard the old saying: "a good craftsman doesn't blame his tools" for his difficulties. Unfortunately, that saying generally isn't properly completed. The reason a good craftsman doesn't blame his tools, is because one of the first things a *good* craftsman does is make sure he has *good* tools. He won't attempt to do a job with poor ones.

Conventional Hand Tools

Hammers are available in a wide variety of sizes, ranging from a light 7-ounce tack hammer to the 20-ounce framing hammer. For the electrician's purposes, a hammer no heavier than 13 ounces is adequate. The curved claw is generally more useful than the rip claw type for electrical work. Wood handles in hammers will eventually work loose, but you can easily tighten them by simply driving the steel wedges in deeper. If one gets too loose, you may need a new wedge. Keep the hammer face clean to promote straight driving and avoid marring finished surfaces.

An electrician needs several screwdrivers and three types of tips. The first tip is sometimes called an electrician's tip or a cabinet tip. Obviously, the name depends on who uses it. Regardless of what the supplier calls it, an electrician should have a #3 and a #6 screwdriver with this tip. A medium and a large standard tip plus a Phillips will be useful; so will a small stubby type. You may not think of a screwdriver as a tool that needs sharpening, but it is. Both flat and Phillips tips become worn and rounded with use. They no longer fit snugly into the screw heads. Worn tips will severely mar screw heads, often making it impossible to withdraw a properly-tightened screw. Flat screwdriver tips can easily be sharpened by filing, but once a Phillips tip is worn, discard it.

A couple of wood chisels kept usefully sharp are helpful additions to the electrician's tool kit. You can use a 1 inch and a 1½ inch to make notches for cable, flexible conduit (Greenfield), or conduit in wood framing. Don't allow your wood chisel to become dull — then it's worse than useless, it's dangerous. A chisel that's too dull to cut wood may bounce back out of the wood and cut you instead.

You can use a hacksaw to cut Greenfield and armored cable (BX), as well as pipe conduit when a standard pipe cutter isn't available. Hacksaw blades vary from 18 teeth per inch to 32 teeth per inch. For electrical work, the general rule is to stay with the finest blade available. A coarse blade will cut faster, but will bind badly in any type of tubing. The finer the blade the more smoothly it'll cut through thin materials.

A *keyhole saw* is the tool for cutting rectangular, or irregular, holes in walls, floors, or ceilings. Usually, two types of blades are available: a general-purpose blade with 11 teeth per inch, and a hacksaw blade with 32 teeth per inch. To cut an opening for a wall box or a recessed ceiling light, first mark the outline of the desired cutout on the wall or ceiling. Then, within that outline, drill a starting hole big enough to get the end of the saw through. Then, cut to the outline and continue around it. Use very *light pressure* on a keyhole saw. They bind and the blade bends easily. Don't be in a hurry. "Light pressure and patience" is the slogan for success with a keyhole saw. If you bend a blade, which is very likely, you can more or less straighten it by hand. But, a bent blade can never fully be straightened, and it'll never again make an exact cut.

Electricians primarily use an adjustable crescent wrench when installing or removing compression couplings on thinwall conduit. For ½- and ¾-inch conduit compression couplings and box connectors, a 6-inch crescent wrench is adequate. As the conduit size increases, so should the wrench size. Someone doing a lot of work with conduit and using a great many compression couplings should have two wrenches. Use one to hold the center of the coupling while tightening the ends with the other. A word of caution to those who sometimes forget — a crescent wrench *isn't* a hammer.

After cutting, electricians use the same reamer to ream conduit that plumbers use to ream galvanized pipe. See Figure 3-1. Reaming conduit after cutting is vital to prevent the burrs left by the pipe cutter

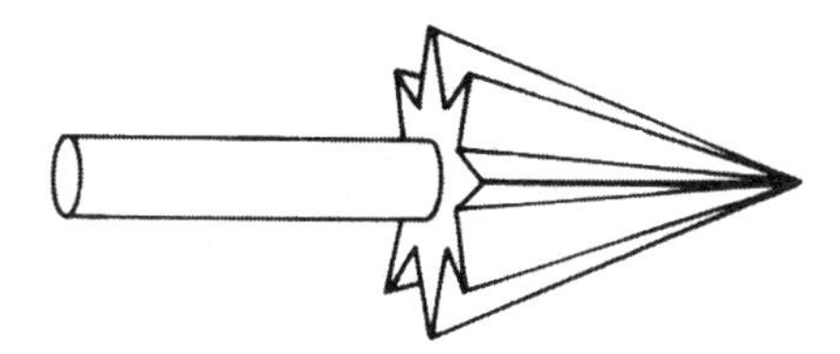

Figure 3-1
Pipe reamer

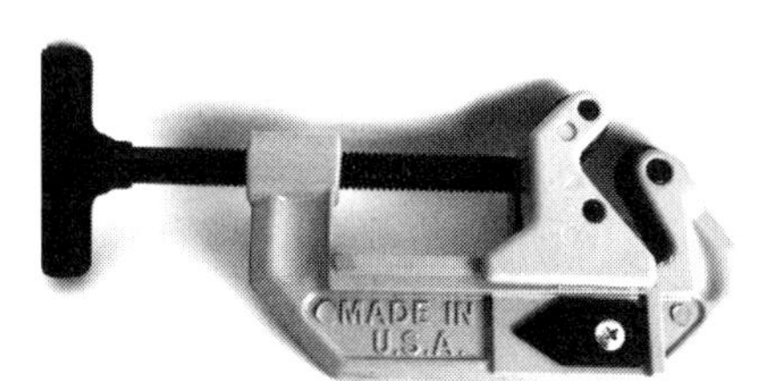

Figure 3-2
Pipe cutter

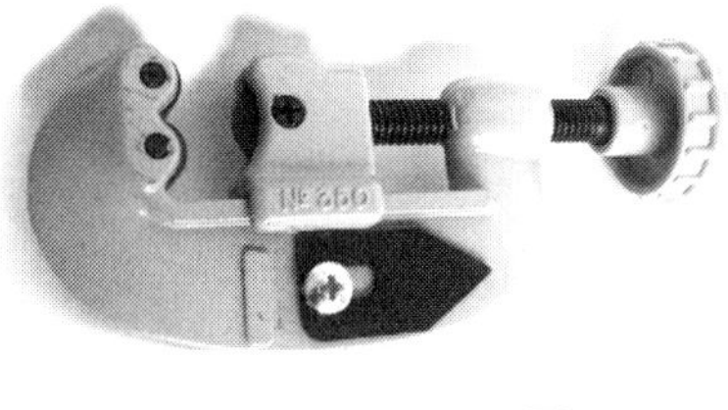

Figure 3-3
Tubing cutter

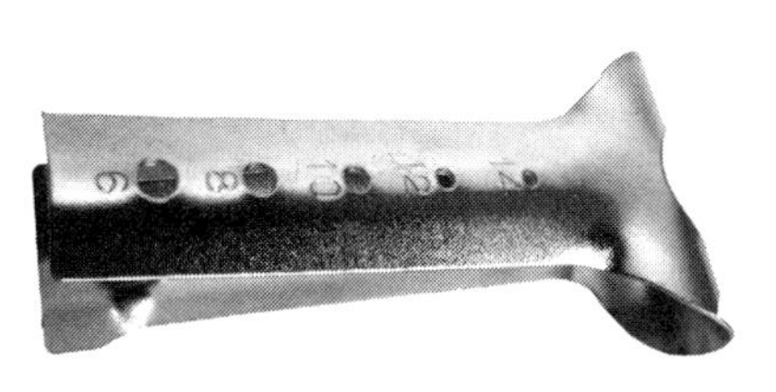

Figure 3-4
Cable stripper

or hacksaw from damaging the insulation on the wire as it's pulled through. We'll discuss this in more detail in Chapter 10.

Flexible conduit, liquidtight, and BX are normally cut with a hacksaw, which will also cut thinwall, rigid, or nonmetallic (PVC) conduits. But a plumber's pipe cutter will do a better job of cutting these. See Figure 3-2. The most important and vulnerable part of the pipe cutter is the cutter wheel itself. In normal use, they'll eventually become dull and need replacement. If abused or used improperly, the wheel may be nicked or warped, which will either produce ragged or crooked cuts, or cause it to stop cutting. If this happens, replace the cutter wheel, and be more careful.

The small pocket tubing cutter that plumbers use to cut small diameter copper pipe (shown in Figure 3-3), will also cut ½-inch steel thinwall conduit. But go easy on the cutter wheel. Feed it slowly and remember that it was designed to cut copper. You're using it to cut steel, and that's a very different material. Also, the small reamer on the side of the tubing cutter will satisfactorily ream copper, but it won't adequately ream ½-inch steel conduit. You'll need a full-size pipe reamer for that.

Electrician's Hand Tools

The majority of residential wiring requires nonmetallic (NM) sheathed cable. Look ahead to Chapter 5 and see the example in Figure 5-1. In order to make connections and splices, you must strip the inside insulated conductors of the exterior plastic protective sheathing. You do this with a *cable stripper*, like the one shown in Figure 3-4. Place the stripper over the cable with the sharpened cutter pressed through the sheathing. As you pull the stripper off the cable, the cutter makes a slit through the sheathing. Strippers are easier to use than a pocketknife, but they're generally poor in quality, and consequently, have a very short service life. They're cheap, so buy them by the dozen, and as soon as one gets dull throw it away and use another. Why no manufacturer has chosen to produce a decent quality cable stripper is a mystery to me.

After stripping the cable sheathing, you have to strip the insulation on the ends of the individual wires before any connections or splices can be made. Do this with a *wire*

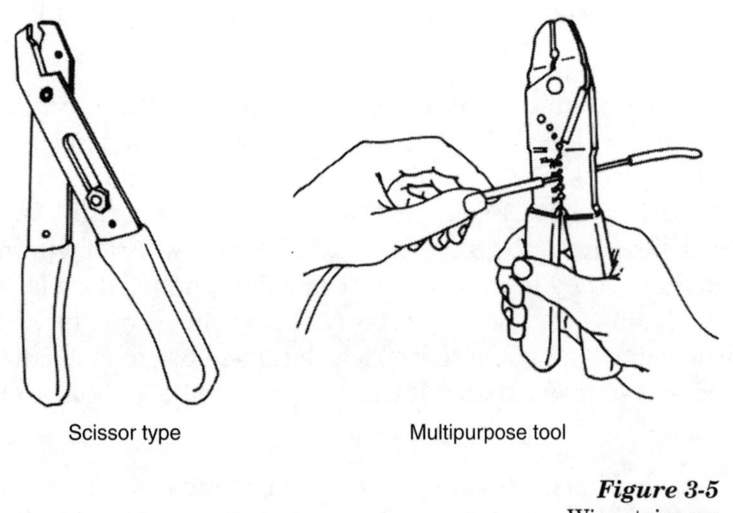

Figure 3-5
Wire strippers

stripper, like the examples in Figure 3-5. Fortunately, wire strippers are much better made than cable strippers. The scissor type is compact, light and will handle the bulk of the average electrician's stripping — ranging in size from #14 AWG to #10 AWG. This tool requires some practice. You have to learn to use just the right amount of pressure to cut through the plastic insulation, but stop before damaging and weakening the wire.

The *multipurpose tool* is a combination wire stripper, crimper and threading die, which will cut small machine screws to desired length and repair the end thread after the cut. I've found this tool inferior to the scissor type as a wire stripper, but I have to admit it has other useful capabilities.

You can use the *needlenose plier* as a wire cutter, but its primary purpose is to bend an end loop in a wire to go around a screw terminal. Like all pliers used for electrical work, be sure to buy a pair with insulated handles. You can trim individual wires with needlenose pliers or the scissor-type wire stripper, but to cut off an NM cable, it's much easier to use a substantial pair of diagonal cutters. Make sure this tool has insulated handles as well.

Cartridge fuses are among the overcurrent protective devices that we'll discuss in Chapter 9. These fuses are most easily inserted and removed with a *fuse puller*, shown in Figure 3-6. Fuse pullers are simply curved-jaw pliers made entirely of nonconductive material. They come in several sizes to match the diameters of the various cartridge fuse cylinders. In a pinch, you can use a pair of metal pliers with insulated handles, but a fuse puller is best for the safety of the fused equipment — and the electrician who's changing fuses.

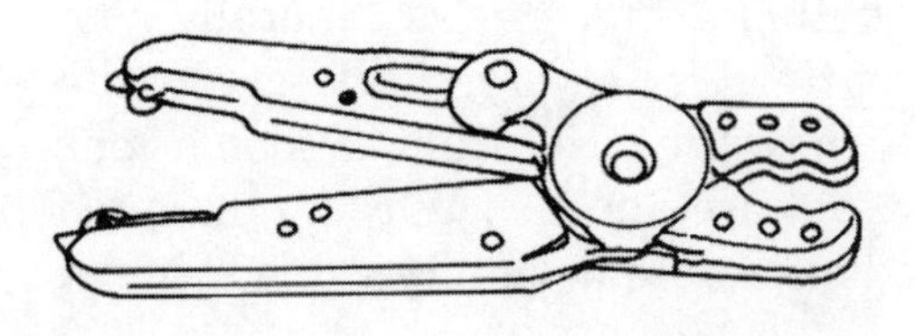

Figure 3-6
Fuse puller

We'll discuss thinwall conduit (among other materials) in Chapter 5, and again in connection with rough wiring in Chapter 10. The process of bending thinwall conduit requires a *conduit bender*. This tool, shown in Figure 3-7, can bend conduit to any shape or angle the job requires. It can also make saddle bends and offsets,

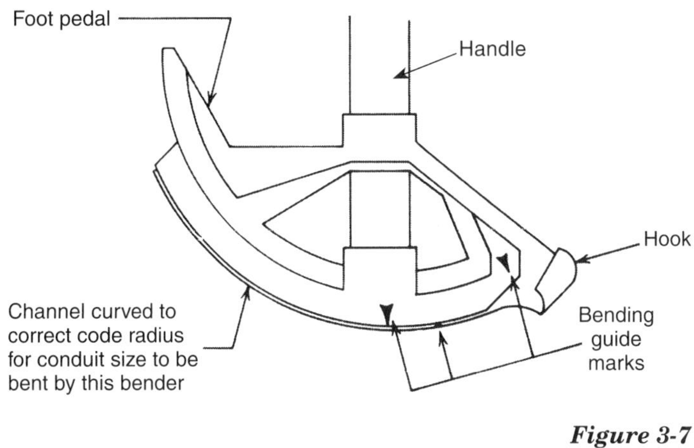

Figure 3-7
Principal parts of a conduit bender

as well as the various complicated shapes sometimes necessary to follow architectural irregularities.

Different conduit sizes require different benders. The conduit bender is a curved channel into which the conduit fits as it's bent. For each conduit size, the curve of the bender is the arc of a circle having the specified minimum radius for that size. You can bend a small conduit size with a bender intended for a larger size, but not the reverse. For example, ½-inch conduit can be bent with a ¾-inch bender. The bend will be to a 5-inch radius rather than the 4-inch minimum allowed for ½-inch conduit, but this will be acceptable to any inspection department. On the other hand, a ¾-inch conduit can't be bent with a ½-inch bender because it won't fit in the tool. And, it wouldn't be acceptable if it *did* fit, so it's just as well that the tool itself prevents such a mistake.

An experienced electrician working regularly with conduit can bend and fit it with amazing accuracy purely from practice and experience. An inexperienced electrician will find a bender such as the Benfield a great help. Look again at Figure 3-7. This manufacturer supplies benders furnished with various calibrating marks cast into the tool. They also provide an instruction manual detailing how to use their benders. By following the instructions carefully, a beginner can do satisfactory work almost immediately.

Figure 3-8
Fish tape

Electrical metallic tubing (thinwall conduit), rigid metal conduit, rigid nonmetallic conduit, flexible conduit (Greenfield), liquidtight flexible conduit, and surface raceways (Wiremold) will be described in detail in Chapter 5. They all have one characteristic in common; they start out as empty protective tubes through which wires are pulled after the protective "plumbing" is in place. Pulling the wiring through any of these materials calls for another specialized electrician's tool, the fish tape, shown in Figure 3-8.

A fish tape is a steel tape, about 1/8-inch wide and about 1/16-inch thick, wound on a reel. The length varies from 25 feet to 100 feet. On the end, the tape is turned back on itself to form a small loop. After bending and installing a run of conduit in the building, pass the fish tape through it from one box to another, attach the required wires to the loop at the end of the tape, and then pull the wires back through the conduit. This process can't be successfully done without a fish tape. We'll discuss pulling wires through conduit in detail in Chapter 10.

Power Tools

The power tool most frequently needed in electrical work is an electric drill. The usual ¼-inch drill turning at 1,750 rpm is satisfactory if there isn't a great deal of work required. It'll handle the drilling you need for modest repair and alteration jobs, but when there's an entire house or other small building to be wired, you should supplement it with a ⅜-inch slow speed and perhaps a ½-inch slow speed as well, depending on the size of the total job.

Most holes drilled for electrical wiring are moderately large in diameter, approximately ¾ to 1 inch, and some are larger. Let's consider for a moment what happens to drill bits as the diameter increases. A ¼-inch drill has a circumference of a little over ¾ inch (0.7854 inch). Turning at the standard 1,750 rpm, its rim speed is approximately 114.5 *feet per minute*. Drilling wood at that speed, the ¼-inch drill gets rather hot, but it can handle it. When the drill diameter is increased to ¾ inch, its circumference becomes more than 2¼ inches (2.3562 inches). The drill, now turning at 1,750 rpm, develops a rim speed of 343.66 feet per minute. At that speed, the tip of a drill bit gets extremely hot almost immediately. The high heat very quickly dulls the bit. When you're drilling a few dozen holes through 1½-inch thick studs, plus a lot more holes through 3 inches of partition top plates, a standard speed drill will go through drill bits almost faster than you can buy them.

That speed of 343.66 feet per minute was the rim speed for a ¾-inch drill. Most holes that an electrician drills are 1 inch. The 1-inch bit has a circumference of 3.14 inches, which at 1,750 rpm, means a rim speed of 458 feet per minute. A slow, variable speed drill with a top rate of 600 rpm brings the rim speeds of ¾- and 1-inch drills down to 117.75 and 157 feet per minute, respectively. At those speeds, the bit life increases to more acceptable levels. A fixed-speed 450-rpm drill motor is much better yet. Using the 1-inch bit, rim speed decreases to 117.75 inches per minute.

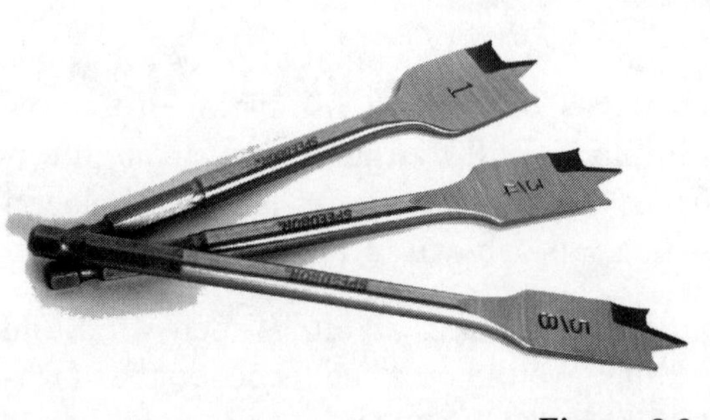

Figure 3-9
Spade bits

Bits used for drilling wood framing members are called either *spade bits* or *speed bits*, shown in Figure 3-9. Electricians should keep a fine metal

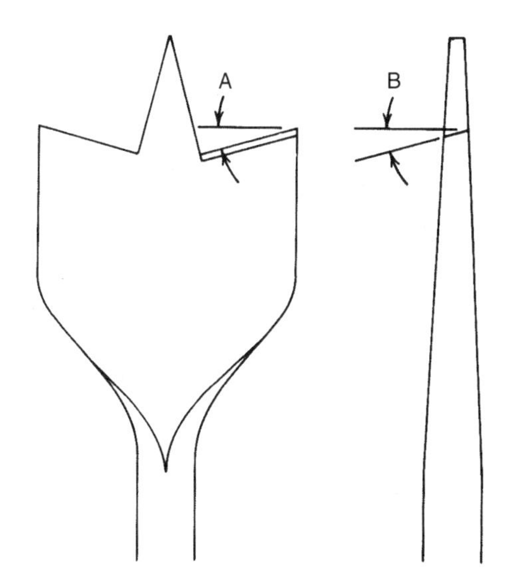

Figure 3-10
Sharpening angles for spade bits

Figure 3-11
Carbide tip masonry bits

Figure 3-12
Star drill

file for touching up the edges of the bits. A sharp drill bit produces clean holes quickly. In contrast, a dull bit will produce ragged holes and take an unreasonable amount of time to do so.

When sharpening spade bits, be careful to maintain the edges at the correct angle. Angle A (shown in Figure 3-10) should slope down from the outside tip toward the center of the drill. And angle B, the angle of the cutting edge (also shown in Figure 3-10), must also be correctly maintained. If either angle varies significantly from its original design, it will cut poorly or not at all. Don't wait for a drill bit to get dull before sharpening it. Touch it up while the correct angle is clearly apparent and it's still cutting well. That way, you'll never have to fight a dull drill.

In addition to making holes for wiring in the building framework, the electrician's other major use for the electric drill is to make holes in masonry. These holes are needed either to insert fasteners or to pass wiring through a masonry wall. In either case, a slow speed drill is an absolute must. Carbide tipped masonry bits, like those shown in Figure 3-11, must turn slowly or they'll overheat and dull in *seconds*. Turn them slowly while applying heavy pressure. They need plenty of pressure to cut well. Press hard for no more than 20 or 30 seconds at a time, and give them a few seconds to cool before leaning on them again. If you're drilling in concrete and dust stops coming out of the hole while you're drilling and pressing as hard as possible, you've hit a piece of rock aggregate. Stop, before you ruin the drill! Get a *star drill*, shown in Figure 3-12, and a hammer to break up whatever is in the way until the masonry bit will cut again.

The other power tool an electrician will find useful is a *saber saw*. Most rectangular- or irregular-shaped holes that are cut using a keyhole saw could be cut more rapidly and accurately with a saber saw. Keep in mind that the saber saw cuts primarily on the up-stroke; so when the preservation of a smooth top surface is important, use a fine-tooth blade. It won't cut as fast as a coarse tooth, but it'll leave a much smoother top surface. Padding the foot of the saw with masking tape will also help to avoid marring an expensive finished surface as the saw passes over it.

Test Instruments

Whether installing a new electrical system, adding to an existing one, or troubleshooting a malfunction, you'll need a few simple test instruments. A basic kit of three inexpensive instruments will handle most problems. These, and a fourth, more expensive one, will handle the average electrician's testing needs. The three basic test instruments are:

1. a test light
2. a multimeter
3. an outlet analyzer

The fourth, a clamp-on ammeter, isn't used as often, but is particularly useful in tracking down circuit loading problems, which we'll discuss later in Chapter 13.

Figure 3-13
Test light

The *test light*, shown in Figure 3-13, is an inexpensive, compact, lightweight, and easy-to-handle device for tracing power through a circuit, as well as testing switches, outlets, fuses or breakers. A word of caution — don't buy a small, cheap test light. Get the best you can find. It'll still be inexpensive and it's worth the cost difference.

A test light consists of two probes connected to each other by a wire, and a neon light placed between them. With one probe on ground and the other on a wire or terminal that should be hot, the test light illuminates, showing that the power is there. No light means no power. If the light illuminates with one probe on each end of a cartridge fuse, then the fuse is blown. If the light *doesn't* show, the fuse is good, providing it lights with one probe at one end of the fuse and the other on ground.

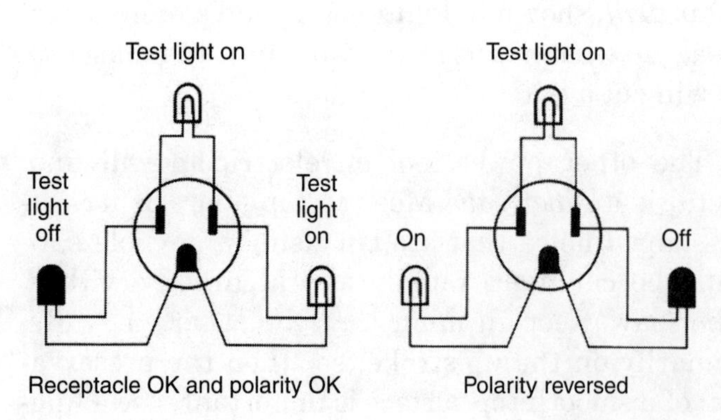

Figure 3-14
Test light used to test receptacle

To test a 120-volt power receptacle with a test light, first push the probes in the two slots. See Figure 3-14. The light should come on. With one probe in the short slot and one in the ground hole it should also light, but with a probe in the long slot and in the ground it shouldn't light. If it lights with a probe in the long slot and in the ground, but it doesn't light with one in the short slot and one in the ground, the polarity is reversed.

To check a single pole switch, put it in the *ON* position. Then, with one probe in the ground, place the other alternately in each of the two terminals. It should light both times. If it doesn't, the switch is bad. With a breaker in the *ON* position, place a probe on the breaker output screw and one on the ground buss in the breaker box, and the test light will go on if the breaker is OK.

The test light will show where there's power and where there isn't, but it doesn't indicate whether the voltage is correct. Most of the time, you'll be looking for 120 volts AC nominal. The 120 volts *nominal* means between 110 and 120 volts. Line voltage from the utility company occasionally rises above or drops below these limits, but only under unusual or emergency conditions.

The instrument you use to read the actual voltage is a *multimeter*; look back to Figure 1-6. As mentioned in Chapter 1, be sure to place the function selector switch on the AC side when looking for AC voltages, and on a scale higher than the voltage you expect to find. When looking for nominal 120-volt readings, use the 250-volt scale. When looking for 240 volts, start on the 500-volt scale, and then switch back to the 250-volt scale. This is to protect the meter movement in case the line voltage should happen to be running high for any reason.

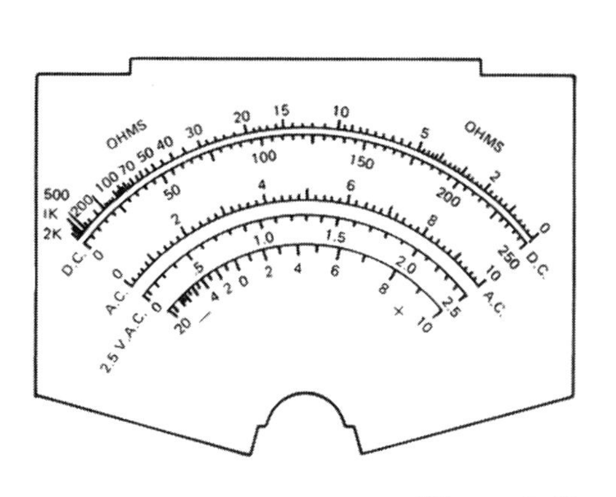

Figure 3-15
Multimeter dial plate

In addition to reading voltages, the multimeter will read resistance, or the lack of resistance, as the case may be. Consequently, it's used to check the continuity of wiring and to trace shorts. The part of the multimeter that reads resistance is called an ohmmeter. To use the ohmmeter, first switch the selector to one of the ohm scales — usually there are three: × 1, × 10, and × 100, meaning that the reading on the scale is to be taken as is (× 1), multiplied by 10 (× 10), or multiplied by 100 (× 100). Note also that the ohms scale (the one at the top of the dial), reads from right to left: 0 is to the right and infinity is to the left. See Figure 3-15. All the voltage scales read the other way, with 0 at the left and the high end at the right.

Remember to check that the meter is functioning properly before taking any readings. Do this by shorting the test probes to each other after turning the selector switch to the ohms scale. The meter should then read 0 ohms, with the needle on 0 at the right side of the upper scale. If it's close, but not right on, adjust it with the *ohms adjust* dial. If you can't adjust it to read 0, probably the battery is low and must be replaced.

You can't take resistance measurements while the power is on. If you try, you'll damage the meter. Use the voltage scales of the multimeter first to make certain the power is off before switching to the ohms scales for resistance readings. A reading of 0 on the meter means that there's no resistance between the two points in which the probes have been placed. For example, in any duplex receptacle box, if one probe is placed on the white wire or common and the other on the bare or ground wire, the meter reads 0 when the circuit is wired properly. With one probe on the black wire and the other on either white common or ground, the meter should stay on infinity (to the far left). This time a 0 reading means a short.

Multimeters range in cost from under $20.00 to around $1,200.00. For building wiring, a very small, very inexpensive one is adequate. The cheaper models are about 4 inches by 2½ inches by slightly over 1-inch thick. They have one great advantage over the big expensive models for our purposes: they're small enough to fit in a shirt pocket, leaving both hands free to climb or crawl into inconvenient places.

The plug-in *outlet analyzer* quickly and easily indicates whether an outlet has been wired properly. See Figure 3-16. When plugged into a live outlet, the way it lights up or doesn't light up will show whether the polarity is correct, and will immediately reveal an open ground or an open common. There are a couple of different versions of this tester, but they all have instructions on them showing how they should be read. Since these units are very inexpensive, cost is no reason for being without one.

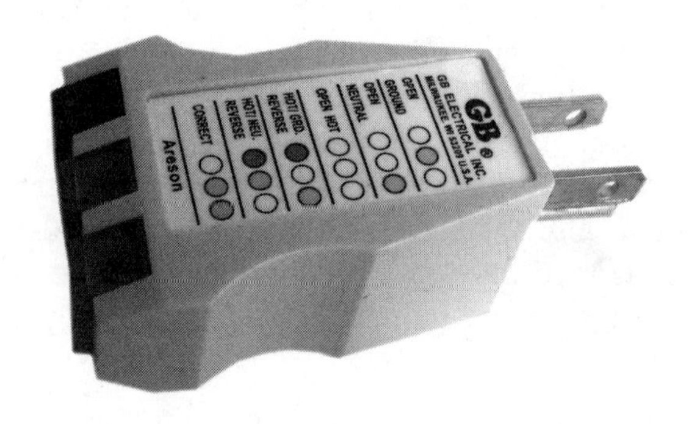

Figure 3-16
Plug-in outlet analyzer

The fourth test instrument is a *clamp-on ammeter*, shown in Figure 3-17. This instrument reads the amperage flowing in a circuit without interfering with the flow. Only the hot wire is placed between the jaws of the meter. The magnetic field produced by the current passing through that wire is translated by the meter into amperage and shown on the dial. Different manufacturers of similar instruments have different dial displays, but all operate by reading the magnetic field created by a current passing through a single conductor. Most clamp-on ammeters are equipped with test probes and can also be used as AC voltmeters. Some have an additional attachment allowing them to be used as ohmmeters.

Figure 3-17
Clamp-on ammeter

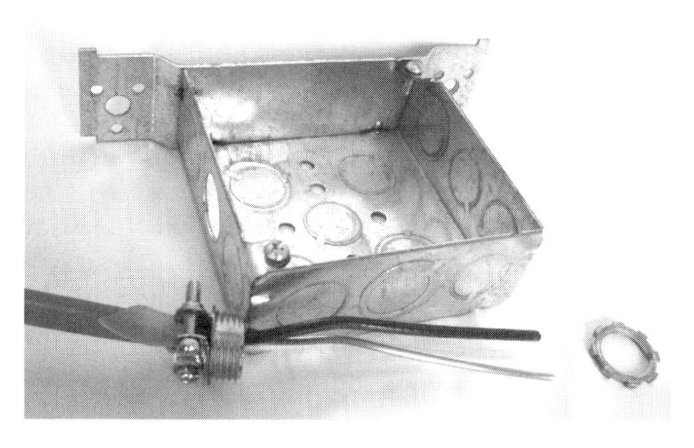

NM cable correctly clamped in connector

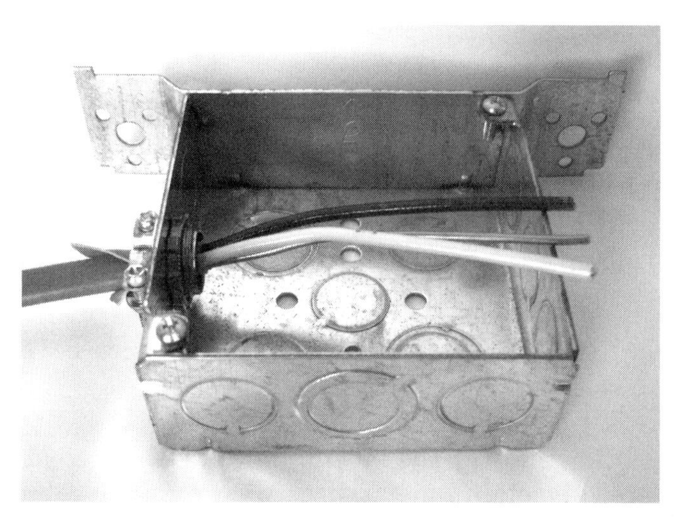

Connector in box

Figure 3-18
Mounting cable in electrical box

Basic Wiring Techniques

Electrical wiring involves the use of some tools that may be unfamiliar, and some techniques and procedures requiring the use of those unfamiliar tools.

To begin, use the cable stripper, shown earlier in Figure 3-4, to slit the sheathing on the ends of NM cable. The sheathing should be slit for a distance of 6 to 8 inches from the end, and then the slit sheathing cut away, along with any paper or fiber filler inside the cable, leaving only the insulated conductors and the bare ground wire. After cutting away the sheathing, observe the insulated conductors carefully to make certain the cable stripper didn't cut through the insulation on the insulated conductors. If it did, you'll have to tape the damaged areas, or cut all the wires off and start over again. When the cable end has been stripped of sheathing, mount it in the electrical box so that the sheathing terminates at the box clamp. See Figure 3-18. The wires inside the box are free of cable sheathing.

Now that the cable has been stripped of sheathing, the ends of the individual conductors must be stripped of insulation in order to make connections. Using the wire stripper (refer back to Figure 3-5), strip the insulated conductors back ½ to ⅝ inch from their ends. A common mistake made by beginners is stripping the cable sheathing too short, and the wire insulation too long. Stripping the insulation too long is inviting trouble. Strip an insulated conductor no more than absolutely necessary to make the required connection, as described below.

One commonly-required connection is the splicing of two or more wires to each other. Make splices in the building power wiring with solderless connectors. Most of your splices will be done with small size wires (#14 up to #10) and will be secured with wire nuts. To make a splice, strip the insulation as noted above, place the wire ends to be spliced inside the wire nut, and twist the wire nut clockwise until the splice is tight and firm. When the splice is completed, no uninsulated wire should extend beyond the end of the wire nut. There are several sizes of wire nuts; their size increases as the number, size, or both, of the wires to be spliced increase.

Figure 3-19
Split bolt connector

When the number of wires or the size of the wires to be spliced has increased beyond the capacity of the largest available wire nut, you must then secure the splices with a split bolt connector. See Figure 3-19. These come in a variety of sizes to handle the larger wire diameters. Note that the wire nut, having an outside shell of insulating material, needs no further insulation. So a splice made with a wire nut is already insulated, but one made with a split bolt connector isn't. A split bolt connector splice must be thoroughly taped with an approved electrical tape. Medical adhesive tape or masking tape will *not* do.

Ultimately, wires must terminate at some device or another, whether it's a switch, a receptacle, a fixture, or at the supply end in the breaker box. Where the wire ends on a terminal screw, it should be wrapped around the screw clockwise in the same direction the screw will turn when being tightened. See Figure 3-20. Unless the screw terminal is clearly marked to allow it, no more than one wire should be attached to a single screw. When two or more wires require a common attachment, splice them together in a wire nut along with a *pigtail*, which will then be used to make the common attachment to the screw terminal. See Figure 3-21.

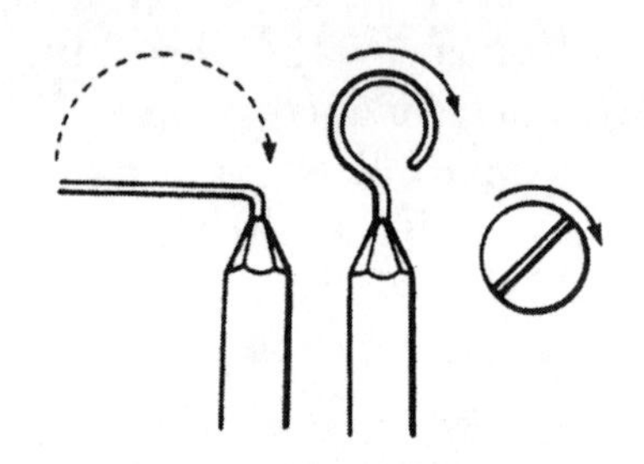

Insulation close to terminal screw

Figure 3-20
Wire properly connected to screw terminal

Another means for making connections to switches and receptacles is the *push-in terminal*, shown in Figure 3-22. Here the wire is simply pushed into a hole in the back of the device, where it's secured by an internal spring-clamp type contact. My enthusiasm for the push-in terminal isn't great, because the area of metal-to-metal contact is very slight. The screw terminal provides considerably more metal-to-metal contact between the wire and the terminal than is possible with the push-in. The primary advantage of using the push-in is the savings in time and labor cost. However, many connection failures and burnouts have occurred at push-in terminals that wouldn't have happened at screw terminals.

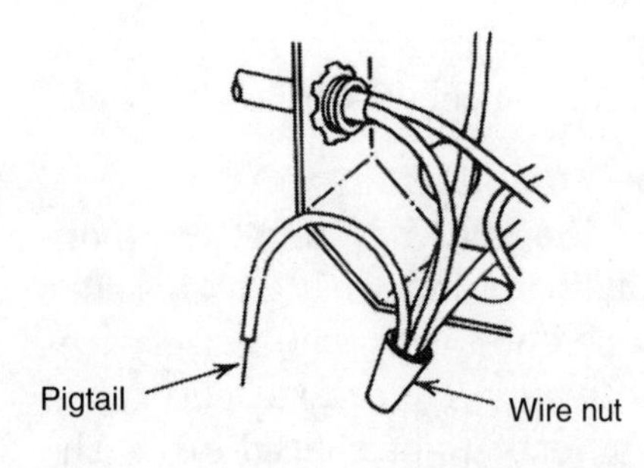

Figure 3-21
Splice with pigtail for terminal attachment

Safety

The primary safety concern in doing electrical work is avoiding electrical shock. Many people who've experienced mild electrical shocks from time to time, or who like to boast of their machismo, have a tendency to view shocks rather lightly. This view isn't very wise.

The smallest circuits used in building wiring have a current-carrying capacity of 15 amperes at 120 volts. The amount of current that will cause serious physical damage varies in

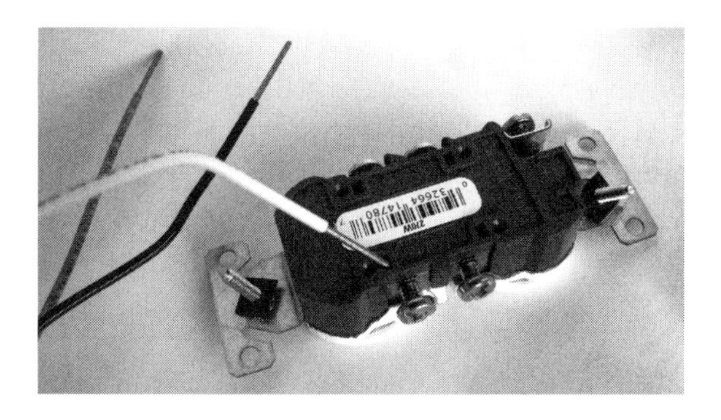

Figure 3-22
White wire in push-in terminal

each individual case, depending on age, general physical condition, and the circumstances under which the body makes contact with the electrical supply. Generally speaking, a current of 0.01 amperes will be unpleasant enough to get a person's attention. The current available in the *smallest* circuit in a building is *1,500 times that amount*. Less than 0.1 ampere is quite likely to be fatal. In other words, the smallest circuit is capable of killing the average person 150 times over. While it shouldn't be feared, this fact should be sufficiently sobering to produce a wholehearted respect for electricity.

Fatalities due to electrical shock result from stoppage of breathing, or interruption of heart function, or both, caused by paralysis of the muscles. The severity of a shock will depend on the resistance of the body to the passage of current, the voltage pushing that current, the path of the current through the body, and the length of time the current flows.

A person's resistance will be the least when they're wet, and contact is made to bare skin. Therefore, stay dry when working on electrical apparatus. Use tools with insulated handles. Wear gloves when working around live conductors. Wear thick, preferably rubber-soled, shoes and don't stand in damp areas to work on live circuits.

Since the major danger from electrical shock is stoppage of the heart or breathing, the most dangerous path for an electrical current through the body is up one arm, across the chest (through the lungs and heart), and down the other arm. So as you're working, be sure that if one hand can touch a live conductor, there's no way the other hand can contact ground — put it in your pocket or put a glove on. A path up one arm, down through the torso, and out one leg or the other isn't as bad as arm to arm; but since it goes through so many vital organs, it's still dangerous.

The next best thing to avoiding an accident is to know what to do in case one occurs — and do it quickly. The most common result of electrical shock is paralysis of the voluntary muscles. For example, a person touches a live conductor, becomes paralyzed, and can't let go. If this happens to a colleague, shut off the power immediately. If you can't, then separate the victim from the live wire or terminal as quickly as possible. *But don't touch him with your bare hands in the process.* If you do, the current passing through him might go through you and paralyze you as well. Throw a dry piece of rope over him and pull him off, or throw a dry shirt or sweater over him and pull him off with the sleeves. If he has a heavy jacket on, pull him off by the jacket.

Once the victim is safely loose from the power supply, immediately check to see if he's still breathing and if his heart is still functioning. If not, administer CPR at once. If you don't know how to give CPR, you need to get certified. The American Heart Association, Red Cross, and many hospitals give CPR lessons to anyone who wants to take them. The lessons take approximately six hours. Some places charge a nominal fee for the course, but in many places, it's free.

It's vital you know CPR in a case of heart failure or lung dysfunction. A few seconds' delay in administering it may mean the difference between life and death. Make sure that someone on every jobsite is trained in this life-saving technique.

Electrical shock can easily be avoided by following a few simple safety rules:

1. Shut off the power before working on wiring.
2. Verify the power is off by testing it with a voltmeter or a test light.
3. Leave a note on the fuse or breaker box warning others to keep the circuit off.
4. Use only screwdrivers, wrenches or pliers with insulated handles.
5. Stand on dry material, preferably insulating material, like wood, concrete, etc.
6. Replace worn or damaged cables on electrical power tools.
7. Short any capacitor to ground before touching terminals. Do this by running a piece of wire from a terminal on the capacitor to ground — but be sure you have gloves on!

In addition to electrical shock, the electrician is exposed to all the other normal hazards around a construction jobsite, and should observe the same safety precautions as everyone else. An electrician can fall or have things fall on him just like anyone else. So, be alert, be careful, and be safe. To modify an old saying among aircraft pilots, "There are old electricians and there are bold electricians, but there are very few old, bold electricians."

STUDY QUESTIONS

1. What hand tools are recommended for cutting an opening for a wall box or recessed ceiling light?

A) Hammer and chisel
B) Drill and hacksaw
C) Keyhole saw and drill
D) Box cutter and hammer

2. How is a fuse puller different from most other hand tools in an electrician's toolbox?

A) It has no insulation on the handles
B) It's usually poorly made, and must be frequently replaced
C) It must be kept sharp, or the jaws may slip off the fuse
D) There's no tool just for pulling fuses; electricians use pliers with insulated handles

3. What's the best method for using a carbide-tipped masonry drill?

A) Use short bursts with a high-speed drill, applying very little pressure
B) Use a slow-speed drill in short bursts, applying heavy pressure
C) Apply steady pressure with the drill at alternately high, then low speed to allow cooling
D) Drill a pilot hole first using a star drill

4. Which instrument will indicate that polarity in a receptacle is reversed?

A) Ammeter
B) Multimeter
C) Outlet analyzer
D) Test light

5. How would you use a test light to check a single-pole switch?

A) With the switch *ON*, put one probe in the ground, and the other alternately in each of the two terminals
B) With the breaker and single-pole switch *ON*, put the probes on both terminals

C) With the switch *ON*, put one probe in the ground, and the other on the breaker output screw
D) Detach the switch from the wall to access the wiring *before* it reaches the switch, then connect the probes to the bare wires at the connections

6. Using a standard multimeter to read the actual AC voltage, where would you place the function selector switch?

A) At any point on the AC side
B) Start at the low end of the DC scale, then switch to AC to protect the meter
C) At the top of the dial, at 0 volts
D) On the AC side, on a scale higher than the voltage you assume

7. What would you be checking on a dial-type multimeter if you're reading a scale from right to left?

A) Resistance
B) Voltage
C) Amperage
D) Microamps

8. What is the proper technique for stripping cable sheathing and wire insulation?

A) Cut away the cable sheathing but leave the fiber filler attached
B) Strip up to 1 to 2 inches of each individual wire insulation
C) Strip NM cable sheathing 6 to 8 inches before attaching to box
D) Strip only ¼ inch of sheathing from NM cable through to the insulation

9. In splicing wires, what connector must you use when the size or number of wires won't fit into the largest wire nut?

A) A push-in connector
B) A split bolt connector
C A screw terminal with a pigtail connector
D A clamp-on connector

10. Which of the following will *not* help to avoid a lethal electrical shock?

A) Keep one hand in your pocket
B) Run a wire to ground from a terminal on any capacitor
C) On circuits carrying current greater than 0.01 amps, disconnect the ground wire
D) Put a note on the fuse box to warn others not to turn the power on

4

CONDUCTORS

While *conductors* might seem like an overly-technical term for *wires*, it isn't actually. A wire is bare metal and a conductor is metal *with* or *without* a covering of some kind. The meanings of these two words *aren't* the same. *NEC* Article 310 deals with conductors. Section 2(A) of that article states clearly that "conductors shall be insulated." This is immediately followed by the confusing statement "Exception: Where covered or bare conductors are specifically permitted elsewhere in this Code."

They have said, then, that a conductor is insulated except when it's covered or when it's bare. In other words, a conductor is insulated except when it's not! Obviously, this needs some explanation.

According to the code, to be considered "insulated," a wire must be fully encased in a material that has been tested and found acceptably resistant to the passage of electrical current at the thicknesses specified for that material in *NEC* Table 310.13(A), shown here in Figure 4-1. In contrast, a covered conductor is a wire sheathed so that either the material used, or the thickness of that material, or both, don't result in a coating of sufficient resistance to the passage of electricity to be considered an electrical insulation. The covering simply acts as protection for the metal inside, not for the people outside.

Wires are manufactured for a variety of purposes and from many different metals, such as brass, copper, silver, aluminum, tungsten, and various steel alloys. However, for use as power wiring in buildings, only copper, aluminum, or copper-clad aluminum are approved, according to *NEC* Section 310.2(B).

Table 310.13(A) Conductor Applications and Insulations Rated 600 Volts

Trade Name	Type Letter	Maximum Operating Temperature	Application Provisions	Insulation	Thickness of Insulation			Outer Covering[1]
					AWG or kcmil	mm	mils	
Fluorinated ethylene propylene	FEP or FEPB	90°C 194°F	Dry and damp locations	Fluorinated ethylene propylene	14–10 8–2	0.51 0.76	20 30	None
		200°C 392°F	Dry locations — special applications[2]	Fluorinated ethylene propylene	14–8	0.36	14	Glass braid
					6–2	0.36	14	Glass or other suitable braid material
Mineral insulation (metal sheathed)	MI	90°C 194°F	Dry and wet locations	Magnesium oxide	18–16[3] 16–10	0.58 0.91	23 36	Copper or alloy steel
		250°C 482°F	For special applications[2]		9–4 3–500	1.27 1.40	50 55	
Moisture-, heat-, and oil-resistant thermoplastic	MTW	60°C 140°F	Machine tool wiring in wet locations	Flame-retardant moisture-, heat-, and oil-resistant thermoplastic		(A)	(A)	(A) None (B) Nylon jacket or equivalent
		90°C 194°F	Machine tool wiring in dry locations. FPN: See NFPA 79.		22–12 10 8 6 4–2 1–4/0 213–500 501–1000	0.76 0.76 1.14 1.52 1.52 2.03 2.41 2.79	30 30 45 60 60 80 95 110	
Paper		85°C 185°F	For underground service conductors, or by special permission	Paper				Lead sheath
Perfluoro-alkoxy	PFA	90°C 194°F 200°C 392°F	Dry and damp locations Dry locations — special applications[2]	Perfluoro-alkoxy	14–10 8–2 1–4/0	0.51 0.76 1.14	20 30 45	None
Perfluoro-alkoxy	PFAH	250°C 482°F	Dry locations only. Only for leads within apparatus or within raceways connected to apparatus (nickel or nickel-coated copper only)	Perfluoro-alkoxy	14–10 8–2 1–4/0	0.51 0.76 1.14	20 30 45	None
Thermoset •	RHH	90°C 194°F	Dry and damp locations		14-10 8–2 1–4/0 213–500 501–1000 1001–2000	1.14 1.52 2.03 2.41 2.79 3.18	45 60 80 95 110 125	Moisture-resistant, flame-retardant, nonmetallic covering[1]
Moisture-resistant thermoset •	RHW[4]	75°C 167°F	Dry and wet locations	Flame-retardant, moisture-resistant thermoset	14–10 8–2 1–4/0 213–500 501–1000 1001–2000	1.14 1.52 2.03 2.41 2.79 3.18	45 60 80 95 110 125	Moisture-resistant, flame-retardant, nonmetallic covering[4]
	RHW-2	90°C 194°F						
Silicone	SA	90°C 194°F	Dry and damp locations	Silicone rubber	14–10 8–2 1–4/0	1.14 1.52 2.03	45 60 80	Glass or other suitable braid material
		200°C 392°F	For special application[2]		213–500 501–1000 1001–2000	2.41 2.79 3.18	95 110 125	

(Continues)

Figure 4-1
Conductor applications and insulations rated 600 volts

Table 310.13(A) *Continued*

Trade Name	Type Letter	Maximum Operating Temperature	Application Provisions	Insulation	Thickness of Insulation			Outer Covering[1]
					AWG or kcmil	mm	mils	
Thermoset	SIS	90°C 194°F	Switchboard wiring only	Flame-retardant thermoset	14–10 8–2 1–4/0	0.76 1.14 2.41	30 45 55	None
Thermoplastic and fibrous outer braid	TBS	90°C 194°F	Switchboard wiring only	Thermoplastic	14–10 8 6–2 1–4/0	0.76 1.14 1.52 2.03	30 45 60 80	Flame-retardant, nonmetallic covering
Extended polytetra-fluoro-ethylene	TFE	250°C 482°F	Dry locations only. Only for leads within apparatus or within raceways connected to apparatus, or as open wiring (nickel or nickel-coated copper only)	Extruded polytetra-fluoroethylene	14–10 8–2 1–4/0	0.51 0.76 1.14	20 30 45	None
Heat-resistant thermoplastic	THHN	90°C 194°F	Dry and damp locations	Flame-retardant, heat-resistant thermoplastic	14–12 10 8–6 4–2 1–4/0 250–500 501–1000	0.38 0.51 0.76 1.02 1.27 1.52 1.78	15 20 30 40 50 60 70	Nylon jacket or equivalent
Moisture- and heat-resistant thermoplastic	THHW	75°C 167°F 90°C 194°F	Wet location Dry location	Flame-retardant, moisture- and heat-resistant thermoplastic	14–10 8 6–2 1–4/0 213–500 501–1000 1001–2000	0.76 1.14 1.52 2.03 2.41 2.79 3.18	30 45 60 80 95 110 125	None
Moisture- and heat-resistant thermoplastic	THW	75°C 167°F 90°C 194°F	Dry and wet locations Special applications within electric discharge lighting equipment. Limited to 1000 open-circuit volts or less. (size 14-8 only as permitted in 410.33)	Flame-retardant, moisture- and heat-resistant thermoplastic	14–10 8 6–2 1–4/0 213–500 501–1000 1001–2000	0.76 1.14 1.52 2.03 2.41 2.79 3.18	30 45 60 80 95 110 125	None
	THW-2	90°C 194°F	Dry and wet locations					
Moisture- and heat-resistant thermoplastic	THWN	75°C 167°F	Dry and wet locations	Flame-retardant, moisture- and heat-resistant thermoplastic	14–12 10 8–6 4–2 1–4/0 250–500 501–1000	0.38 0.51 0.76 1.02 1.27 1.52 1.78	15 20 30 40 50 60 70	Nylon jacket or equivalent
	THWN-2	90°C 194°F						
Moisture-resistant thermoplastic	TW	60°C 140°F	Dry and wet locations	Flame-retardant, moisture-resistant thermoplastic	14–10 8 6–2 1–4/0 213–500 501–1000 1001–2000	0.76 1.14 1.52 2.03 2.41 2.79 3.18	30 45 60 80 95 110 125	None
Underground feeder and branch-circuit cable — single conductor (for Type UF cable employing more than one conductor, see Article 340.)	UF	60°C 140°F 75°C 167°F[6]	See Article 340.	Moisture-resistant Moisture- and heat-resistant	14–10 8–2 1–4/0	1.52 2.03 2.41	60[5] 80[5] 95[5]	Integral with insulation

Figure 4-1 *(continued)*
Conductor applications and insulations rated 600 volts

Table 310.13(A) Continued

Trade Name	Type Letter	Maximum Operating Temperature	Application Provisions	Insulation	Thickness of Insulation			Outer Covering[1]
					AWG or kcmil	mm	mils	
Underground service-entrance cable — single conductor (for Type USE cable employing more than one conductor, see Article 338.)	USE	75°C 167°F	See Article 338.	Heat- and moisture-resistant	14–10 8–2 1–4/0 213–500 501–1000 1001–2000	1.14 1.52 2.03 2.41 2.79 3.18	45 60 80 95[7] 110 125	Moisture-resistant nonmetallic covering (See 338.2.)
	USE-2	90°C 194°F	Dry and wet locations					
Thermoset	XHH	90°C 194°F	Dry and damp locations	Flame-retardant thermoset	14–10 8–2 1–4/0 213–500 501–1000 1001–2000	0.76 1.14 1.40 1.65 2.03 2.41	30 45 55 65 80 95	None
Moisture-resistant thermoset	XHHW[4]	90°C 194°F 75°C 167°F	Dry and damp locations Wet locations	Flame-retardant, moisture-resistant thermoset	14–10 8–2 1–4/0 213–500 501–1000 1001–2000	0.76 1.14 1.40 1.65 2.03 2.41	30 45 55 65 80 95	None
Moisture-resistant thermoset	XHHW-2	90°C 194°F	Dry and wet locations	Flame-retardant, moisture-resistant thermoset	14–10 8–2 1–4/0 213–500 501–1000 1001–2000	0.76 1.14 1.40 1.65 2.03 2.41	30 45 55 65 80 95	None
Modified ethylene tetrafluoro-ethylene	Z	90°C 194°F 150°C 302°F	Dry and damp locations Dry locations — special applications[2]	Modified ethylene tetrafluoro-ethylene	14–12 10 8–4 3–1 1/0–4/0	0.38 0.51 0.64 0.89 1.14	15 20 25 35 45	None
Modified ethylene tetrafluoro-ethylene	ZW	75°C 167°F 90°C 194°F 150°C 302°F	Wet locations Dry and damp locations Dry locations — special applications[2]	Modified ethylene tetrafluoro-ethylene	14–10 8–2	0.76 1.14	30 45	None
	ZW-2	90°C 194°F	Dry and wet locations					

[1] Some insulations do not require an outer covering.
[2] Where design conditions require maximum conductor operating temperatures above 90°C (194°F).
[3] For signaling circuits permitting 300-volt insulation.
•
[4] Some rubber insulations do not require an outer covering.
[5] Includes integral jacket.
[6] For ampacity limitation, see 340.80.
[7] Insulation thickness shall be permitted to be 2.03 mm (80 mils) for listed Type USE conductors that have been subjected to special investigations. The nonmetallic covering over individual rubber-covered conductors of aluminum-sheathed cable and of lead-sheathed or multiconductor cable shall not be required to be flame retardant. For Type MC cable, see 330.104. For nonmetallic-sheathed cable, see Article 334, Part III. For Type UF cable, see Article 340, Part III.

Figure 4-1 *(continued)*
Conductor applications and insulations rated 600 volts

Wire Materials

While aluminum is considerably less expensive than copper, the cost difference is absolutely its *only* advantage. In many other respects, aluminum is significantly less desirable. Most importantly, it has a much higher resistance to electrical current than copper. This means that a circuit wired in aluminum will always require a larger gauge wire than the same circuit wired in copper.

All metals, when exposed to air and moisture, form oxides. A common metal oxide is the rust that forms on iron and steel. For some metals, such as copper or brass, this oxidation process is self-limiting. Either of these metals will form a thin surface oxide coating that effectively seals the metal beneath from further oxidation until that surface film is removed. In the case of aluminum and steel, the oxidation of the metal isn't self-limiting. Once aluminum starts to oxidize, it'll continue indefinitely — converting the metal to aluminum oxide, a fine white powder. As long as aluminum is covered by insulation, it's protected from direct contact with the air and oxidation. However, the end of an aluminum wire that's been stripped of its insulation to make a connection is now exposed and subject to oxidation. Once stripped, aluminum wire should be treated with an antioxidant, and even then, you should periodically inspect it for signs of white aluminum oxide powder being formed. If you find any, clean the aluminum with a fine abrasive, and re-treat it with the antioxidant.

With copper-clad aluminum, the copper surface film protects the aluminum within from oxidation as long as that copper film remains intact, but this is precisely the problem. The copper film is too easily scraped. And, again, stripping the insulation leaves areas of bare aluminum subject to oxidation.

In addition to the wire sizing and oxidation difficulties, aluminum has other disadvantages. Its coefficient of expansion is considerably greater than that of copper, meaning it expands and contracts more with changes in temperature. This expansion and contraction of the metal may result in loose connections at odd and out-of-the-way points in the circuit. At *best*, these loose connections fail electrically, meaning the circuit or some part of it goes dead; at *worst*, a loose connection becomes a high resistance point that causes the wire to heat and either arc to a terminal, or melt off insulation, allowing arcing between wires. This arcing can and has caused many fires.

A further disadvantage of aluminum is its brittleness. For example, copper is far more plastic than aluminum. It's also more resistant to bending and rebending stresses, and resists crystallizing or breaking better than aluminum. Overall, the consensus of people in the electrical field is that the best thing to do with aluminum wire is *nothing*. Avoid using it if possible. It's simply not worth the cost savings.

Terminal and Device Markings

Due to the important differences in the properties of the two approved metals, copper and aluminum, the terminals and devices to which wires may be attached won't necessarily accommodate both. Some will take copper only, some will take copper or copper-clad aluminum but not bare aluminum, and some will take any of the three: copper, copper-clad, or bare aluminum. In order for the electrician to avoid connecting aluminum to terminals that won't accept it, a system of markings has been adopted to forewarn him.

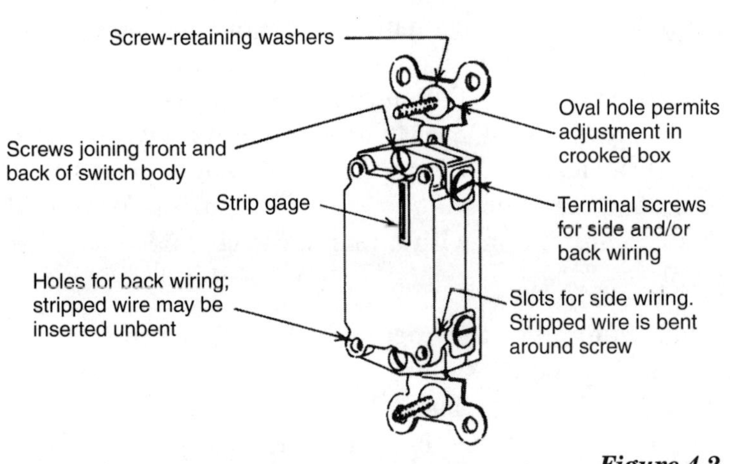

Figure 4-2
Switch with both screw and push-in terminals

Receptacles and switches rated for 20 amperes or less must be marked *Co/Alr* if aluminum may be connected to them. Unless otherwise marked, only copper or copper-clad aluminum may be connected — unless they're marked *copper only*. In any case, only copper or copper-clad aluminum may be used with the push-in type pressure terminals, commonly found on switches and duplex receptacles. See Figure 4-2. Bare aluminum, where permitted, must always be connected to a screw-type terminal.

Receptacles rated at 30 amperes and up must be marked *Al-Cu*, if aluminum may be connected to them. Unless otherwise marked, only copper or copper-clad aluminum may be connected.

It's imperative that aluminum only be connected to terminals or screws made of metals that are compatible with it. Look carefully for the marking *Co/Alr* on devices rated at 20 amperes or less, and the mark *Al-Cu* on those rated at 30 amperes or over. If these markings can't be found on a particular device, do *not* under any circumstances connect aluminum conductors to it.

Gauge number	Diameter (mils)	Cross section Circular mils	Square inches
0000	460.0	212,000.0	0.166
000	410.0	168,000.0	.132
00	365.0	133,000.0	.105
0	325.0	106,000.0	.0829
1	289.0	83,700.0	.0657
2	258.0	66,400.0	.0521
3	229.0	52,600.0	.0413
4	204.0	41,700.0	.0328
5	182.0	33,100.0	.0260
6	162.0	26,300.0	.0206
7	144.0	20,800.0	.0164
8	128.0	16,500.0	.0130
9	114.0	13,100.0	.0103
10	102.0	10,400.0	.00815
11	91.0	8,230.0	.00647
12	81.0	6,530.0	.00513
13	72.0	5,180.0	.00407
14	64.0	4,110.0	.00323
15	57.0	3,260.0	.00256
16	51.0	2,580.0	.00203
17	45.0	2,050.0	.00161
18	40.0	1,620.0	.00128
19	36.0	1,290.0	.00101
20	32.0	1,020.0	.000802
21	28.5	810.0	.000636
22	25.3	642.0	.000505
23	22.6	509.0	.000400
24	20.1	404.0	.000317
25	17.9	320.0	.000252
26	15.9	254.0	.000200
27	14.2	202.0	.000158
28	12.6	160.0	.000126
29	11.3	127.0	.0000995
30	10.0	101.0	.0000789
31	8.9	79.7	.0000626
32	8.0	63.2	.0000496
33	7.1	50.1	.0000394
34	6.3	39.8	.0000312
35	5.6	31.5	.0000248
36	5.0	25.0	.0000196
37	4.5	19.8	.0000156
38	4.0	15.7	.0000123
39	3.5	12.5	.0000098
40	3.1	9.9	.0000078

Figure 4-3
Table of AWG numbers and nominal diameters: #4/0 to 40

Wire Size

In order to provide a uniform standard of reference, nearly all wire manufactured in the U. S. is sized to conform to the American Wire Gauge (AWG). The standard sizes manufactured in accordance with this gauge are identified by gauge numbers that range from #40 up to #4/0 (0000). Yes, you read it correctly. The smallest size, #40, is the largest number. The numbers decrease as the wire thickness increases up to #0, then #2/0 (00), #3/0 (000), and finally the largest AWG size, #4/0 (0000). See Figure 4-3.

In building power wiring, we aren't concerned with any sizes smaller than #14, because *NEC* Table 310.5 (Figure 4-4) tells us that #14 copper or #12 aluminum are the smallest sizes permissible. Smaller size wires are permitted for low-voltage signal, alarm, and control circuits, such as intercoms, thermostats, bells, buzzers, sprinkler timers, or burglar alarms. And, you can also use smaller sizes for various exposed flexible cords. For fixture wire ampacity, operating temperature, minimum wire size, etc., see *NEC* Article 402.

Table 310.5 Minimum Size of Conductors

Conductor Voltage Rating (Volts)	Minimum Conductor Size (AWG) Copper	Aluminum or Copper-Clad Aluminum
0–2000	14	12
2001–8000	8	8
8001–15,000	2	2
15,001–28,000	1	1
28,001–35,000	1/0	1/0

From the *National Electrical Code* © 2007, NFPA

Figure 4-4
Minimum conductor size

The standard AWG wire sizes most commonly used in building power wiring are shown in actual size in Figure 4-5. Note that from #14 all the way up to #2, only even numbers appear. The odd numbered sizes aren't made for electrical wiring.

Wire size 4/0 AWG certainly isn't the largest wire made. 4/0 is merely the large end of the American Wire Gauge sizes. Much larger wires are commonly made, and they're identified by their cross-sectional areas as measured in thousands of circular mils. This designation is abbreviated MCM. The next electrical wire size larger than

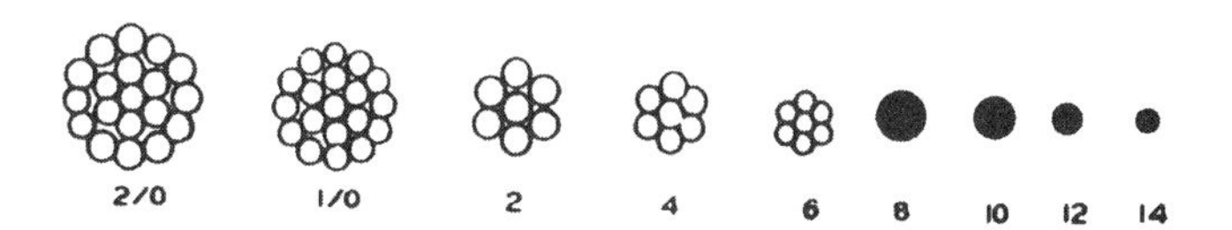

Figure 4-5
AWG wire gauges, actual size

4/0 is 250 MCM. From there on, the numbers in the designation increase as the wire size increases because the numbers now equal the cross-sectional area of the wire.

This change in numbering systems is evident in *NEC* Table 310.16, shown in Figure 4-6. Notice that the wire sizes are given in both the far left and far right columns. From the top of the column, the numbers decrease as wire size increases, then halfway down the page, the sequence changes. The number 0000 is followed by 250 then 300, and so on. At that point, AWG numbers change to MCM.

Wire Construction

All wires used in building power wiring sized from #14 up to #8 are solid wires, meaning that a single round strand of metal makes up the full thickness of the wire. Look again at Figure 4-5. All wires larger than #8 are stranded; that means a large wire is made up of several smaller wires twisted together. Size #8 AWG marks the dividing line. It's available in either solid or stranded, and it's used in building wiring both ways.

Sizes #14, #12, and #10 are also available in stranded form. However, in the permanent parts of a building electrical system, they're not customarily used in this form. Solid wire in these sizes is easier to strip without damaging the metal, and it makes more positive connections when spliced, and when connected to terminals.

The reason for changing to stranded wire at #6 is that it and larger sizes in a solid rod would be stiff and unwieldy to handle. Splices in these larger wire sizes are made using pressure clamps, such as the split bolt connectors shown in Figure 3-19 in the previous chapter. Connections at switches, breakers, or outlets are made in clamp or set screw terminals.

Insulations

Within buildings, as well as on the way to and from them, power wiring is exposed to a wide variety of environmental conditions. Temperatures may vary from subfreezing cold to extreme heat. While one area is absolutely dry, another may be moderately damp, while yet another is soaking wet. These differences in temperature and humidity might vary from one place to another in and around a building, or they may vary in a particular spot with the seasons of the year, or from day to day, or even from hour to hour.

Table 310.16 Allowable Ampacities of Insulated Conductors Rated 0 Through 2000 Volts, 60°C Through 90°C (140°F Through 194°F), Not More Than Three Current-Carrying Conductors in Raceway, Cable, or Earth (Directly Buried), Based on Ambient Temperature of 30°C (86°F)

Size AWG or kcmil	Temperature Rating of Conductor [See Table 310.13(A).]						Size AWG or kcmil
	60°C (140°F)	75°C (167°F)	90°C (194°F)	60°C (140°F)	75°C (167°F)	90°C (194°F)	
	Types TW, UF	Types RHW, THHW, THW, THWN, XHHW, USE, ZW	Types TBS, SA, SIS, FEP, FEPB, MI, RHH, RHW-2, THHN, THHW, THW-2, THWN-2, USE-2, XHH, XHHW, XHHW-2, ZW-2	Types TW, UF	Types RHW, THHW, THW, THWN, XHHW, USE	Types TBS, SA, SIS, THHN, THHW, THW-2, THWN-2, RHH, RHW-2, USE-2, XHH, XHHW, XHHW-2, ZW-2	
	COPPER			ALUMINUM OR COPPER-CLAD ALUMINUM			
18	—	—	14	—	—	—	—
16	—	—	18	—	—	—	—
14*	20	20	25	—	—	—	—
12*	25	25	30	20	20	25	12*
10*	30	35	40	25	30	35	10*
8	40	50	55	30	40	45	8
6	55	65	75	40	50	60	6
4	70	85	95	55	65	75	4
3	85	100	110	65	75	85	3
2	95	115	130	75	90	100	2
1	110	130	150	85	100	115	1
1/0	125	150	170	100	120	135	1/0
2/0	145	175	195	115	135	150	2/0
3/0	165	200	225	130	155	175	3/0
4/0	195	230	260	150	180	205	4/0
250	215	255	290	170	205	230	250
300	240	285	320	190	230	255	300
350	260	310	350	210	250	280	350
400	280	335	380	225	270	305	400
500	320	380	430	260	310	350	500
600	355	420	475	285	340	385	600
700	385	460	520	310	375	420	700
750	400	475	535	320	385	435	750
800	410	490	555	330	395	450	800
900	435	520	585	355	425	480	900
1000	455	545	615	375	445	500	1000
1250	495	590	665	405	485	545	1250
1500	520	625	705	435	520	585	1500
1750	545	650	735	455	545	615	1750
2000	560	665	750	470	560	630	2000
	CORRECTION FACTORS						
Ambient Temp. (°C)	For ambient temperatures other than 30°C (86°F), multiply the allowable ampacities shown above by the appropriate factor shown below.						Ambient Temp. (°F)
21–25	1.08	1.05	1.04	1.08	1.05	1.04	70–77
26–30	1.00	1.00	1.00	1.00	1.00	1.00	78–86
31–35	0.91	0.94	0.96	0.91	0.94	0.96	87–95
36–40	0.82	0.88	0.91	0.82	0.88	0.91	96–104
41–45	0.71	0.82	0.87	0.71	0.82	0.87	105–113
46–50	0.58	0.75	0.82	0.58	0.75	0.82	114–122
51–55	0.41	0.67	0.76	0.41	0.67	0.76	123–131
56–60	—	0.58	0.71	—	0.58	0.71	132–140
61–70	—	0.33	0.58	—	0.33	0.58	141–158
71–80	—	—	0.41	—	—	0.41	159–176

* See 240.4(D).

Figure 4-6
Allowable ampacities of insulated conductors

In addition to the variations in temperature and humidity, power wiring may be exposed to many different types of oils, greases, chemicals, tars, gases, fumes, or liquids. Some of these might have a corrosive effect on the metal in the conductors, while others might attack the insulation or sheathing used to protect the conductors. In order to deal safely with these different conditions, a variety of insulations have been developed. Each provides a high degree of protection from one or more of the unfavorable conditions that the wiring might encounter. No individual insulation is suitable for all conditions.

Since the purpose of a residential building is to provide an environment suitable for human habitation, extreme conditions of temperature, humidity, or exposure to corrosive chemicals aren't normally found. The extremes, particularly in terms of corrosive materials, occur in industrial structures and occasionally in commercial buildings. However, a qualified electrician, regardless of his area of specialization, should be familiar with the general range of available materials and have an overall understanding of their properties and uses.

Look back at Figure 4-1. *NEC* Table 310.13(A) recognizes 29 different types of insulated conductors with a 600-volt rating for use in general wiring. Although the average electrician won't use more than maybe a half a dozen of these, for reference purposes, they're all listed here. Many, as you'll quickly see, are for very specialized applications.

All of the types starting with the code letter "R," such as RHH, and RHW, are rubber. Rubber was once the primary insulation material, but it's now largely supplanted by plastics. "R" insulations are currently found on main building feeders or on some major subfeeders, but not used on general branch circuit wiring. You'll rarely find any of these insulations used in residential work.

The ones you'll use constantly will be from the "T" or thermoplastic group. These are THHN, THHW, THW, THW-2, THWN, THWN-2 and TW. Of these, TW and THW are used most frequently.

TW is the standard insulation type used in NM cables, which we'll discuss in detail in Chapter 5. Since the vast majority of residential wiring is done with NM cables, it follows that TW is the insulation that residential electricians work with daily. It fits just about any situation. It's moderately heat-resistant — up to 140 degrees F; it's moderately moisture-resistant; and it's moderately priced compared with some of the more highly-resistant materials, whose properties aren't needed in residential applications.

When you're using wiring with either rigid or flexible conduit, you'll likely be using THW rather than TW. Due to its higher operating temperature, combined with moisture resistance, THW fits a wider range of situations; therefore, it's routinely used to replace TW in conduit runs.

When you need a higher temperature resistance, switch to THHN. THW is good up to 90 degrees C (194 degrees F) only for special applications, while THHN is good to 90 degrees C for general purpose wiring. Another reason for utilizing THHN might be for its insulation thickness. For example, on a #14 wire using THW, the insulation has to be 30 mils thick. With THHN it only needs to be 15 mils thick.

As we'll see in Chapter 10, the matter of insulation thickness becomes a primary concern when pulling wiring through any of the various types of conduit. According to *NEC* Chapter 9, Table 1, you're permitted to fill no more than 40 percent of the cross-sectional area of a conduit with multiple conductors. In the case of a single conductor, the fill percentage can go as high as 53 percent; if there are only two conductors, the fill can go up to 31 percent; however, when there are over two conductors being pulled through a conduit, the fill can't exceed 40 percent. A *conductor* includes both the wire and its associated insulation. So, the thinner the insulation, the more space there'll be left for wires.

To define the number of conductors permitted in the various types of conduits, the code repeatedly refers to the percentages of fill specified in *NEC* Chapter 9, Table 1. In residential work, we seldom pull either very many or very large wires, so this limitation isn't usually a problem. If you intend to do considerable conduit work, refer to Chapter 9, Annex C, Tables C1 through C12(A) for full details of all the *NEC* requirements.

UF is the sheathing for cables consisting of two or more conductors plus a ground wire. It's used in underground feeds for decorative lighting, and for supplying power to outdoor utility outlets near and on the outside of buildings. It's also used where exposed wiring is needed.

In these outdoor applications, the UF-type insulation forms the cable sheathing, which protects individual conductors insulated with TW. We'll cover this in more detail in the next chapter.

When the electrical power supply reaches a building via an underground service entrance, the buried wires connecting the transformer to the building meter box will probably be underground service entrance cable or USE. (We'll be covering the Service Entrance in Chapter 9.) This material is similar to UF in terms of moisture and temperature resistance; however, a thinner layer of USE will do the same insulating job as a much thicker layer of UF. The *NEC* specifically permits the use of UF on underground feeders, but just as specifically forbids its use on underground service entrances for which USE is clearly approved.

The extensive group of high-temperature-rated plastic insulations in *NEC* Table 310.13(A), shown in Figure 4-1, are for use in special applications, mostly in dry locations. In any event, none of them has become at all popular for use in general building power wiring.

Markings

3/0 AWG TYPE MTW OR THHN OR THWN (UL) 600V GASOLINE AND OIL RESISTANT

Figure 4-7
Typical code-required conductor markings

In order for there to be no doubt, question, or confusion concerning wire size, insulation type, approved maximum voltage rating, or responsible manufacturer, all individual conductors and all factory-assembled cables must be properly marked, per *NEC* Section 310.11. Properly marked means that at intervals of not more than 24 inches, all individual conductors and all assembled cables must be durably marked, showing wire size in AWG or MCM, whichever applies. See Figure 4-7. We'll discuss this more in the next chapter.

Marked at intervals of not more than 40 inches, conductors and assembled cables will show:

a. Proper letter designation for the insulation type being used, such as TW, THW, THHN.

b. Maximum voltage for which that insulation type and thickness is approved.

c. Name or trademark of manufacturer.

Color Code

In addition to the identifying markings described above, the *NEC* also requires that the insulation on individual conductors be color coded so that conductors being used for specific purposes may be identified instantly.

NEC Section 250.119 states that the grounding conductor may be bare, covered, or insulated. When covered or insulated, it must be colored green or green with one or more yellow stripes.

NEC Section 200.6(A) states that for sizes AWG 6 or smaller, an insulated grounded conductor must be identified by a continuous white or gray outer finish, or by three continuous white stripes on other than green insulation along its entire length. Wires finished white or gray, but with colored threads showing, are permitted. With mineral-insulated metal-sheathed cable, mark the ends at the time of installation.

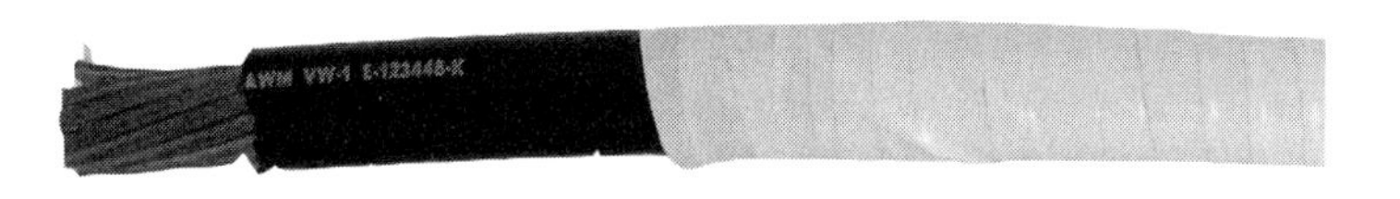

Figure 4-8
Black conductor reidentified with white tape

NEC Section 200.6(B) states that, for sizes larger than AWG 6, if the insulation isn't colored along its entire length per (A) above, then it must be marked at its ends (see Figure 4-8).

Ungrounded (hot or power) conductors, per *NEC* Section 310.12(C), are not allowed to be white, grey, or green — since those colors have been specifically assigned to common and ground. The colors most frequently used for the ungrounded leads are black, red, and blue, in that order. As we've just seen, when white or grey aren't available, it's permissible to reidentify another color with white paint or tape, and use the wire as a common. The reverse is also permissible. Perhaps black, red, or any other power color isn't available, but white or grey is. In that case, white or grey may be reidentified with black paint or tape *everywhere* the wire is accessible. Then it can be used as a power wire.

Ampacities

Amperage, as explained in Chapter 1, is the measure of the rate of flow of an electrical current. The *NEC* imposes strict amperage limitations on conductors used in building wiring, based on both the wire sizes measured in AWG or MCM, and the insulation type that covers the wire. These limitations are clearly shown in *NEC* Table 310.16 (Figure 4-6).

In this table, the wire sizes are listed twice, in both the far left and the far right columns. Temperature ratings and insulation types are also listed twice across the top of the table. You'll find four columns of each above the heading *Copper*, and four more above the heading *Aluminum or copper-clad aluminum*. Listed in the columns below those headings are the permissible amperages for the various wire sizes and insulation types.

As an example, let's look up the allowable amperage for #12 AWG copper wire with TW insulation. Go down the left column of wire sizes to the number 12. The column to the right of the wire sizes is titled *Types TW, UF*. In that column, in line with wire size 12, is the number 25. Notice that size 12 wire has an asterisk. A note at the bottom of the table refers the reader to *NEC* Section 240.4(D). There, the code states that #12 copper shall be rated at 20 amperes. Any number in the table *not* followed by an asterisk is the actual amperage allowed for that wire size and insulation.

Let's look at another example. #8 copper wire with TW insulation is rated at 40 amperes. The same #8 wire with THW insulation (next column to the right) is good for 50 amperes. However, #8 *aluminum* wire with THW insulation is only good for 40 amperes (fifth column to the right from the wire sizes).

Remember, the *NEC* lists minimum requirements. It's always permissible to exceed these code minimums, but never to fall short of them. #12 copper wire with TW insulation is good for 20 amperes; so it's permissible to use it for a 15-ampere circuit, if desired. On the other hand, #14 copper with the same TW is only good for 15 amperes — so don't try to get away with using it for a 20-ampere circuit.

STUDY QUESTIONS

1. Which metals are approved for use as power wiring in buildings, per the *NEC*?

A) Copper, steel, copper-clad aluminum
B) Aluminum, brass, copper
C) Copper, aluminum, steel
D) Copper, aluminum, copper-clad aluminum

2. What advantage does using aluminum wire as a conductor have over copper wire?

A) It is more economical
B) It is not as prone to oxidation
C) It has a higher coefficient of expansion
D) It doesn't have any advantage over copper

3. Which conductor material and terminal type are *incorrectly* paired?

A) Bare aluminum; screw-type terminal
B) Bare aluminum; push-in type terminal
C) Copper-clad aluminum; push-in type terminal
D) Copper; screw-type terminal marked *Al-Cu*

4. What is the smallest size of aluminum wire used for power wiring installed in a building?

A) #2 (00)
B) #4/0
C) #12
D) #14

5. For what purpose are wires smaller than #14 copper or #12 aluminum permitted in building power wiring?

A) Storage closet lights
B) Lights on garage door operators
C) LED lighting
D) Low voltage alarms and controls

6. What is the largest solid wire size used in building electrical wiring?

A) #10
B) #8
C) #6
D) #4

7. Which of the following gauges of wire are available in *both* stranded and solid form?

A) #6 and larger
B) #8 only
C) #14, #12, #10 and #8
D) All wire is available in either form

8. What insulation is normally used in NM cable?

A) TW
B) THW
C) THHN
D) THWN

9. In a conduit carrying three conductors, up to what percentage of the cross-sectional area may the conductors occupy?

A) 31
B) 40
C) 53
D) 60

10. Which is a permitted coloring for an insulated grounded conductor for sizes AWG 6 or smaller?

A) Three continuous white stripes on green insulation
B) White with one or more yellow stripes
C) Black
D) White or gray its entire length

5

GROUPED CONDUCTORS

With the exception of the lead connecting the ground bus in the breaker box to the grounding electrode system, all wire runs in building power wiring must contain at least two conductors plus a ground path. That ground path might be an interconnected system of grounding wires, such as is used in NM type cable, or it might be provided through a metallic protective casing, such as electrical metallic tubing, usually called conduit.

Since all power wiring runs must consist of at least two conductors plus ground, a number of different methods for grouping wires have been approved to provide for the varying degrees and types of protection required under different operating and environmental conditions. The code clearly defines the uses and limitations of each approved method.

Nonmetallic-Sheathed Cable — Types NM, NMC and NMS (*NEC*® Article 334)

For a number of years, the majority of residential power wiring has been done with nonmetallic-sheathed cable. It's referred to as NM cable, and defined in the code as a factory assembly of two or more insulated conductors having an outer sheath of moisture-resistant, flame-retardant, nonmetallic material. A typical example of this cable is shown in Figure 5-1.

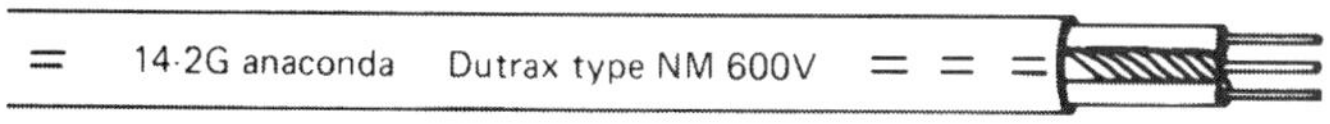

14-2G anaconda Dutrax type NM 600V

Figure 5-1
NM cable

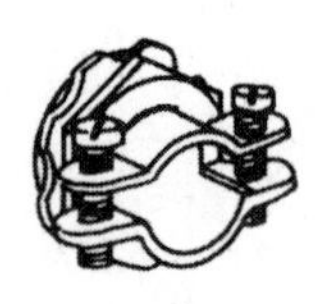

Metal

Plastic

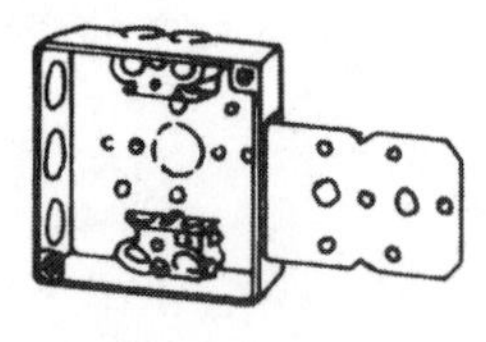

Internal within box

Figure 5-2
NM cable box connectors

The insulation used on the individual conductors is Type TW; the outer sheathing is a flame-retardant and moisture-resistant plastic; the grounding conductor in this cable is uninsulated, although wrapped with paper. Note in Figure 5-1 that the sheathing carries markings indicating wire size (14), number of wires (2), the presence of the ground wire (G), the name of the manufacturer (Anaconda), the cable type (NM), and the maximum voltage approved for this cable (600V). It's required that all of this information appear on the sheathing of NM cable, repeated at intervals not exceeding 24 inches. NM and NMC cables come in both 2-wire and 3-wire with ground. Wire sizes run from #14 to #2 AWG.

The difference between Type NM and NMC is in the outer sheathing of NMC. Besides being flame-retardant and moisture-resistant, it's fungus- and corrosion-resistant. NMS also contains signal or control wires.

The *NEC* permits Types NM and NMC cables to be used in single- or multi-family dwellings and other structures not over three floors above grade. While the *NEC* permits the use of NM and NMC for stores, offices, or other commercial or industrial buildings as long as they're not over three stories, it's good to find out what local codes say about this subject. Many local codes require that all wiring in structures other than residential be metal-protected.

Type NM is permitted for both exposed and concealed runs in normally dry interior locations. This means that it can also be passed through the voids in concrete block walls, as long as the walls are normally dry. It can't be used where it'll be exposed to corrosive fumes or be embedded in concrete or plaster. For example, you can't cut a channel in a masonry wall or a concrete floor, place NM cable in the channel, and then seal it in place by covering it with plaster or concrete.

You can use NMC anywhere that NM is permitted. It can also be used in damp areas or in the presence of corrosive fumes. Unlike NM, it can be embedded in a shallow chase in masonry, concrete or adobe. But neither NM nor NMC can be used for service entrance cables.

Both NM and NMC cables must be secured to metal boxes with cable clamps. The cable clamps can be metal, plastic or they can be an integral part of the box itself, all shown in Figure 5-2. Clamps aren't required when plastic boxes are used. Nonmetallic cables must be secured to the structure within 12 inches of the point where they emerge from a box, and every 4½ feet thereafter, until within 12 inches of another box. This can be accomplished using staples or straps — staples like the one shown in Figure 5-3 are most common.

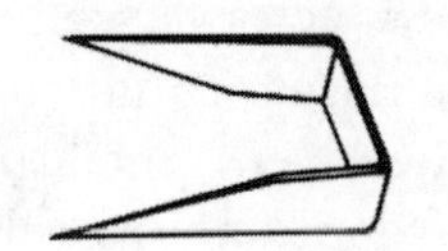

Figure 5-3
Staple for NM cable

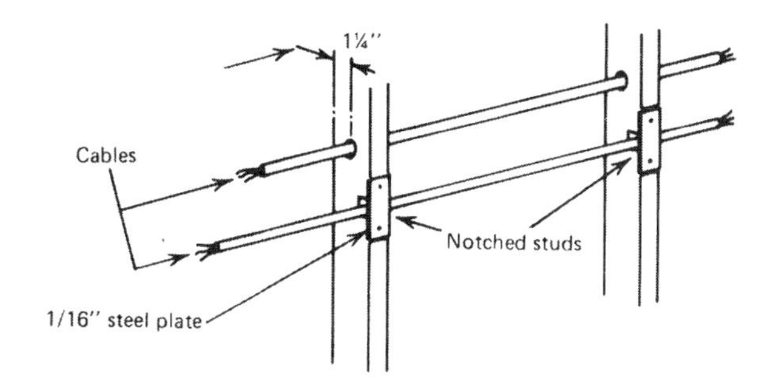

Figure 5-4
Running NM cable horizontally through studs as required by Code

Where a cable could be exposed to physical damage, it must be protected by guard strips or conduit. Where it passes through a floor, conduit is required, and that conduit must extend at least 6 inches above the floor. When cables pass through studs, joists, or rafters, as shown in Figure 5-4, the hole for the cable must be at least $1\frac{1}{4}$ inches from the nearest edge of the wood. Or, where the wood is notched for the cable, a $\frac{1}{16}$-inch-thick steel cover plate must be used to protect the cable from drywall or paneling nails.

When running cables through attics, a distinction is made between what's termed an accessible and an inaccessible attic. See Figure 5-5. If there's a permanent ladder or stairway leading to the attic, it's considered accessible. If there's merely a scuttle hole in the ceiling, making it necessary to bring a ladder to gain access, it's classified as *inaccessible*. In an inaccessible attic, it's only necessary to protect cables on top of the joists within 6 feet of the scuttle hole by flanking them with guard strips. In an accessible attic, all cables running across the joists, plus any cables within 7 feet above the joists, will require guard strips.

Service Entrance Cable — Type SE and USE (*NEC* Article 338)

Type SE cable has a flame-retardant and moisture-resistant cover. While USE can be used underground, it isn't flame-retardant. In SE and USE cables containing two or more conductors, one conductor can be uninsulated. Don't use this type for interior branch circuits.

When all conductors are insulated, you can use Type SE in interior wiring. It should be installed to the same requirements as Type NM cable.

Underground Feeder Cable — Type UF (*NEC* Article 340)

Type UF cable is similar to NM cable because it consists of insulated conductors plus an uninsulated ground wire bound together inside a plastic sheathing, shown in Figure 5-6. The insulation for these

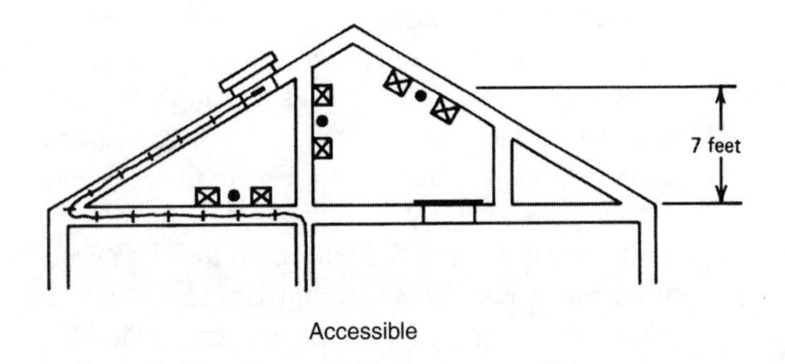

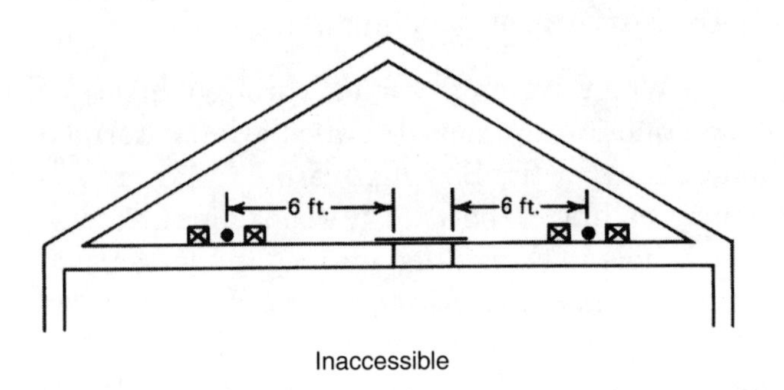

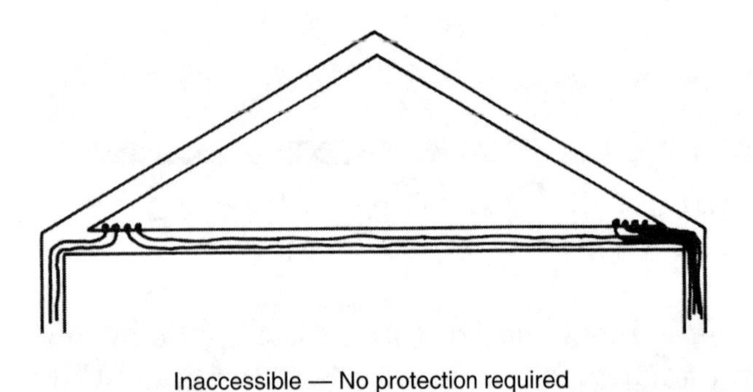

Figure 5-5
Protection of NM cable in attics

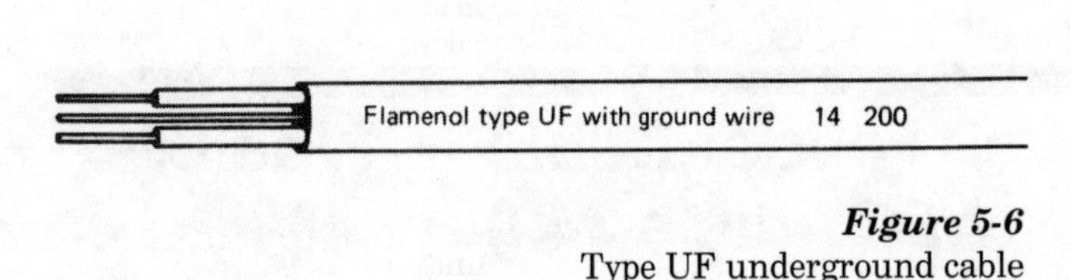

Figure 5-6
Type UF underground cable

conductors is TW, which is also used on NM. The exterior covering is flame-retardant; moisture-, fungus- and corrosion-resistant, and is suitable for direct burial in the ground. A sunlight-resistant type is also available, making it suitable for general outdoor use, whether buried or exposed. The sunlight-resistant type is specifically marked. The other markings on UF cables resemble those required on NM. The wire size is given, along with cable type, maximum voltage and manufacturer name. It's available in wire sizes ranging from #14 to #2 AWG.

The major visible difference in the cable is that the outside covering of the NM cable is a comparatively thin sheathing, but UF is actually a solid plastic extrusion in which the insulated conductors and the uninsulated ground wire are completely embedded.

UF cable is approved for direct burial in the ground, as long as it's buried at least 24 inches deep. If a 2-inch-thick concrete layer is placed above the cable, the depth of burial can be reduced by 6 inches. If there's a building slab above the cable, the slab is considered adequate protection and no additional burial depth is necessary. However, prior to pouring the slab, the cable should be placed under a thin layer of sand, so it won't be embedded in the finished slab. *NEC* Table 300.5, shown in Figure 5-7, gives necessary depth requirements for buried wiring.

In addition to underground use, UF cable can be used in any of the interior applications where NM or NMC cables are approved. In those cases, the UF installation must conform to the same requirements as NM or NMC, especially regarding protection from physical damage. UF can't be used as a service entrance cable or be embedded in concrete.

Table 300.5 Minimum Cover Requirements, 0 to 600 Volts, Nominal, Burial in Millimeters (Inches)

	Type of Wiring Method or Circuit									
Location of Wiring Method or Circuit	Column 1 Direct Burial Cables or Conductors		Column 2 Rigid Metal Conduit or Intermediate Metal Conduit		Column 3 Nonmetallic Raceways Listed for Direct Burial Without Concrete Encasement or Other Approved Raceways		Column 4 Residential Branch Circuits Rated 120 Volts or Less with GFCI Protection and Maximum Overcurrent Protection of 20 Amperes		Column 5 Circuits for Control of Irrigation and Landscape Lighting Limited to Not More Than 30 Volts and Installed with Type UF or in Other Identified Cable or Raceway	
	mm	in.	mm	in.	mm	in.	mm	in.	mm	in.
All locations not specified below	600	24	150	6	450	18	300	12	150	6
In trench below 50-mm (2-in.) thick concrete or equivalent	450	18	150	6	300	12	150	6	150	6
Under a building	0 (in raceway only)	0	0	0	0	0	0 (in raceway only)	0	0 (in raceway only)	0
Under minimum of 102-mm (4-in.) thick concrete exterior slab with no vehicular traffic and the slab extending not less than 152 mm (6 in.) beyond the underground installation	450	18	100	4	100	4	150 (direct burial) 100 (in raceway)	6 4	150 (direct burial) 100 (in raceway)	6 4
Under streets, highways, roads, alleys, driveways, and parking lots	600	24	600	24	600	24	600	24	600	24
One- and two-family dwelling driveways and outdoor parking areas, and used only for dwelling-related purposes	450	18	450	18	450	18	300	12	450	18
In or under airport runways, including adjacent areas where trespassing prohibited	450	18	450	18	450	18	450	18	450	18

Notes:
1. Cover is defined as the shortest distance in millimeters (inches) measured between a point on the top surface of any direct-buried conductor, cable, conduit, or other raceway and the top surface of finished grade, concrete, or similar cover.
2. Raceways approved for burial only where concrete encased shall require concrete envelope not less than 50 mm (2 in.) thick.
3. Lesser depths shall be permitted where cables and conductors rise for terminations or splices or where access is otherwise required.
4. Where one of the wiring method types listed in Columns 1–3 is used for one of the circuit types in Columns 4 and 5, the shallowest depth of burial shall be permitted.
5. Where solid rock prevents compliance with the cover depths specified in this table, the wiring shall be installed in metal or nonmetallic raceway permitted for direct burial. The raceways shall be covered by a minimum of 50 mm (2 in.) of concrete extending down to rock.

Figure 5-7
Minimum cover requirements for underground installations

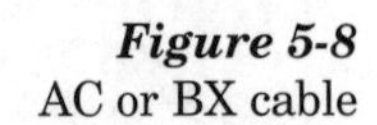

Figure 5-8
AC or BX cable

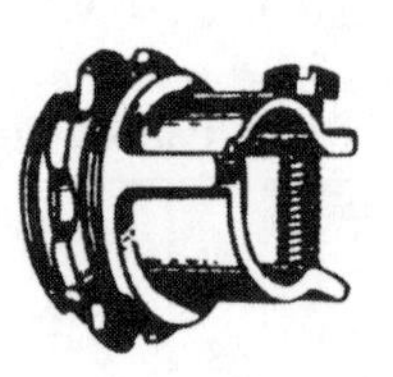

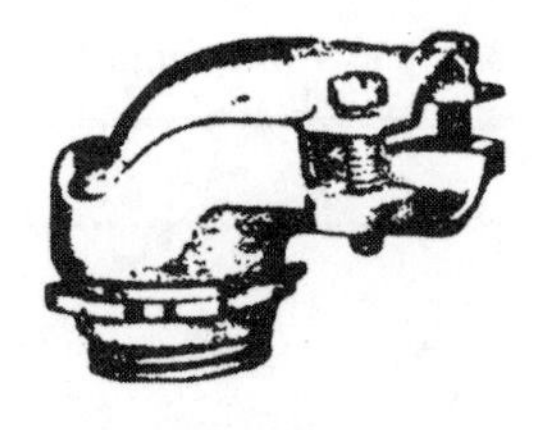

Cable connectors for armored cable

Figure 5-9
Box connectors for BX cable

Armored Cable — Type AC (*NEC* Article 320)

AC cable is more commonly known as *BX* cable. It consists of a fabricated assembly of insulated conductors cased in a flexible spiral metal sheathing, as shown in Figure 5-8. The flexible metal armor of AC cable is accepted as a grounding conductor, so the separate grounding conductor included in NM and UF cables is omitted. However, an internal bonding strip is included in AC cables to supplement the effectiveness of the spiral armor as its grounding conductor. Where AC cables are attached to boxes, you must use approved bushings and connectors. See Figure 5-9.

AC cable displays the usual marking information. Wire size, number of wires, cable type, maximum approved voltage, and maker's name are found on a tag attached to the coil of an AC cable. The wire sizes available in AC cable range from #14 to #1. Both 2- and 3-conductor cables are available. The designation ACT indicates that the conductors are insulated with a thermoplastic insulation, and ACL indicates that a lead covering is applied over the conductor assembly, greatly improving its moisture resistance.

AC cable can be used anywhere that NM is permitted; it can also be used for underplaster extensions and be embedded in plaster finishes on masonry. It must be secured with staples or straps within 12 inches of each box, and at intervals not exceeding 4½ feet in runs between boxes. Bends must be made so that the cable armor isn't damaged, and the inside radius of any bend must be at least five times the diameter of the cable being bent. Like NM cable, you can fish AC through the voids in concrete block walls when those walls are normally dry. When walls are subject to considerable dampness, or are below grade, you should use Type ACL.

When used in attics, the limitations regarding protection for AC cable are the same as for NM. In inaccessible attics, protection is only required within 6 feet of the scuttle hole. However, in accessible attics,

Figure 5-10
Flexible metal conduit — Greenfield

Table 1 Percent of Cross Section of Conduit and Tubing for Conductors

Number of Conductors	All Conductor Types
1	53
2	31
Over 2	40

Figure 5-11
Percent of cross section of conduit and tubing for conductors

all cables passing over the ceiling/floor joists must have guard strips. Any cable above the floor level, up to 7 feet, also needs guard strips. Look back to Figure 5-5.

Flexible Metal Conduit — Type FMC (*NEC* Article 348)

Flexible metal conduit (FMC) is also a spiral metal armor with a similar exterior appearance to Type AC cable. See Figure 5-10. FMC, often referred to in the trade as *Greenfield*, differs from AC cable in that it doesn't contain wires until the electrician adds them. You can't use FMC smaller than ½-inch electrical trade diameter as a permanent part of the building wiring, although 3/8-inch trade size can be used in some cases to carry power to motors, and to connect an outlet box to a light fixture.

This material is just an empty armor through which the necessary wires are pulled. Because of this, there are no markings stating the number of wires, insulation type, and voltage rating, etc. on the outside. According to *NEC* Section 348.22, the number of conductors permitted in FMC should never exceed the percentage fill specified in *NEC* Chapter 9, Table 1, shown in Figure 5-11.

FMC is commonly used concealed in stud or concrete block walls, but it can't be used in any wet locations. Although it's a type of metal conduit, it's only acceptable as a grounding means if both the conduit and the fittings that are being used are approved for grounding. Otherwise, only lengths up to 6 feet are permitted as a grounding means. If the total length of the ground return path is longer than that, a separate ground wire must be added.

Figure 5-12
Greenfield box connectors

Figure 5-12 shows examples of box connectors used with FMC. Box connectors may thread inside the spiral of the material, or connect two pieces of flex with a coupling that threads into the inside of both.

Like both NM and AC cables, FMC must be secured within 12 inches of each outlet box and every 4½ feet from there on, until within 12 inches of the next box. Between boxes, a run of FMC shouldn't contain more than four quarter bends. That means no more than 360 degrees in bends (a quarter bend equals 90 degrees). Also, these bends must have sufficient sweep so that the armor isn't damaged — meaning an inside radius of at least five times the diameter of the armor being bent.

Liquidtight Flexible Metal Conduit — Type LFMC (*NEC* Article 350)

Liquidtight is another type of spiral flexible metal armor, similar to Greenfield, except that the spiral turns are a little tighter, and it has a plastic outer jacket that's both liquidtight and sunlight-resistant. An example of LFMC is shown in Figure 5-13. Like Greenfield, liquidtight carries no markings for wire size, insulation, or voltage rating because it comes without wires; they're pulled through as needed. The number of wires of different sizes and insulation types that can be pulled must agree with Table 1 of *NEC* Chapter 9. Look back to Figure 5-11.

While the *NEC* permits the use of liquidtight in a wide range of applications, it's expensive. Consequently, its use in residences is quite limited. It's often used in short runs for wiring outdoor air conditioning equipment or for wiring disposals under kitchen sinks.

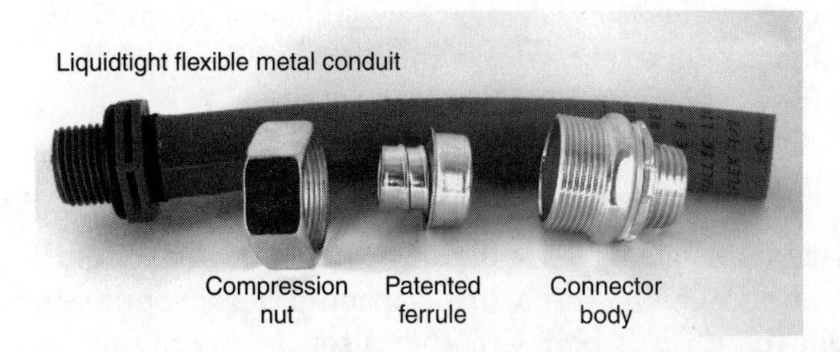

Figure 5-13
Liquidtight flexible metal conduit and compression box connectors

Liquidtight flexible conduit is always attached to boxes by means of special liquidtight compression connectors, also in Figure 5-13. If liquidtight is installed on or in walls, it must be secured with staples or straps every 4½ feet to within 12 inches of the boxes, the same as other flexible wiring materials.

The *NEC* also recognizes liquidtight flexible *nonmetallic* conduit, Type LFNC. In residential work, it can be used in the same places as the metallic type — except its use outdoors is limited to moderate climates. It can become brittle in extreme cold. Conduit lengths are limited to no more than 6 feet. A grounding conductor can be installed either inside or outside the conduit, but when outside it must be no more than 6 feet in length.

Rigid Metal Conduit — Type RMC (*NEC* Article 344)

While the *NEC* refers to, and permits, the use of both ferrous and nonferrous metals as rigid metal conduit, in practice, galvanized pipe is generally used — the same type galvanized pipe commonly found in water supply and gas lines. As always, the use of pipe smaller than ½-inch trade size isn't allowed.

Since this material is plain, everyday steel pipe, it can be cut and threaded the same as water or gas pipe. When it's cut, all cut ends must be reamed to remove any rough edges. If any rough edges are left, they're likely to tear the insulation when insulated conductors are pulled through. Threading is done with the same type of threader and dies that a plumber would use on the same pipe. Standard American Pipe threads are to be used throughout.

Red brass RMC can be used in direct burial and for swimming pool installations. Aluminum RMC can be used when the environment is considered suitable. Rigid aluminum encased in concrete or in contact with earth must have approved supplementary corrosion protection.

The number of wires of specific sizes and insulation types that can be pulled through pipe conduit of the various trade pipe sizes must conform to *NEC* Table 1 of Chapter 9. Look again at Figure 5-11.

Between boxes or fittings, the number of bends allowed is again limited to no more than the equivalent of four quarter bends, or 360 degrees. This is to facilitate the plumbing, removal, or repulling of conductors. The minimum radius of bends for various conduit sizes is listed in *NEC* Chapter 9, Table 2, shown in Figure 5-14. When space doesn't allow for bends of the radii required in the table, you can use fittings called *pulling ells*. See Figure 5-15. Whenever a pulling ell is used, the count on bends can be started again.

Pulling ells are made with the side, long end, or both sides removable. This makes it possible to get into them even when they're located in unlikely and difficult-to-access spaces. In addition to ells, there are tees and straight conduit bodies. Any of these fittings can also be used with FMC, which we've discussed, or with electrical metallic tubing (EMT), which we'll discuss next.

Table 2 Radius of Conduit and Tubing Bends

Conduit or Tubing Size		One Shot and Full Shoe Benders		Other Bends	
Metric Designator	**Trade Size**	**mm**	**in.**	**mm**	**in.**
16	½	101.6	4	101.6	4
21	¾	114.3	4½	127	5
27	1	146.05	5¾	152.4	6
35	1¼	184.15	7¼	203.2	8
41	1½	209.55	8¼	254	10
53	2	241.3	9½	304.8	12
63	2½	266.7	10½	381	15
78	3	330.2	13	457.2	18
91	3½	381	15	533.4	21
103	4	406.4	16	609.6	24
129	5	609.6	24	762	30
155	6	762	30	914.4	36

From the *National Electrical Code* © 2007, NFPA

Figure 5-14
Radius of conduit and tubing bends

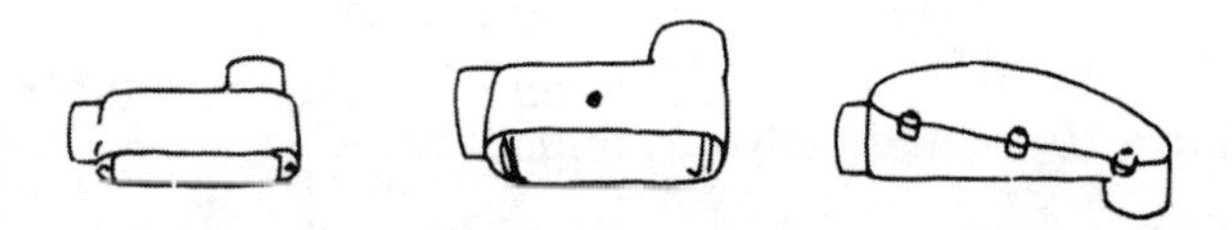

Figure 5-15
Conduit fittings (pulling ells) used with rigid, EMT, and flexible metal conduits

Table 344.30(B)(2) Supports for Rigid Metal Conduit

Conduit Size		Maximum Distance Between Rigid Metal Conduit Supports	
Metric Designator	**Trade Size**	**m**	**ft**
16–21	½–¾	3.0	10
27	1	3.7	12
35–41	1¼–1½	4.3	14
53–63	2–2½	4.9	16
78 and larger	3 and larger	6.1	20

From the *National Electrical Code* © 2007, NFPA

Figure 5-16
Spacing between supports for rigid metal conduit

Rigid metal conduit must be secured to supporting material within 3 feet of each box or fitting, and every 10 feet thereafter until within 3 feet of another box or fitting. For conduit sizes of 1 inch and greater, the spacing between supports increases with the conduit size, in accordance with *NEC* Table 344.30(B)(2); see Figure 5-16.

In the past, rigid metal conduit was used far more commonly than it is today. In many small residential and commercial buildings, the only rigid metal conduit used, if any, will be for the mast supporting the overhead service entrance. Electrical metallic tubing is used more today because it's so much easier to handle.

Electrical Metallic Tubing, Type EMT (*NEC* Article 358)

EMT or *thinwall* conduit is the most widely used nonflexible conduit. Except in situations where it'll be subject to

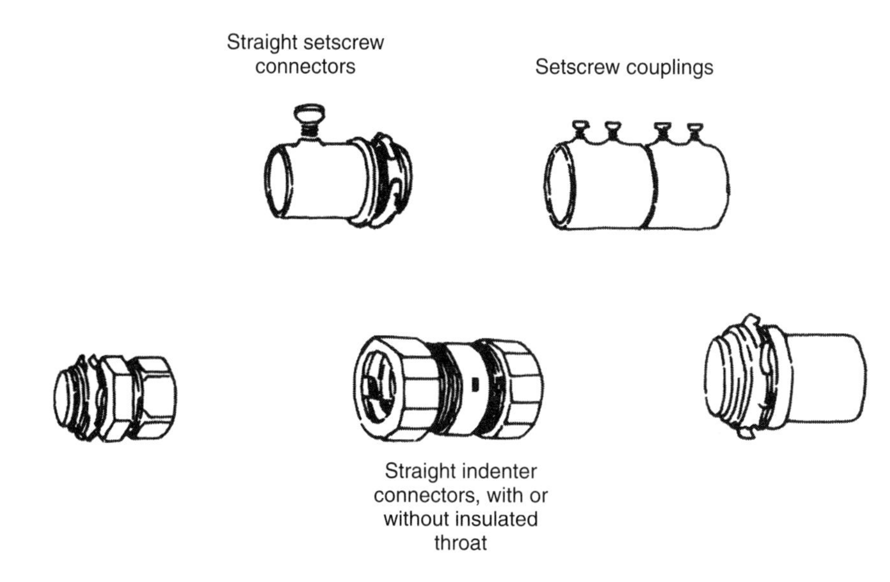

Figure 5-17
EMT box connectors and couplings

severe physical damage, it's used in virtually all places where rigid conduit can be used. It can be used either exposed or concealed in or on walls, floors or ceilings. With proper protection and fittings, it can also be embedded in concrete, buried in the ground, or used in areas subject to severe corrosion.

It's preferred over rigid conduit because it's much lighter and easier to handle. It's also easier to cut and ream. Bending to follow fairly complex forms is quite easy with the smaller diameters. And threading isn't only unnecessary, it can't be done. Use either compression-type connectors and couplings, or setscrews to make connections between successive lengths of conduit or between lengths of EMT and boxes. See Figure 5-17.

Refer once again to *NEC* Chapter 9, Table 1 (Figure 5-11) for the number of wires of different sizes and insulation types that can be used in EMT. There can't be more than a total of four quarter bends in a single run between boxes. More than that and you won't be able to pull the required wiring through the conduit. The minimum radius for bends in EMT (in the various available trade sizes) is the same as for rigid conduit. You can find them in *NEC* Chapter 9, Table 2 (Figure 5-14).

The same fittings — pulling ells, tees, and straights — used with rigid are also used with EMT. As with rigid, you can start counting bends again after any fitting or box. EMT must be securely attached to supporting structures within 3 feet of a box and every 10 feet thereafter. The material is supplied in 10-foot lengths. This insures that there are no full lengths supported only by couplings at both ends, as this would likely be too weak for safety.

Since EMT is a continuous metal armor made up of parts that are in solid, firm electrical contact with each other, it also constitutes an acceptable continuous ground return path. This makes a separate ground wire unnecessary.

Rigid Polyvinyl Chloride Conduit — Type PVC (*NEC* Article 352)

Although other materials are approved for this purpose, most of the rigid nonmetallic electrical conduit presently used is polyvinyl chloride (PVC). Other approved materials are asbestos cement, fiberglass epoxy, or high-density polyethylene.

Nonmetallic conduit can be used above ground or underground, or be concealed in walls, ceilings, or floors. It can also be used in wet locations and has excellent resistance to many highly-corrosive chemicals. Since PVC is designed for connection to couplings, fittings and boxes using solvent-type cement, it's absolutely waterproof. This makes it ideal for the many really wet areas in industrial and commercial structures.

NEC Article 352 doesn't limit the conductors allowed in a single rigid conduit by number, but instead limits the percentage of the cross section that can be filled. Again, you'll find that percentage in Figure 5-11.

As always, no conduit smaller than ½-inch trade size is permitted. Requirements for supporting rigid nonmetallic conduit are listed in *NEC* Table 352.30. Rigid nonmetallic conduit must be attached within 3 feet of each box; see Figure 5-18.

Conduit Size	*Support Spacing*
½ to 1 inch	3 feet
1¼ to 2 inch	5 feet
2½ to 3 inch	6 feet
3½ to 5 inch	7 feet
6 inch	8 feet

Figure 5-18
Support spacing for rigid nonmetallic conduit

Every length of nonmetallic conduit must be marked at least every 10 feet, giving the manufacturer's name and the type of material, so that the electrician (and the inspector) can be sure that the approved material is being used.

Surface Metal Raceways (*NEC* Article 386)

Occasionally in residential applications, and quite often in offices and commercial applications, it's necessary or desirable to surface-mount electrical wiring. That mounting must be visually acceptable to a degree that exposed NM cable or exposed EMT is not. For this purpose, complete systems of raceways, boxes and fittings, both metallic and nonmetallic, designed for surface mounting, are available under the registered trade name *Wiremold.*

Surface raceways are permitted only in dry areas. Figure 5-19 shows a variety of the fittings, couplings, connectors and boxes that are available. These allow considerable flexibility in the design and installation of surface-mounted wiring. The joints between fittings in the metal Wiremold systems are a type of compression connection. The *NEC* accepts these metal-to-metal contacts as an approved ground path, and doesn't require an additional ground wire inside the raceway.

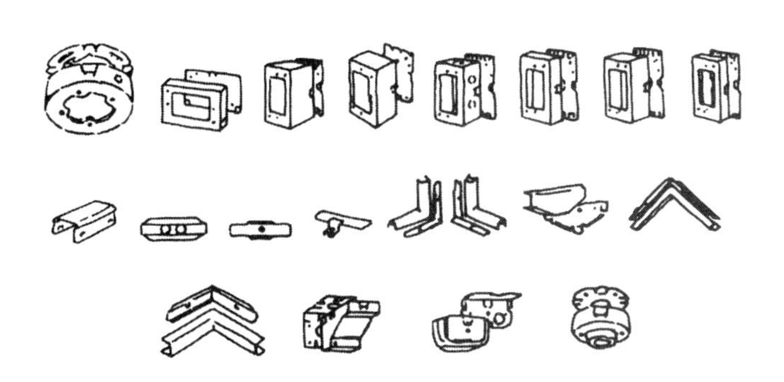

Figure 5-19
Wiremold fittings

There are several different sizes of Wiremold metal raceways, intended to handle different numbers of wires of various sizes and insulations. A table showing these raceways, and their conductor capacities, is in Chapter 10, Figure 10-19.

Surface Nonmetallic Raceways (*NEC* Article 388)

Like their metallic counterparts, you can only use surface nonmetallic raceways in dry areas. They can't be concealed, or exposed to physical damage. They can pass crossways through dry walls or a partition, as long as the section inside is unbroken. The number of conductors permitted in this type of raceway is:

Wire size	*THHN*	*TW*
14 AWG	10	6
12 AWG	7	5
10 AWG	5	4

Of course, a grounding conductor must be included when using nonmetallic raceways.

Multioutlet Assembly (*NEC* Article 380)

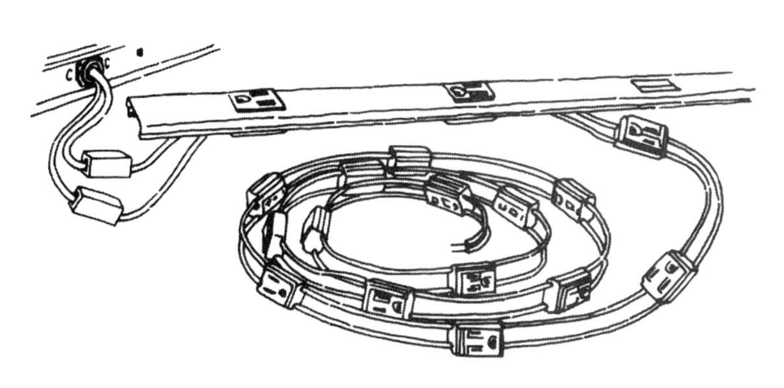

Figure 5-20
A multioutlet assembly (Wiremold)

The multioutlet assembly, shown in Figure 5-20, is a surface-mounted raceway in which outlets are placed at intervals of either six or 12 inches. The raceway contains the necessary wiring to serve the installed outlets.

These assemblies are available in lengths of three, five and six feet. Suitable connectors are also available to provide for the attachment of multioutlet assemblies to existing wall outlet boxes, conduit, flexible conduit or Wiremold.

STUDY QUESTIONS

1. What's the difference between NM and NMC cable?

A) NMC is metal protected
B) NMC is used only in nonresidential wiring
C) NMC contains signal control wires
D) NMC is fungus- and corrosion-resistant

2. When is an attic considered *accessible*?

A) There's easy access from the dwelling area through the scuttle hole in the ceiling
B) There's a permanent ladder or stairway leading to the attic
C) Two-thirds of the attic has 6 feet of headroom
D) There's a minimum 30 inches of clear headroom above the attic access opening

3. Under what circumstances is UF cable approved for direct burial in the ground?

A) If it's embedded in concrete
B) If it's used as service entrance cable
C) If it's buried at least 24 inches deep
D) If it's buried 6 inches deep and covered with a 2-inch-thick concrete layer

4. When would you use ACL rather than BX cable?

A) In inaccessible attics
B) For underplaster extensions
C) When a separate grounding conductor is needed
D) In very damp walls or areas below grade

5. When is it necessary to provide a separate grounding means for Greenfield?

A) In runs longer than 6 feet
B) When the conductors exceed 31 percent of the cross-sectional area of the conduit
C) When you use ¼-inch-diameter trade size
D) In all applications

6. Which material is *not* approved for use as rigid metal conduit?

A) Galvanized steel
B) Red brass
C) Aluminum
D) Copper-clad aluminum

7. Why is EMT preferred over rigid metal conduit in residential wiring?

A) It comes in a variety of lengths
B) The minimum radius for bends is larger
C) It's lighter and easier to handle
D) Quarter bends between boxes are unlimited

8. At what distances must EMT be fastened to supporting structures?

A) Within 3 feet of each box and every 10 feet thereafter
B) Within 3 inches of each box and every 4½ feet thereafter
C) Within 12 inches of each box and every 10 feet thereafter
D) Within 12 inches of each box and every 4½ feet thereafter

9. Which material is *not* approved for use as rigid nonmetallic electrical conduit?

A) PVC
B) LFNC
C) HDPE
D) Fiberglass epoxy

10. Where are surface metal raceways permitted for use?

A) In both concealed or exposed areas
B) In either wet or dry areas
C) Where they are out of sight and inconspicuous
D) In dry areas only

6 ELECTRICAL BOXES

All receptacles, switches, and fixtures must be installed using appropriate electrical boxes. *NEC*® Section 300.15 requires that all connections of conductors to devices, which include receptacles, switches, and fixtures, be made in boxes. Some light fixtures are supplied with junction boxes as an integral part of the fixture. Those that aren't supplied must be attached to boxes that are part of the power wiring system. *NEC* Section 300.15 further states that all connections or splices of conductors to each other must be made in boxes, or *fittings*. Fittings, also called conduit bodies, were mentioned in Chapter 5 when we discussed rigid metal conduit and liquidtight flexible metal conduit.

The number of conductors that must meet and connect in, or pass through, a particular box varies greatly, as does the number of devices mounted there. The method used to install a box in a wall, and the amount of space available for it, also varies. The *NEC* has restrictions on the number of conductors that can be brought into a box of any given size. It also has requirements on how they may be brought in, and how the box may be installed. Because of the broad range of requirements and restrictions regarding the use of electrical boxes, there are now several different types available in both metal and plastic to meet these needs.

Metal Boxes

Metal boxes can be used with any wiring system: NM cable, rigid or EMT conduit, flexible conduit, BX cable, or liquidtight flexible conduit. There are two basic shapes: rectangular boxes, to conform to the standard sizes of switches and receptacles, and octal or round boxes, for mounting ceiling or wall light fixtures.

When a box is needed solely as a joining or junction point for wiring, you can use either a rectangular or an octal box.

Single Gang Boxes

The most commonly-used single gang type is called a *handy box*. The ordinary use for this box is as the mounting place for a single switch or a duplex outlet, but other devices may be mounted in it as well. We'll look at its other uses later in the book.

Figure 6-1 shows some of the different handy boxes available. While at first glance they may look very similar, each is significantly different from the others. All of the boxes illustrated are fabricated from

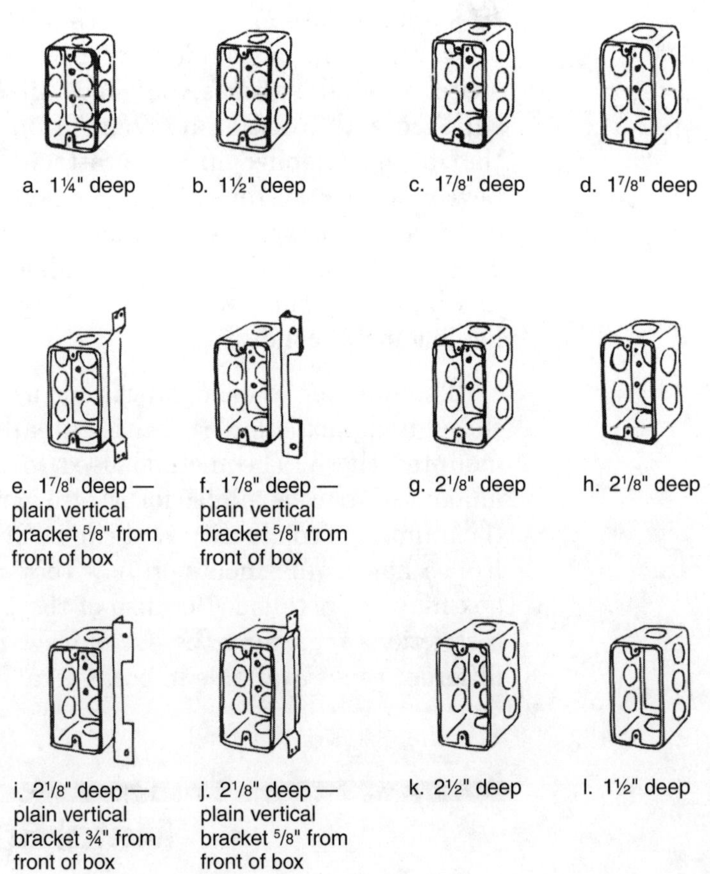

Figure 6-1
2⅛" x 4" metal handy boxes

Figure 6-2
Removing knockout

stamped sheet steel. The circles or disks on the sides, bottom and ends are partially cut out to be easily removable. They're called *knockouts* because wherever a cable or conduit is to be connected to the box, one of these disks is literally knocked out to provide an opening for mounting the appropriate connector. See Figure 6-2. Only the knockouts for openings that will be used may be removed. You can't remove ones that won't have wires through them, and leave the holes open. The small front flanges on the top and bottom are spaced and threaded to accept the mounting screws for standard switches or outlets.

Look again at Figure 6-1. The boxes shown in (a), (b), and (c) all have the same number of knockouts, and all are the same ½-inch trade size. The only difference among these boxes is their depth. The reason for this is that the number of conductors that can be brought into a box, according to *NEC* Article 314, is limited by its cubic volume. Since the length and width of a single gang box are limited by the standard dimensions of switches and outlets, the only way you can increase the number of conductors you can put in a box is by using a deeper box.

Box (d) and box (c) in Figure 6-1 have the same dimensions; only the number of knockouts is different. The knockouts in box (d) are larger because they're made for ¾-inch trade size connectors.

Boxes (e) and (f) are just as deep as boxes (c) and (d), but they're provided with side brackets for easy attachment to wall studs. The bracket on box (e) allows it to be mounted with the box front at any desired distance forward of the stud, while box (f) assures a constant measurement. Your choice depends on whether you need flexibility or uniformity.

Box (g) and box (h) are deeper than the previous ones, meaning they're approved for more conductors. Box (g) is stamped for ½-inch trade-size knockouts, while box (h) has ¾-inch holes. The two types of mounting brackets that were used on boxes (e) and (f) are used again on boxes (i) and (j), which are deeper boxes, both with ½-inch knockouts throughout.

Box (k) is the deepest box of all, but is otherwise similar to (a), (b), (c) and (g). However, box (l) is a new type; it's an *extension* box. Its back has been removed and a pair of screw holes made that line up with those on the front flanges of any of the other handy boxes. This allows it to be mounted on top of another box, greatly increasing the internal cubic volume available for wires.

In wood frame construction, the majority of handy boxes are mounted on wall studs, using brackets similar to those shown in Figure 6-1. Other methods of installing handy boxes will be discussed in Chapter 10 in connection with rough wiring.

Boxes with conduit knockouts

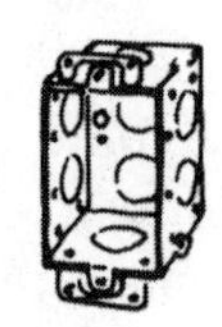
Plaster ears

"Loxbox" support welded on each side — plaster ears

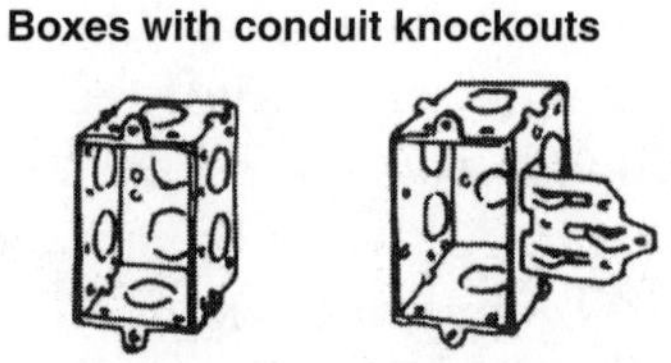
Six nail holes each side — side leveling ridges

Eagle Claw "NL" bracket, ½" from front of box

Eagle Claw vertical bracket, ½" offset

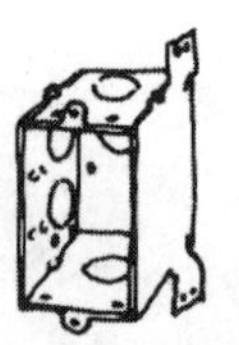
Plain vertical bracket, ¼" offset

Boxes with nonmetallic sheathed cable clamps

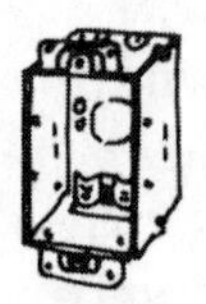
Plaster ears

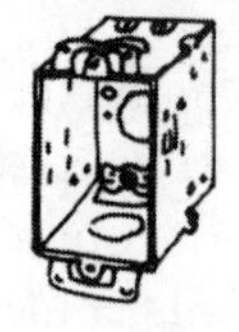
Plaster ears

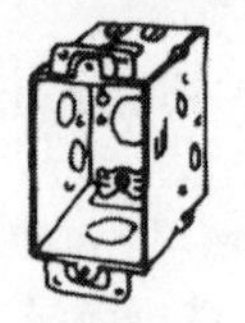
Plaster ears — leveling bosses

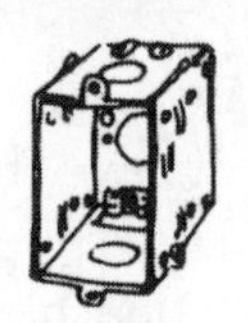
Six nail holes each side — side leveling ridges

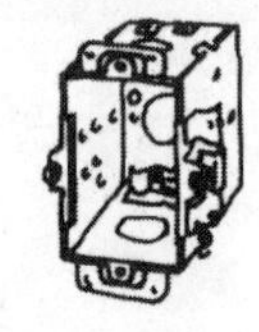
"Loxbox" support welded on each side — plaster ears

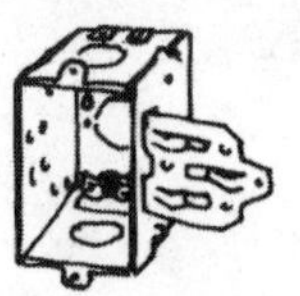
Eagle Claw "NL" bracket, ½" from front of box

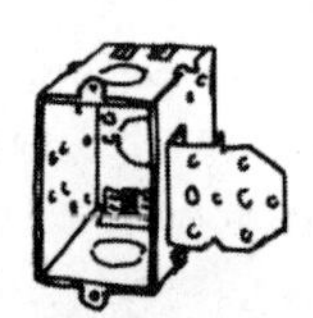
Plain "NL" bracket, ¼" from front of box

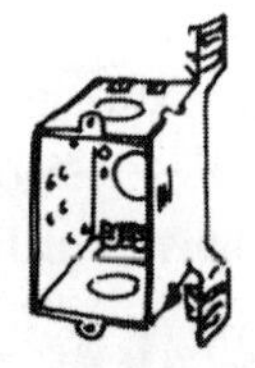
Eagle Claw ½" offset VB bracket

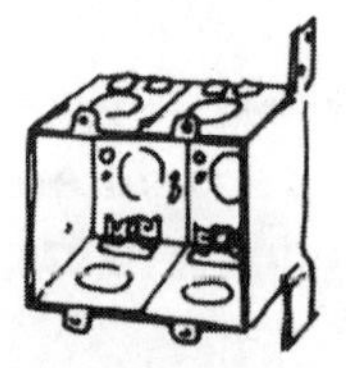
Plain vertical bracket, ¼" offset

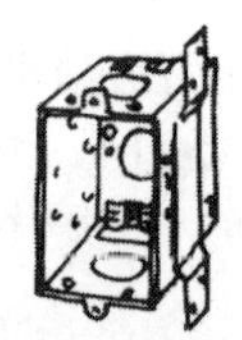
Plain vertical bracket, 3/8" offset

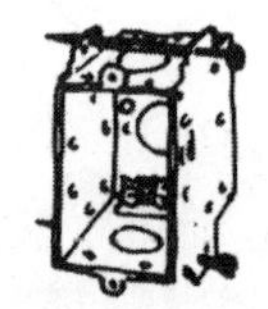
2½" deep — two 16-penny nails in side

Figure 6-3
3" x 2" square corner switch boxes

Figure 6-3 shows examples of the square corner device box. These single gang boxes are also made of sheet steel. Many have removable sides, allowing two or more to be joined together to accommodate combinations of switches, or switches and outlets. These boxes are 3 inches high instead of 4 inches like the handy box. Because they're not as high, the top and bottom front flanges, threaded for switch or receptacle mounting, are on the outside — not the inside.

Square corner boxes are available in a wide variety of knockout and mounting bracket configurations. Due to their reduced height, they're generally deeper than handy boxes to allow adequate internal volume for conductors. Some boxes are only provided with knockouts for conduit or NM connectors. Others have the round knockouts as well as internal

NM cable clamps. The internal cable clamps make installations using NM cable easy, since the separate cable clamps usually required in all knockout holes are omitted.

Another single gang box that can be connected to others to form various multiple combinations is the bevel corner type, shown in Figure 6-4. These are also supplied with internal cable clamps for NM cable and have a selection of possible mounting methods.

4-inch Square Boxes

When two switches, two outlets, or a switch combined with an outlet must be placed in a single box, use a 4-inch square or a 4 by 4 box. See Figure 6-5. You can also use this box in a situation where only a single device is required but the box is also a junction for more conductors than the single gang boxes will take.

4-inch square boxes are available with many different arrangements of knockouts and mounting brackets, as well as several depths, ranging from 1¼ up to 2⅛ inches, to accommodate various numbers and sizes of conductors. Some have all ½-inch knockouts, and

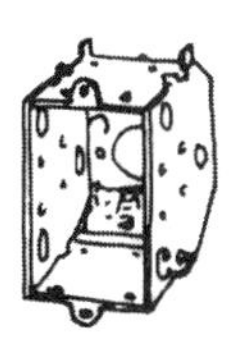

External nail brackets — leveling bosses

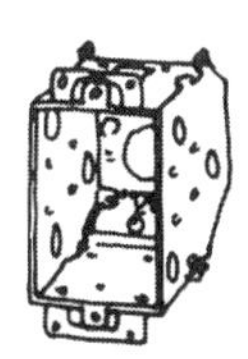

Plaster ears — external nail brackets — leveling bosses

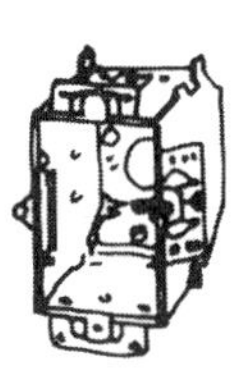

"Loxbox" support welded on each side — plaster ears

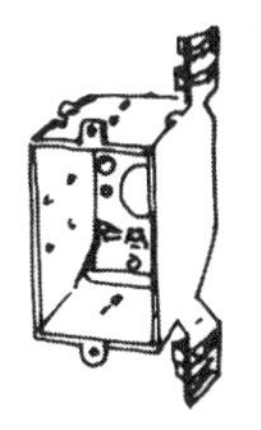

Eagle Claw "VB" bracket — ⅜" offset

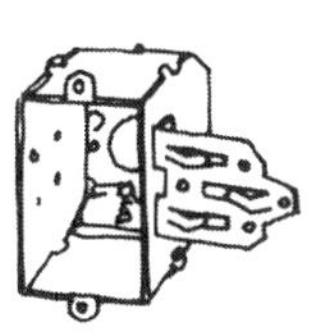

Eagle Claw "NL" bracket

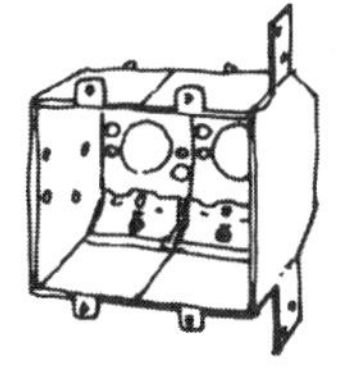

Plain vertical offset bracket

Two 16-penny nails in external brackets — leveling bosses

Figure 6-4
Bevel corner metal device boxes

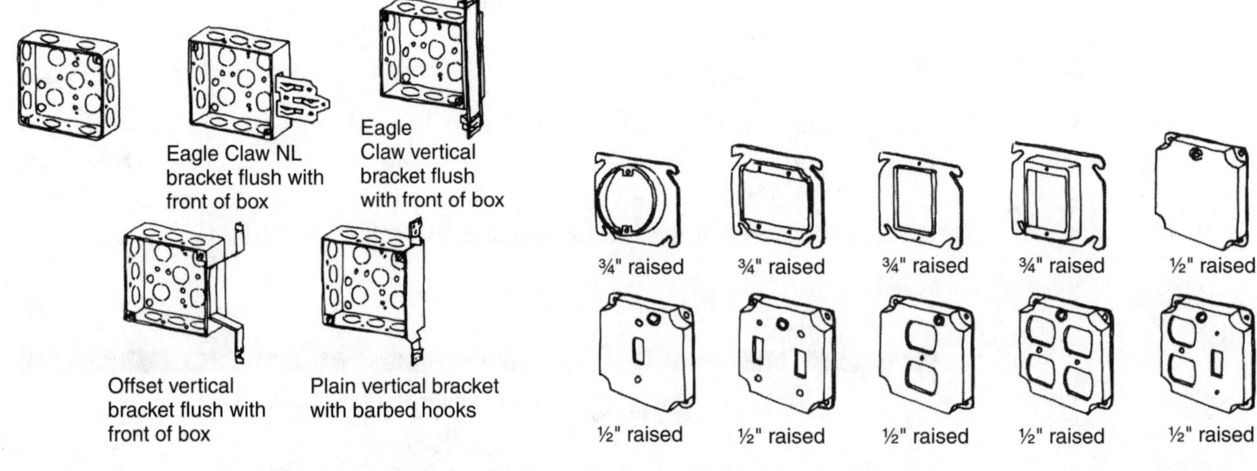

Figure 6-5
4" square metal boxes

Figure 6-6
4" square box covers and adapter plates

others have all ¾-inch, while many have different combinations of ½- and ¾-inch knockouts. You can also get boxes with all 1-inch knockouts, or a combination of all three knockout sizes.

Many 4-inch square boxes have corner-mounting flanges to allow them to accept any of the various cover plates shown in Figure 6-6. Others have two pairs of flanges, exactly the same as the single pair on a handy box. This permits the direct attachment of two switches or two outlets to the box without the necessity of an intermediate cover plate.

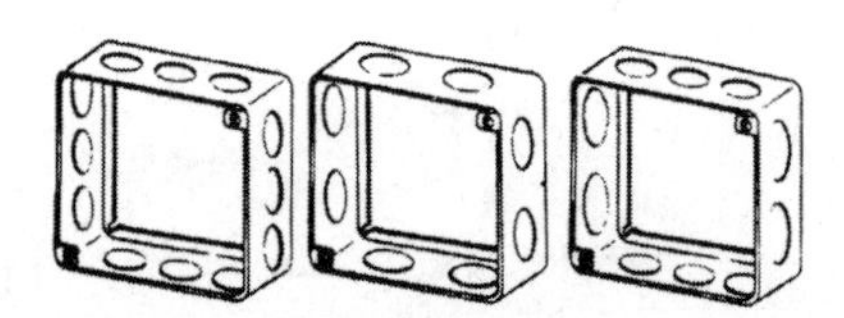

Figure 6-7
4" square box extension rings, 1½" deep

You can add extension rings to 4-inch square boxes if the number of conduits, cables, or wires that must meet in a single box becomes too large. Any of the rings shown in Figure 6-7 can be mounted in front of one of the regular 4-inch square boxes by screwing through the slide-on slots in the back of the ring. Knockouts on these rings are all ½-inch, all ¾-inch, or a combination of the two.

Multiple Gang Boxes

You can obtain larger boxes with space for three, four, five, or even six devices (switches or outlets) when required, as shown in Figure 6-8. While a three-gang box requirement is common, a four-, five- or six-gang box isn't as common, so they might be difficult to find. These larger boxes are stamped for a combination of ½- and ¾-inch knockouts, and supplied with the appropriate cover plates for mounting the devices.

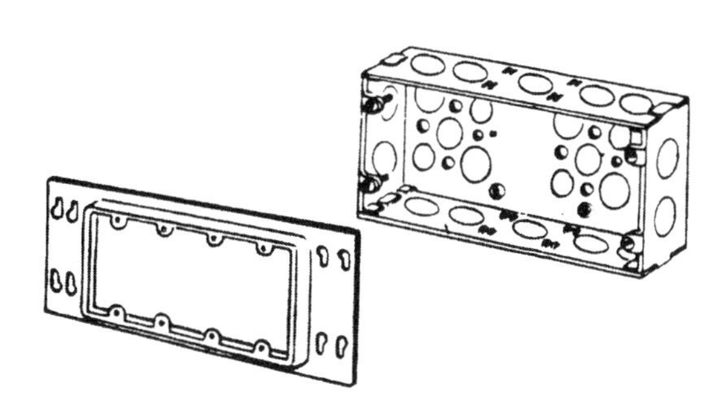

Figure 6-8
Multiple gang box for mounting switches

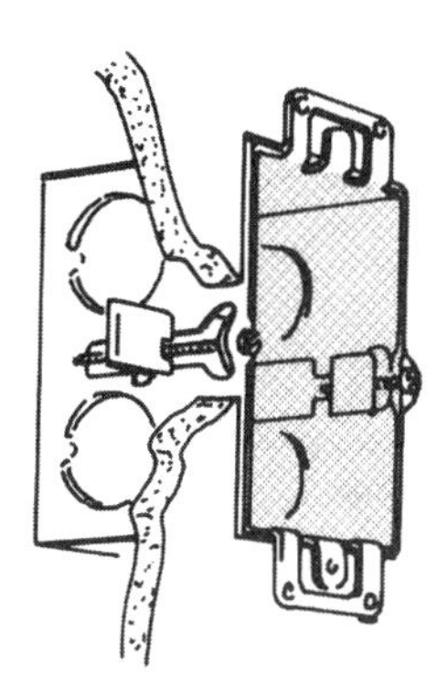

Figure 6-9
Cut-in box

Cut-In Boxes

A variation of the square corner boxes shown back in Figure 6-3 is a box equipped with adjustable clamps on the sides, like the one shown in Figure 6-9. This box is called a cut-in box. It doesn't need to be attached to the wall studs, but is installed directly in the drywall. The plaster ears hold it on the outside while the adjustable clamps are tightened against the back side of the plasterboard. This type of box is particularly useful in alteration and renovation work when switches or outlets must be added to walls that have already been finished on both sides.

Another helpful feature of these boxes is that they can be ganged by simply removing sides and screwing boxes to each other. Thus, two, three or more boxes can be ganged together as needed, and then the entire composite unit installed in the wall using adjustable clamps on the ends.

Octagon Boxes

Octagon (octal) boxes for the support of ceiling or wall light fixtures are either 3 or 4 inches across, with two threaded flanges spaced to conform to the standard used in fixture mounting straps or on the fixture bases themselves. Octagon boxes, shown in Figure 6-10, vary in depth and in knockout sizing. Some have ½-inch knockouts, some ¾ inch, and the others are mixed.

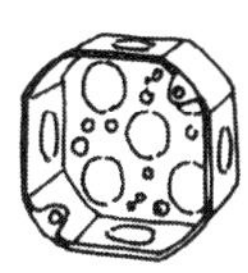

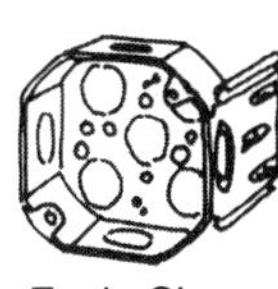

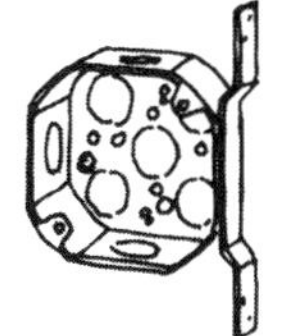

Figure 6-10
Metal octagon boxes

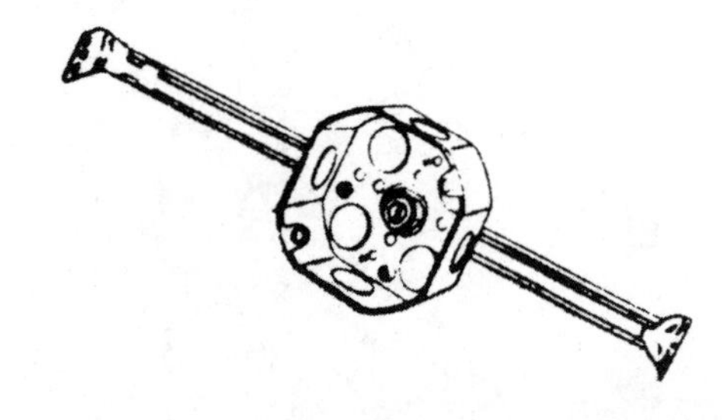

Figure 6-11
Bar hanger with box

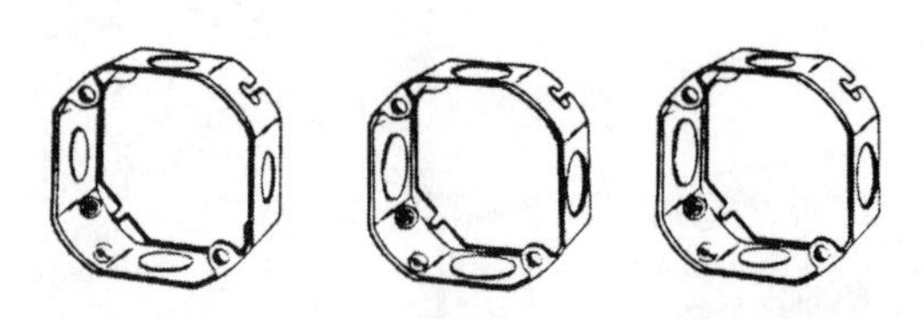

Figure 6-12
Extension rings for octagon boxes

Octagon boxes equipped with mounting brackets may be attached directly to ceiling joists. However, the location of ceiling light fixtures, as dictated by a homeowner or visual requirements, often leaves the electrician without a joist in the proper place. In that case, the electrician can either place a block between joists, or attach the box to a bar hanger, which is then attached to the joists. Figure 6-11 shows a bar hanger with box attached.

Like other boxes, you can deepen octagons using extension rings to provide additional space for a large number of conductors. Attach octagon extensions to the standard octagon boxes by screwing them onto the front, the same as rectangular extensions. The knockouts vary between ½ and ¾ inch as needed. Figure 6-12 shows ½-inch-deep octagon extension rings.

Weatherproof Boxes

According to *NEC* Article 100, weatherproof means "constructed or protected so that exposure to the weather will not interfere with successful operation." Naturally, all boxes that are used outdoors, for floodlights, security lights, convenience outlets, and switches, must be weatherproof. Some boxes in wet indoor locations also need to be weatherproof.

Figure 6-13
Weatherproof box

Weatherproof metal boxes are made of aluminum castings instead of sheet steel like other metal boxes. Instead of knockouts, they contain threaded holes where standard box connectors can be directly threaded. See Figure 6-13. All unused threaded connector holes must be tightly closed with threaded plugs to keep them watertight.

Weatherproof rectangular boxes are available in one, two, and three gang sizes. A selection of waterproof covers allows for the various mounting arrangements of waterproof switches, single and duplex receptacles, and mixed combinations of switches and receptacles.

The receptacles used in weatherproof boxes are the same as those used for normal indoor wiring. A gasket placed between the rectangular cover plate and the box, as well as the gaskets inside the individual receptacle covers, provides the necessary weatherproofing. You can use outdoor switches built into a weatherproof cover plate, or you can use indoor-type switches placed under gasketed weatherproof lids.

Weatherproof boxes come either with or without mounting lugs for wall attachment. Figure 6-14 shows boxes with mounting lugs. When the box isn't directly attached to a wall, it can be supported by two or more conduits threaded wrench tight into the box, and then the conduits are attached to the wall, as shown in Figure 6-15. If the box isn't over 100 cubic inches in size, does *not* contain receptacles or switches, and doesn't support a light fixture, then the conduits may be attached to the wall as far as three feet away from the box they support. However, if the box *does* contain switches or receptacles, or it supports a fixture,

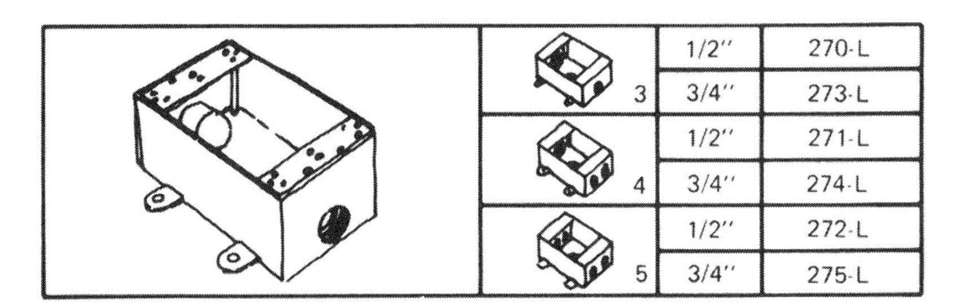

3	1/2"	270-L
3	3/4"	273-L
4	1/2"	271-L
4	3/4"	274-L
5	1/2"	272-L
5	3/4"	275-L

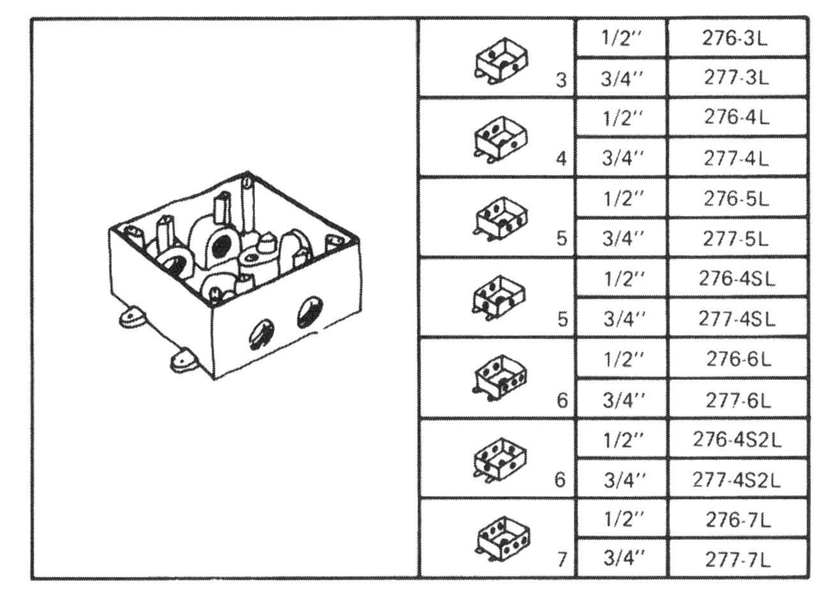

3	1/2"	276-3L
3	3/4"	277-3L
4	1/2"	276-4L
4	3/4"	277-4L
5	1/2"	276-5L
5	3/4"	277-5L
5	1/2"	276-4SL
5	3/4"	277-4SL
6	1/2"	276-6L
6	3/4"	277-6L
6	1/2"	276-4S2L
6	3/4"	277-4S2L
7	1/2"	276-7L
7	3/4"	277-7L

Figure 6-14
Weatherproof boxes with mounting lugs

Figure 6-15
Weatherproof box supported by conduits on the wall

Figure 6-16
Round weatherproof box for floodlight

then the conduits holding it in place must be attached to the wall within one inch of the box. When a weatherproof box is supported only by conduit buried in the ground, such as a box supplying power to outdoor shrub lights, it must be supported by two conduits no more than 18 inches high.

There are also round weatherproof boxes, like the one in Figure 6-16. These come either with or without lugs for wall or soffit mounting. Like rectangular boxes, they can be supported solely by two or more conduits, as long as the conduits are attached to the building as described earlier. Since most round weatherproof boxes are commonly used as supports for outdoor flood and security lights, the 18-inch maximum support spacing for the conduits usually applies.

Plastic Boxes

Plastic boxes are referred to in the *NEC* as nonmetallic boxes. They're made either of a thermosetting polyester or PVC. Plastic boxes are also approved for use with nonmetallic sheathed cable (NM cable), discussed in Chapter 5, which is the wiring method used for most residential work and other buildings not more than three stories above grade. In these wiring jobs, electricians use enormous quantities of plastic boxes. They're light and easily installed; and since they don't require grounding or box connectors, they can be wired considerably faster than conventional metal ones.

However, you most definitely *cannot* use plastic boxes with metal conduits (rigid or EMT), Greenfield, liquidtight flexible metal conduit, or armored cable. All these require metal boxes.

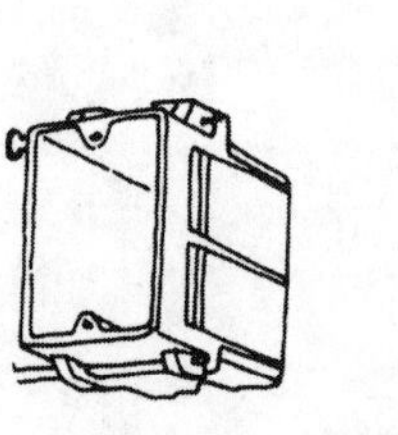

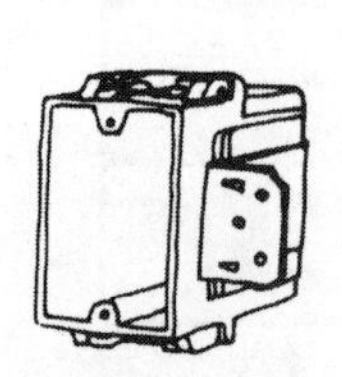

Figure 6-17
Single-gang plastic boxes

Like metal boxes, plastic boxes are manufactured in two shapes: rectangular for switches and receptacles, and round for fixture mounting. Rectangular boxes come in one-, two-, three- and four-gang sizes. Single boxes, like those shown in Figure 6-17, vary in depth from 2 to 2½ inches, and all are equipped for direct stud mounting. The models fitted with nails or *V* brackets can be installed to fit any wall surface material: ¼-inch wood paneling, ½-inch drywall, ⅝-inch drywall, or plaster. The right-angled side-bracket types are set for either ¼- or ½-inch wall surfacing materials.

Two- and three-gang rectangular boxes are available in either direct nailing or bracket types. Some three- and four-gang models are supplied with bar hangers. In my experience, a three gang without a bar hanger is more or less manageable, but a four gang really needs the additional support. Two-, three- and four-gang plastic boxes are shown in Figure 6-18.

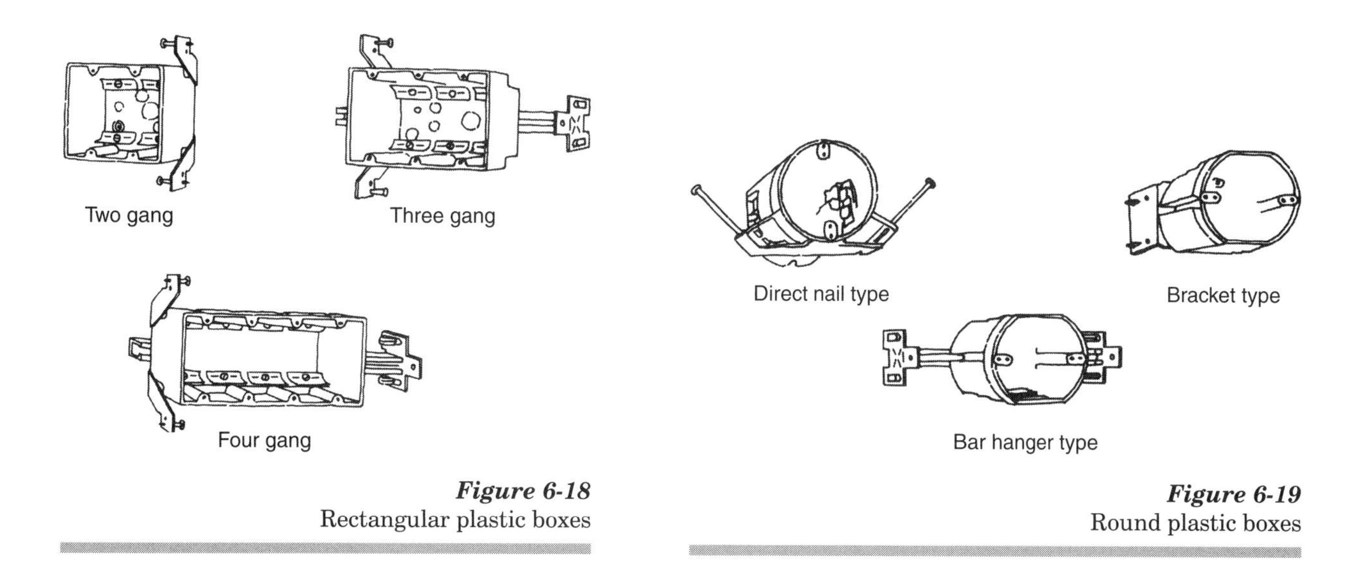

Figure 6-18
Rectangular plastic boxes

Figure 6-19
Round plastic boxes

Fasten round plastic fixture boxes with the nails supplied with the box, an attached bracket, or an adjustable bar hanger. These fastening methods are shown in Figure 6-19. Since boxes for ceiling fixtures require accurate placement, the bar hanger makes mounting fairly easy. When the fixture is rather heavy, the bar hanger provides better support than the nail-on or side-bracket boxes with one-sided attachments.

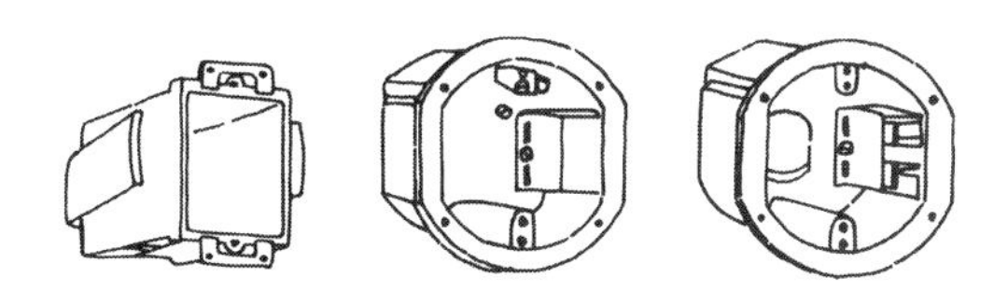

Figure 6-20
Plastic single-gang and round cut-in boxes

Cut-in boxes, both round and single gang (see Figure 6-20), are available in plastic, but the singles can't be attached to each other to make multiples like the metal cut-in boxes. Rounds can be installed directly in a ceiling without requiring any attachment to ceiling joists. But before you mount a fixture on such a box, make certain that the sheetrock or plaster on the ceiling can support its weight.

National Electrical Code Regulations Regarding Boxes

Let's look at some of the code regulations that come up most frequently when installing wiring in electrical boxes.

Number of Conductors Permitted in a Box

According to *NEC* Section 314.16, the number of conductors allowed varies depending on the size of the box, as well as the wire size of the conductors. *NEC* Table 314.16(A), shown in Figure 6-21, gives the

Table 314.16(A) Metal Boxes

Box Trade Size			Minimum Volume		Maximum Number of Conductors* (arranged by AWG size)						
mm	in.		cm³	in.³	18	16	14	12	10	8	6
100 × 32	(4 × 1¼)	round/octagonal	205	12.5	8	7	6	5	5	5	2
100 × 38	(4 × 1½)	round/octagonal	254	15.5	10	8	7	6	6	5	3
100 × 54	(4 × 2⅛)	round/octagonal	353	21.5	14	12	10	9	8	7	4
100 × 32	(4 × 1¼)	square	295	18.0	12	10	9	8	7	6	3
100 × 38	(4 × 1½)	square	344	21.0	14	12	10	9	8	7	4
100 × 54	(4 × 2⅛)	square	497	30.3	20	17	15	13	12	10	6
120 × 32	(4¹¹⁄₁₆ × 1¼)	square	418	25.5	17	14	12	11	10	8	5
120 × 38	(4¹¹⁄₁₆ × 1½)	square	484	29.5	19	16	14	13	11	9	5
120 × 54	(4¹¹⁄₁₆ × 2⅛)	square	689	42.0	28	24	21	18	16	14	8
75 × 50 × 38	(3 × 2 × 1½)	device	123	7.5	5	4	3	3	3	2	1
75 × 50 × 50	(3 × 2 × 2)	device	164	10.0	6	5	5	4	4	3	2
75 × 50 × 57	(3 × 2 × 2¼)	device	172	10.5	7	6	5	4	4	3	2
75 × 50 × 65	(3 × 2 × 2½)	device	205	12.5	8	7	6	5	5	4	2
75 × 50 × 70	(3 × 2 × 2¾)	device	230	14.0	9	8	7	6	5	4	2
75 × 50 × 90	(3 × 2 × 3½)	device	295	18.0	12	10	9	8	7	6	3
100 × 54 × 38	(4 × 2⅛ × 1½)	device	169	10.3	6	5	5	4	4	3	2
100 × 54 × 48	(4 × 2⅛ × 1⅞)	device	213	13.0	8	7	6	5	5	4	2
100 × 54 × 54	(4 × 2⅛ × 2⅛)	device	238	14.5	9	8	7	6	5	4	2
95 × 50 × 65	(3¾ × 2 × 2½)	masonry box/gang	230	14.0	9	8	7	6	5	4	2
95 × 50 × 90	(3¾ × 2 × 3½)	masonry box/gang	344	21.0	14	12	10	9	8	7	4
min. 44.5 depth	FS — single cover/gang (1¾)		221	13.5	9	7	6	6	5	4	2
min. 60.3 depth	FD — single cover/gang (2⅜)		295	18.0	12	10	9	8	7	6	3
min. 44.5 depth	FS — multiple cover/gang (1¾)		295	18.0	12	10	9	8	7	6	3
min. 60.3 depth	FD — multiple cover/gang (2⅜)		395	24.0	16	13	12	10	9	8	4

*Where no volume allowances are required by 314.16(B)(2) through (B)(5).

Figure 6-21
Maximum number of conductors in metal boxes

conductor maximums for many of the most commonly-used round, octal, square, and single-gang rectangular boxes. The permissible number of conductors is directly related to the cubic volume of the box. For example, a box 4 inches square by 1¼ inches deep (see line 4 in the table) may house 9 conductors at wire size #14, but only 8 at size #12. By increasing the depth to 2⅛ inches, the number of size #14 wires permitted increases to 15.

Not only does *NEC* Section 314.16 specify the number of conductors permitted in a box, it also defines how they're to be counted. As you'll see below, *NEC* Section 314.16 has an elaborate procedure for arriving at the total conductors in a box:

Table 314.16(B) Volume Allowance Required per Conductor

Size of Conductor (AWG)	Free Space Within Box for Each Conductor	
	cm³	in.³
18	24.6	1.50
16	28.7	1.75
14	32.8	2.00
12	36.9	2.25
10	41.0	2.50
8	49.2	3.00
6	81.9	5.00

Figure 6-22
Volume allowance per conductor

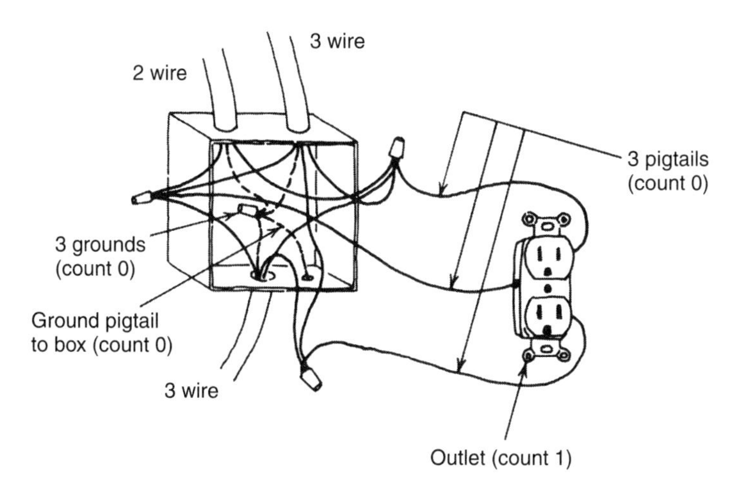

Figure 6-23
Box with cables and outlet showing conductor count

1. An insulated wire entering a box and terminating inside the box in a splice or at a terminal on a device, such as a switch or receptacle, counts as one conductor.
2. With one or more cable clamps present, a single volume allowance, per *NEC* Table 314.16(B), shown in Figure 6-22, is required per fitting type based on the largest conductor present.
3. A conductor both originating and terminating inside the box (a pigtail) isn't counted.
4. For each strap holding one or more devices, make a double allowance, per *NEC* Table 314.16(B), based on the largest connected conductor.
5. Where one or more grounding conductors enter a box, make a single allowance, per *NEC* Table 314.16(B), based on the size of the largest one.
6. A conductor that originates outside a box and simply runs through it and out again without being cut or connected to anything counts as one conductor.

Let's see how this works in an example. Look at Figure 6-23. The box shown has two 3-wire and one 2-wire cables entering, plus three grounds, four pigtails, and a duplex receptacle. The conductor count for this box is:

a. Circuit conductors	8
b. Three grounds (count as only 1)	1
c. Receptacle	1
d. Pigtails	0
Total	*10*

If the conductors in the example above are #14 AWG, according to *NEC* Table 314.16(A), there are three choices of box size permissible:

1. 4 x 2 1/8 round/octagonal
2. 4 x 1½ square
3. 3¾ x 2 x 3½ masonry

Entrance of Conductors into Boxes

Requirements for conductors entering boxes are given in *NEC* Section 314.17. Using the appropriate connectors, like those in Figure 6-24, protects conductors entering metal boxes from abrasion. Notice that all of these connectors completely fill the knockout holes into which they're fitted, as required per *NEC* Section 314.17(A).

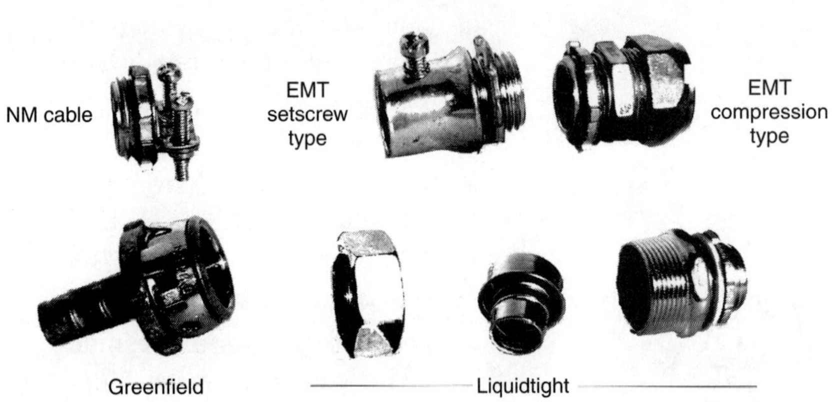

Figure 6-24
Box connectors

Greenfield and liquid-tight connectors protect the wiring by screwing into the inside of the cut ends of the spiral armor. With BX cable, the wiring is protected by a fiber bushing that slips inside the armor end after cutting. EMT is only protected if you properly ream the tube ends after cutting; so, inadequate reaming is extremely dangerous.

NM cable can be brought into plastic boxes by simply passing it through a cable knockout. Since plastic boxes don't present the same abrasion danger as metal, you don't need connectors. However, the cable sheathing must extend at least ¼ inch inside the box, and the cable must be fastened to the wall framing within 8 inches of the box.

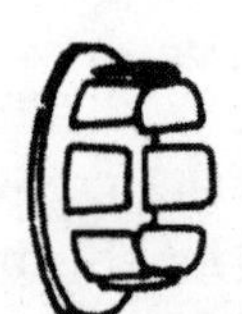

Figure 6-25
Steel knockout plug

Unused Openings All unused box openings must be closed. If by accident or error, you open the wrong knockout in a metal box, the hole must be plugged using a steel knockout plug like the one in Figure 6-25. Unfortunately, plugs are only available for metal boxes. If you knock out the wrong hole in a plastic box, your only option is to use another box.

Flush Devices Flush devices are primarily switches and receptacles. Their boxes must be completely enclosed on the back and sides, and properly support the device, per *NEC* Section 314.19. A box with no back, like the one shown in Figure 6-1, box (l), couldn't be used by itself as a flush device box; nor could you use any of the gangable boxes with removable sides with one side missing (refer back to Figure 6-3).

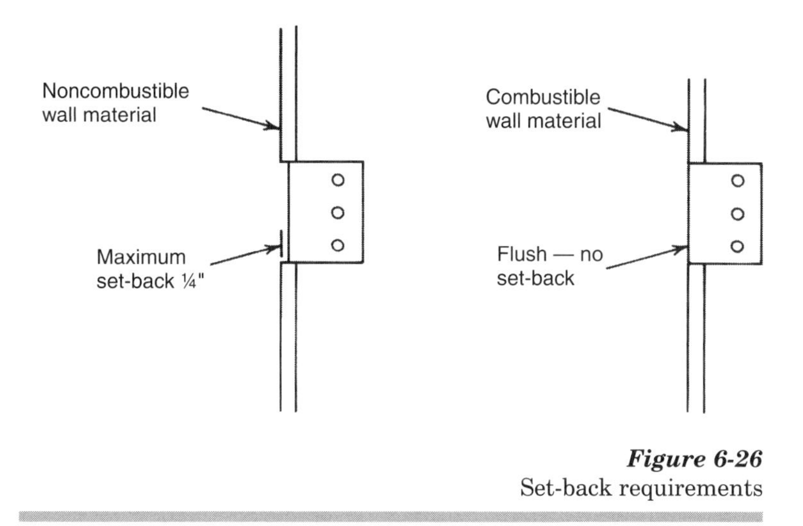

Figure 6-26
Set-back requirements

Wall Box Set-Back

NEC Section 314.20 says that where boxes are set in walls or ceilings of concrete, plaster, sheetrock, or other noncombustible material, the front of the box shouldn't be set more than ¼ inch back from the wall surface, as shown in Figure 6-26. However, if the wall is wood paneling or other combustible material, the box front must be flush with the surface — no setback is allowed.

Surface Repairs

If plaster or plasterboard wall surfaces break around a box, repair them so that no gaps or openings greater than ⅛ inch remain around the edges, per *NEC* Section 314.21. In most cases, this requirement is unnecessary. Any gaps around boxes larger than the code requires closing would be unsightly and the homeowners would insist on having them repaired.

Box Accessibility

NEC Section 314.29 requires you to ensure that boxes, conduit bodies, and handhole enclosures are accessible. What it means by "accessible" and what you may think is accessible may not be the same thing. The code says that boxes must be installed so that the wiring inside can be reached without making a hole in a wall or a ceiling, or cutting any framing out of the way.

In other words, a box can be installed in a location where it will take all your athletic ability to reach it, as long as you can remove the cover and access the wiring without any damage to the building. However, a junction box can be placed behind the plasterboard facing of a living room wall, as long as you provide an access door so you can reach it without damaging the wall. If you make an installation like that, be sure it's in an inconspicuous location or you'll be dealing with some very unhappy customers.

Now that we know what kinds of conductors we'll use to connect devices to each other, and what types and sizes of boxes we'll use to house these devices, it's time to find out exactly how switches and outlets should be wired.

STUDY QUESTIONS

1. What governs the length and width of a handy box?

A) The number of conductors that must be brought in
B) The dimensions of standard switches and outlets
C) The number and size of the knockouts needed
D) *NEC* limitations on internal cubic volume

2. What is the main difference between square corner device boxes and handy boxes?

A) They have external NM cable clamps
B) They're generally shallower than handy boxes
C) The top and bottom front mounting flanges are inside
D) They're 3 inches high instead of 4 inches

3. The largest multiple gang box comes in a size large enough to accommodate how many devices?

A) 10
B) 8
C) 6
D) 4

4. What can you use to center-mount a box for a ceiling light fixture?

A) An extension ring
B) A bar hanger
C) Threaded drywall toggle bolts
D) Adjustable mounting brackets

5. How are weatherproof boxes different from other boxes?

A) Weatherproof boxes are made only of steel
B) Only special receptacles and switches can be used with weatherproof boxes
C) Instead of knockouts, weatherproof boxes have threaded holes
D) They are generally deeper to allow for the weatherproof gasket

6. Which of the following is allowed for use with plastic boxes?

A) NM
B) NM, EMT
C) NM, Greenfield
D) NM, liquidtight, BX

7. In totaling the conductor count in a metal box, how would you count 3 pigtails?

A) They would count as 0 conductors
B) They would count as 1 conductor
C) They would count as 3 conductors
D) You can't have 3 pigtails in one box

8. What type of box connectors do you need with plastic boxes?

A) Screw-in connectors
B) Compression connectors
C) Thermosetting polyester or PVC connectors
D) No box connectors are needed with plastic boxes

9. What must you do if you knock out one too many holes in a metal box?

A) Use another new box
B) Use a closed-end connector in the empty hole
C) Fill the unused hole with a steel plug
D) One extra hole is permitted in locations not exposed to dampness

10. What is a flush device?

A A box that is set-back 3/8 inch into the wall
B) A device to smooth out burrs in conductors
C) A connector that is flush with the box
D) A switch or receptacle

7

WIRING SWITCH CIRCUITS AND OUTLETS

In all buildings — residential, commercial, or industrial — most lights, some appliances and machinery, and the switched half of many convenience outlets are controlled by switches or switch circuits that are integral parts of the building power wiring, not parts of the devices being switched. This lets you place switches at the most convenient operating locations, at considerable distances from the device they control.

Switch Types

In order to accomplish various switching functions, you may need several types of switches, see Figure 7-1.

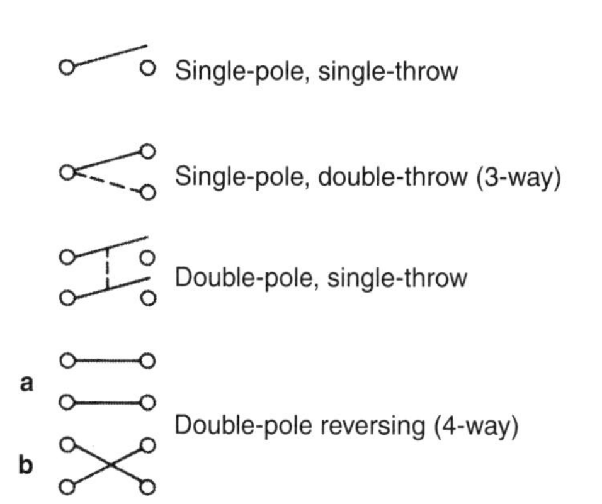

Figure 7-1
Types of switches

Single-Pole, Single-Throw Switch

This basic *on-off* switch is the most common. Internal mechanisms vary somewhat from manufacturer to manufacturer, but the result is the same. Whether accomplished with a toggle or rocker, the two terminals on the case are internally connected to each other or disconnected.

Single-Pole, Double-Throw

The type of switch used in building wiring (as opposed to machinery controls or electronic equipment) is more often referred to as a *3-way switch*. See Figure 7-2. This switch has three terminal screws and its handles aren't marked *on* and *off*. The dark colored screw, known as the "hot" or common terminal, connects to the power. This type is an *either-or* not an *on-off* switch. It connects terminal *A* to either *B* or *C* — never both, and never neither. Thus, it's never *off*. There are various ways of connecting and using this switch.

Double-Pole, Single-Throw

This is a second type of *on-off* switch. This time, two pairs of terminals are simultaneously connected or disconnected. In building wiring, this switch is used on all 240-volt circuits to control both hot lines at once, as required by *NEC*® Article 404. The double breakers required for range, dryer, 240-volt water heater, and split-wired appliance outlets are internally double-pole single-throw switches. The fused disconnecting mechanisms required for small central air conditioning systems are also double-pole single-throws.

In addition, the ground-fault circuit-interrupter (GFCI) discussed in connection with code requirements in Chapter 8 is also a double-pole single-throw, since it disconnects both hot and common sides of a 120-volt circuit simultaneously.

Double-Pole Reversing Switch

This is another *either-or* switch. More commonly known as a *4-way switch*, also shown in Figure 7-2, it's used in conjunction with two 3-ways in the multiple switch circuits. The modern 4-way switch has four terminal screws: one pair for input from the 3- or 4-way switch that feeds it and one pair for output to the 3- or 4-way switch that it feeds. Internally, it connects the two terminals at one end of its case, either straight through to the corresponding two at the other end, as in Figure 7-1 (a), or it connects diagonally across in an *X* formation, as in Figure 7-1 (b).

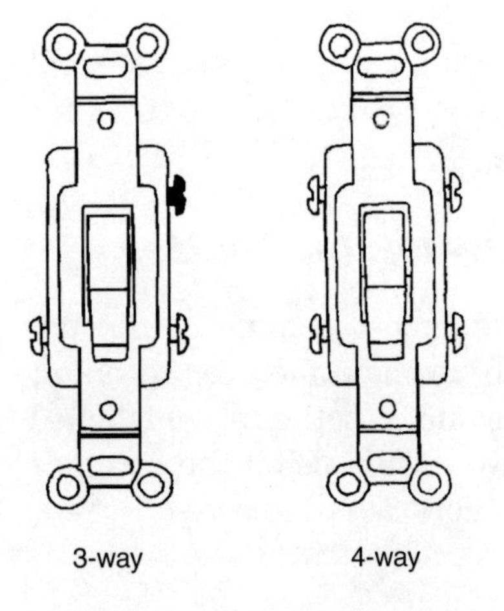

Figure 7-2
Switches

Dimmer Switches

In incandescent lighting circuits, dimmer switches have been popular for residential for many years and even longer for special effect lighting. Although there are many types of specialized dimmers available, the most commonly-used dimmers are the inexpensive rotary type, which I'll discuss here. They're supplied with either wire pigtails or screw or push-in terminals, such as those found on single-pole, 3- or 4-way switches.

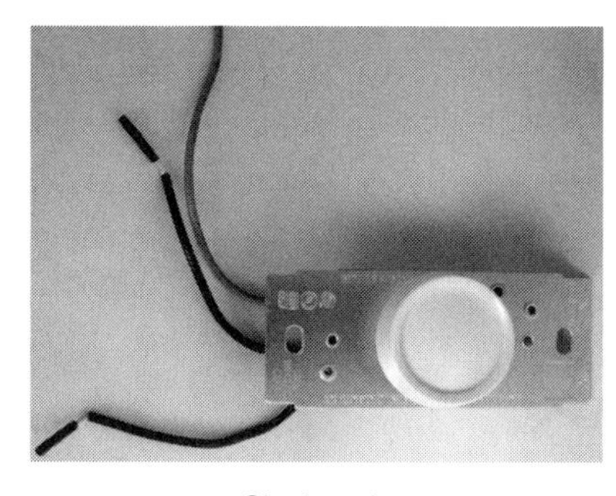

Single pole

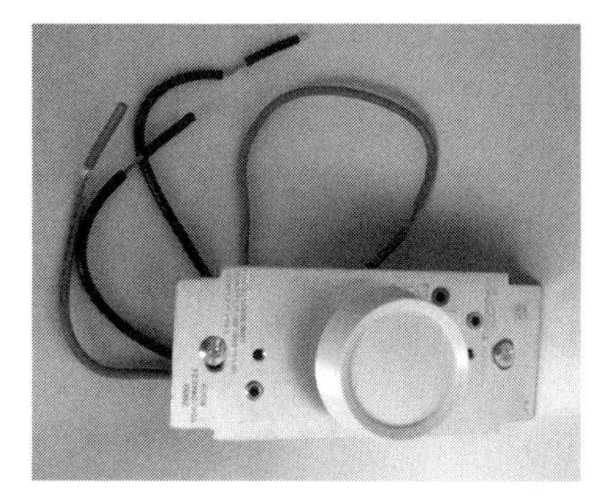

3-way

Figure 7-3
Dimmer switches

Two types of dimmers are available: the dimmer combined with a single-pole switch, or the dimmer combined with a 3-way switch. The single-pole will have two black pigtails for connections; the 3-way will have one black and two reds; see Figure 7-3. The 3-way might be substituted for either one of the standard 3-way switches in any of the multiple switch circuits. The single-pole dimmer may be used to replace any standard single-pole switch. Connections to both types are made by splicing directly to the wire pigtails provided.

> ***REMEMBER:*** Dimmer switches are for use with incandescent lights *only*. Fluorescent lights can also be dimmed, but a different and more expensive dimmer is required. A special dimming ballast must replace the standard ballast supplied with the fixture. Consequently, fluorescent dimmers are rarely used.

The average, inexpensive incandescent dimmer will handle loads of up to 600 watts. For most residential or small office applications, this wattage is generally plenty of capacity. However, just to be sure, make it a routine to total the load that the dimmer must handle. If the load is more than 600 watts, you'll find it necessary to use a somewhat more expensive, higher capacity unit.

Pilot Light Switches

These switches are commonly used on lighting circuits that have been set up so that the light or lights they control aren't visible from the switching location. For example, there might be a light in the attic with a switch on the floor below, or there could be a light in the basement with a switch on the floor above. Pilot lights are commonly used for outdoor lighting circuits such as driveway post lights, decorative shrubbery lights, or security light systems. In all cases, the pilot light acts as a reminder that something not visible from the switch location is turned on.

However, the pilot light indicates *only* that a circuit has been turned on. This *by no means* indicates that the device at the end of the line is operating. The pilot light is wired in parallel with the load being switched. Let's look back at the basic series and parallel circuits illustrated in Figure 1-10 in Chapter 1 to remind you that devices wired in parallel operate independently of each other. Thus, the pilot light at the switch could be operating

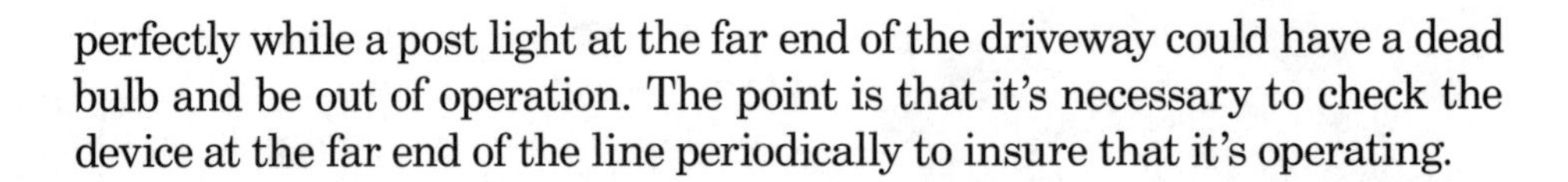

perfectly while a post light at the far end of the driveway could have a dead bulb and be out of operation. The point is that it's necessary to check the device at the far end of the line periodically to insure that it's operating.

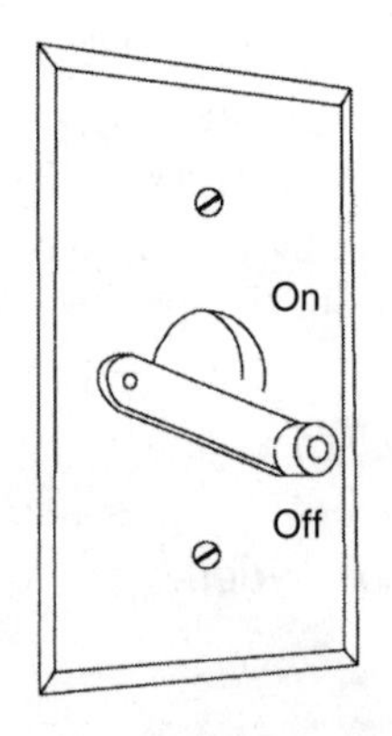

Figure 7-4
Weatherproof switch

Waterproof Switches

Often, it's convenient to control outdoor devices with outdoor switches. Such switches must be waterproof; see Figure 7-4. Waterproof switches are available in single-pole or 3-way types. This permits control of a circuit from one or two outdoor locations, or from one outdoor location plus one or more inside. These switches are made for mounting on either standard or waterproof boxes. They're also supplied with a waterproofing gasket to seal the joint between the switch faceplate and the box.

Photo-Control Switches

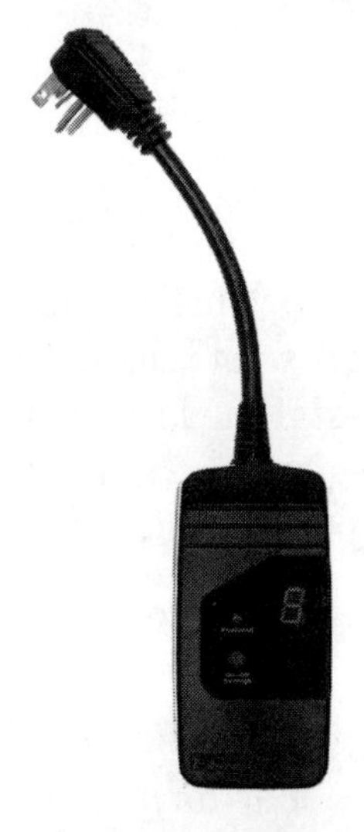

Figure 7-5
Photo-control switch

There are various requirements for lighting that are necessary throughout the night, but there aren't any for daylight. Circuits needed under these conditions can be controlled using timers. However, as seasons change, and the beginning and end of Daylight Savings Time modifies, day and night hours change considerably. This means that timers will require frequent resetting — a big nuisance.

An alternate method for the control of such circuits is the use of photo-control switches, shown in Figure 7-5. These devices contain a photoelectric sensor; actually, it's a small photovoltaic cell that produces a slight voltage sufficient to hold the switch open as long as there's enough light to activate it. When night falls and the light fails, the cell goes dead, the switch closes, and the lights go on. Naturally, the sensor unit must be adequately shielded from any customary night lighting in the area. If not, it'll sense the light, activate the photovoltaic cell, and open its switch — shutting off its light circuit.

Photo-controls are commonly supplied with a threaded fitting that'll enter a ½-inch standard box knockout or will thread directly into a standard waterproof outdoor box. Connections are normally made by splicing directly to wire pigtails on the photo-control. A photo-control is available that doesn't require any wiring at all. It simply consists of a threaded base that screws into a standard light bulb socket above which is installed a standard bulb socket controlled by an internal light-activated switch.

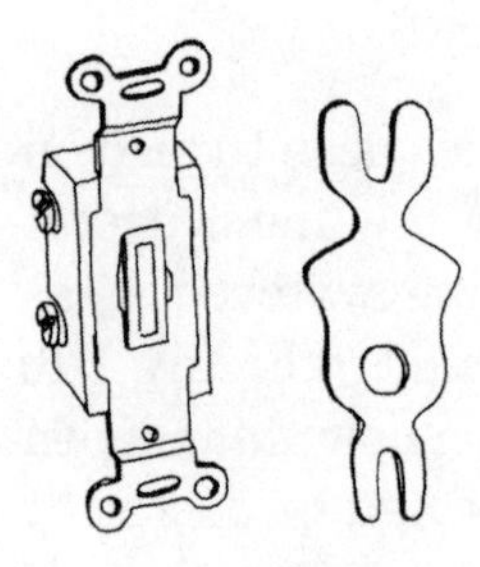

Figure 7-6
Key switch

Key Switches

When children are present or for various reasons you want a safer environment, tamper proof yet accessible switches might be useful. This need can be met with a key switch, like the one in Figure 7-6. The key switch is simply a single-pole switch with the toggle removed and replaced with a key slot. It may be activated only when the proper key is inserted. Electrical connections are the same as for any other single-pole switch.

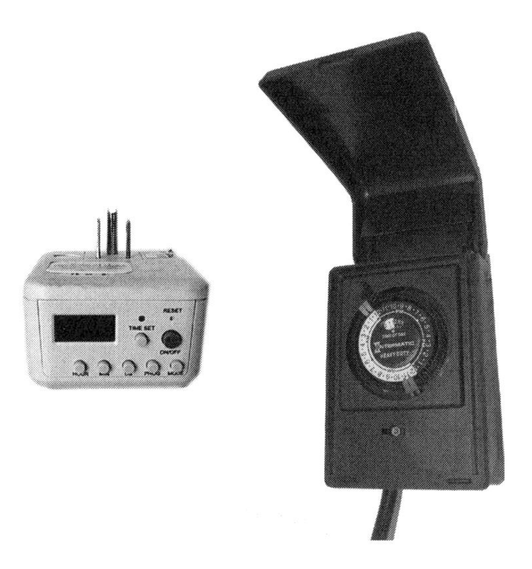

Figure 7-7
Digital timer and timer switch

Pool Filter Systems, Hot Tubs, Sprinkler Systems, Night Lighting

These devices are usually controlled by timer switches, like the one shown in Figure 7-7. Additionally, timers are being included in an increasing number of space and water heaters, as well as air conditioning, in order to conserve energy.

Timed switching units contain a small clock operating on the 60-cycle line current. The clock turns a plate calibrated to 24 hours. Control tabs are attached to the plate that'll open or close a power switch at whatever times have been set on the plate. A digital timer is also available requiring no wiring at all. It easily plugs into a standard wall socket and has a two-prong plug socket on its side to which can be connected whatever needs to be timed. It can be set to turn on and off twice. Larger units that fit into a standard 2 x 4 box can be set to cycle up to 24 times for those who need multiple on-off settings.

Wiring is connected to screw terminals. Inside the cover of each timer switch is a wiring diagram that illustrates the proper connections for that switch. Standard screw head color-coding of brass heads for power and silver heads for common will be used. However, which screws are for input and which are for output varies with each manufacturer. When installing one of these, assume that all else has already failed and read the instructions *first*.

Wiring Switch Circuits

All building wiring runs are done as discussed in Chapter 5, with grouped conductors in one form or another: either with factory-preassembled cables or with some type of conduit through which the needed wires can be pulled. In preassembled cables, the individual wires are already color-coded: for power, either black or black and red; for common, white; for ground, bare, green or green with yellow stripes. In conduit, the electrician must code them correctly when he pulls his wires.

As mentioned earlier, most switches have disconnecting means on the power side of the line only; the common return, or grounded side of the line, remains connected.

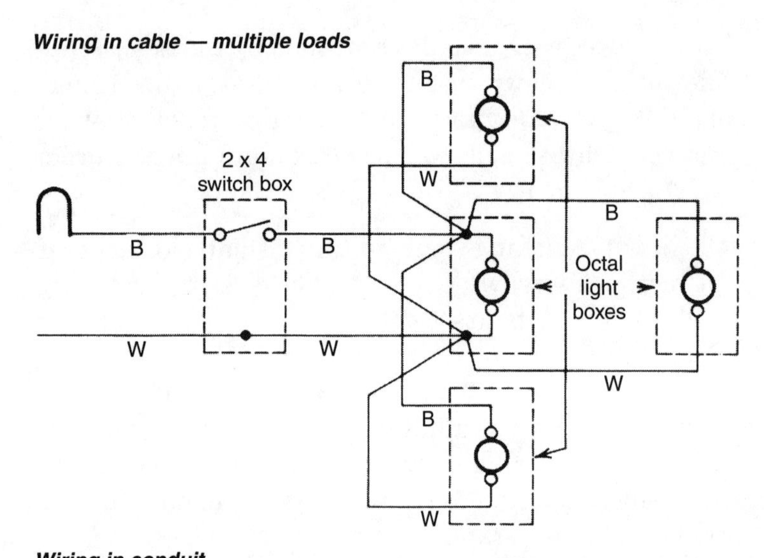

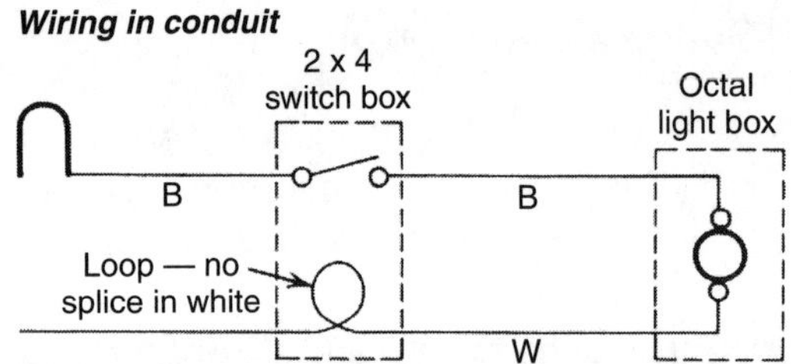

Figure 7-8
Wiring of single-pole switch circuits — power in at switch box

Wiring of Single-Pole Switch Circuits

This switch has two screw terminals, two push-in terminals, or both. Power enters through one and leaves from the other. Internally, the switch either connects or disconnects these two terminals. So it makes no difference which one is the input or output.

The basic parts of a single-pole switch circuit are a switch, a load, and the wires to connect them. The incoming feeder consists of a power or hot wire (usually black, red or blue), a common grounded return (white or grey), and a safety grounding path which may be a grounding wire (cable or a continuous metal sheathing, like EMT).

> ***IMPORTANT!*** The common or grounded return (white or grey) wire goes to all loads, such as lights, outlets, or using devices. It *doesn't* go to switches.

In Figure 7-8, the wiring of a single-pole switch circuit varies depending on where the unswitched power comes in. When the feeder enters the circuit at the switch box, in NM or other cable, the connections will be black (B) power in to one terminal on the switch, then black power out of the other switch terminal, and on to the load. White (W) common incoming splices to white common outgoing. Neither white goes to the switch. Outgoing white connects to the load, shown in the diagram called *wiring in cable* in Figure 7-8. When the single-switch controls multiple loads, the additional wires are spliced onto the switch output in whatever order happens to be convenient; see the diagram labeled *wiring in cable — multiple loads.*

Changing from cable to conduit, to wire the same circuit with either single or multiple loads, only one simple change is made. Now, look at the diagram labeled *wiring in conduit* in Figure 7-8. In the switch box, the white common merely passes through on its way to the load. Since it attaches to no device or other wires in that box, it doesn't need to be cut at all. As a standard procedure when working in conduit, any wire that simply passes through a box on its way to somewhere else shouldn't be cut and spliced. Just leave a loop a couple of inches in diameter in the bottom of the box, and continue it on to its destination. In this case, the white is looped and passes on to the load.

Often, unswitched power enters the circuit at a ceiling light, but it passes onto a controlling switch before it gets to the light. This is common because, often in residential applications, it's convenient to distribute power through the building via the attic. Since the code permits residential buildings and others not over three stories above grade to be wired in cable, a problem appears with insulation color-coding. All two-conductor cables contain one black power wire, one white common wire, and a bare ground. To take only power to a switch and back, two black wires are needed, since both wires are on the power side. This seems to violate *NEC* Section 200.7, which states that only conductors with gray or white covering or marking at the termination, or with three continuous white stripes on insulation colored other than green, may be used unless *NEC* Section 200.7(B) or (C) permit it.

NEC Section 200.7(C)(2) then states:

> *"Where a cable assembly contains an insulated conductor for single-pole, 3-way, or 4-way switch loops, and the conductor with white or gray insulation or a marking of three continuous white stripes is used for the supply to the switch but not as a return conductor from the switch to the switched outlet. In these applications, the conductor with white or gray insulation or with three continuous white stripes shall be permanently reidentified to indicate its use by painting or other effective means at its terminations and at each location where the conductor is visible and accessible."*

When wiring in cable (Figure 7-9), the incoming black supply line from the power source is spliced to white in the cable connecting to the switch box. There, both black and white connect to the switch with black returning power from the switch to the load. Meanwhile, white common from the power source is stopped at the load, because as noted earlier, it never goes to switches. When the switch is to control multiple loads, they may as required, be connected to the first one.

Wiring in cable

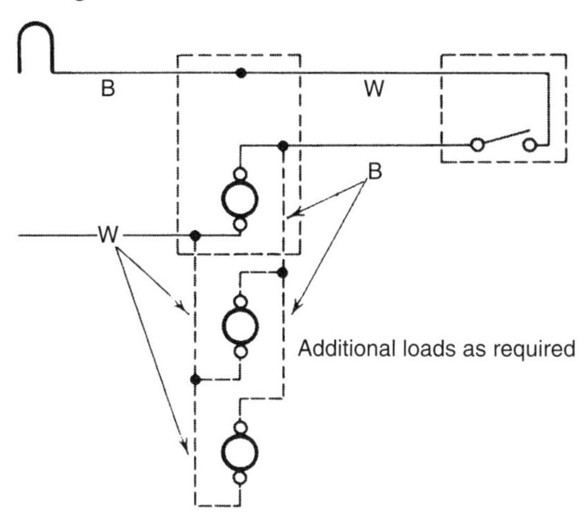

Wiring in conduit

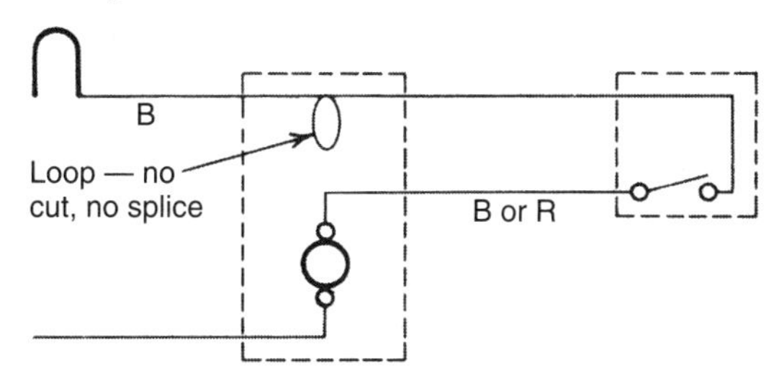

Figure 7-9
Wiring of single-pole switch circuits — power in at load box

To wire the same circuit in conduit (Figure 7-9), the color change from black to white in the power input line to the switch isn't only unnecessary, it's *not permitted*. Since the unswitched power line only passes through the light box to get to the switch, it doesn't need to be cut — it merely loops and passes through. Since the two wires into and out of the switch are both power wires, both must be black, or any other proper power color.

Wiring of 3-Way Switch Circuits

When there's a need to control a load from two different locations, two 3-way switches are used. Controlling a light in a stairway from either the top or the bottom is a common use.

The 3-way switch isn't an *on-off* switch, so when power is supplied to terminal *A*, either terminal *B* or terminal *C* will be live at all times. See Figure 7-10. Two 3-way switches are wired so that the *B* and *C* terminals on one switch are connected to the *B* and *C* terminals on the other by two wires called *travelers*. Power enters a 3-way switch circuit via the *A* terminal of one switch, and it passes through either terminal *B* or *C* to one of the travelers, shown in Figure 7-11 (a). At the second switch, it either goes on through and out the corresponding *A* terminal, or it stops because the second switch is turned toward the dead rather than the live traveler. See the examples in Figure 7-11 (b) and (c).

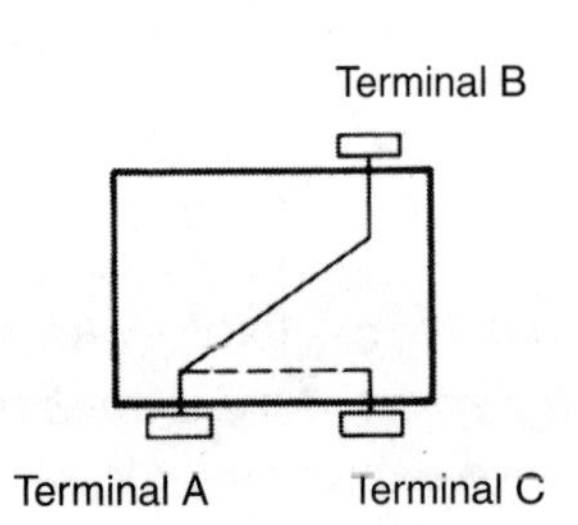

Figure 7-10
3-way switch

Two 3-way switches and the load they control can be wired in cable in any of five variations depending on the electrical sequence in which the parts are arranged, and the point at which power enters the sequence. Figure 7-12 shows a floor plan in which the part sequence places the two switches together, with the load at one end. Regardless of where unswitched power enters this sequence, it's wired using a 3-wire cable between the switches — the other cables to and from the switches are 2-wire type.

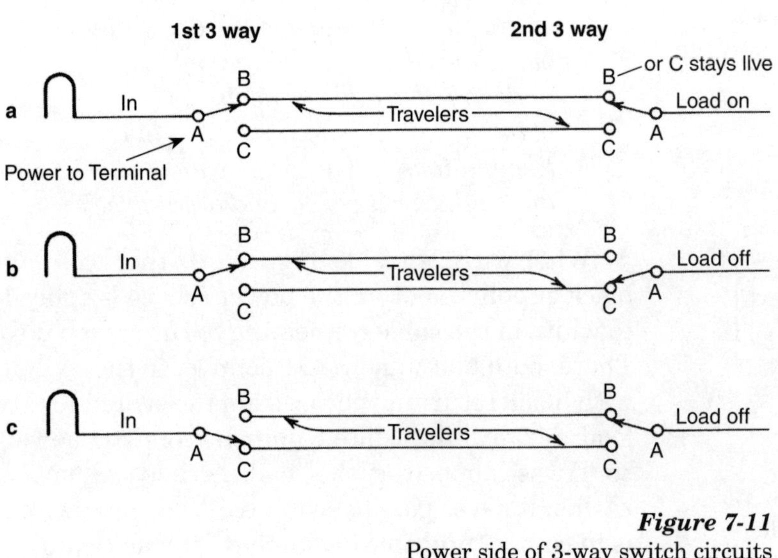

Figure 7-11
Power side of 3-way switch circuits

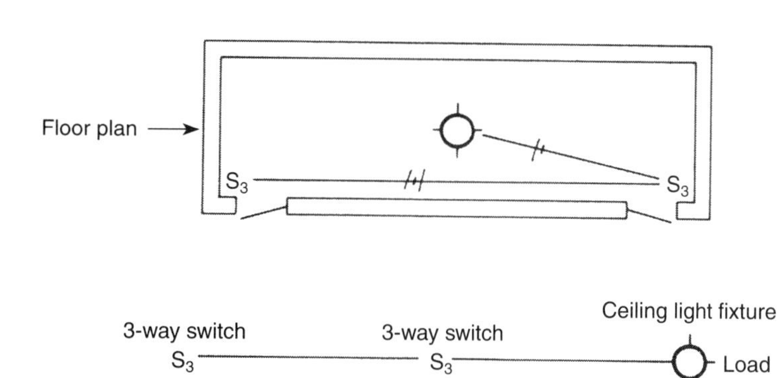

Part sequence A

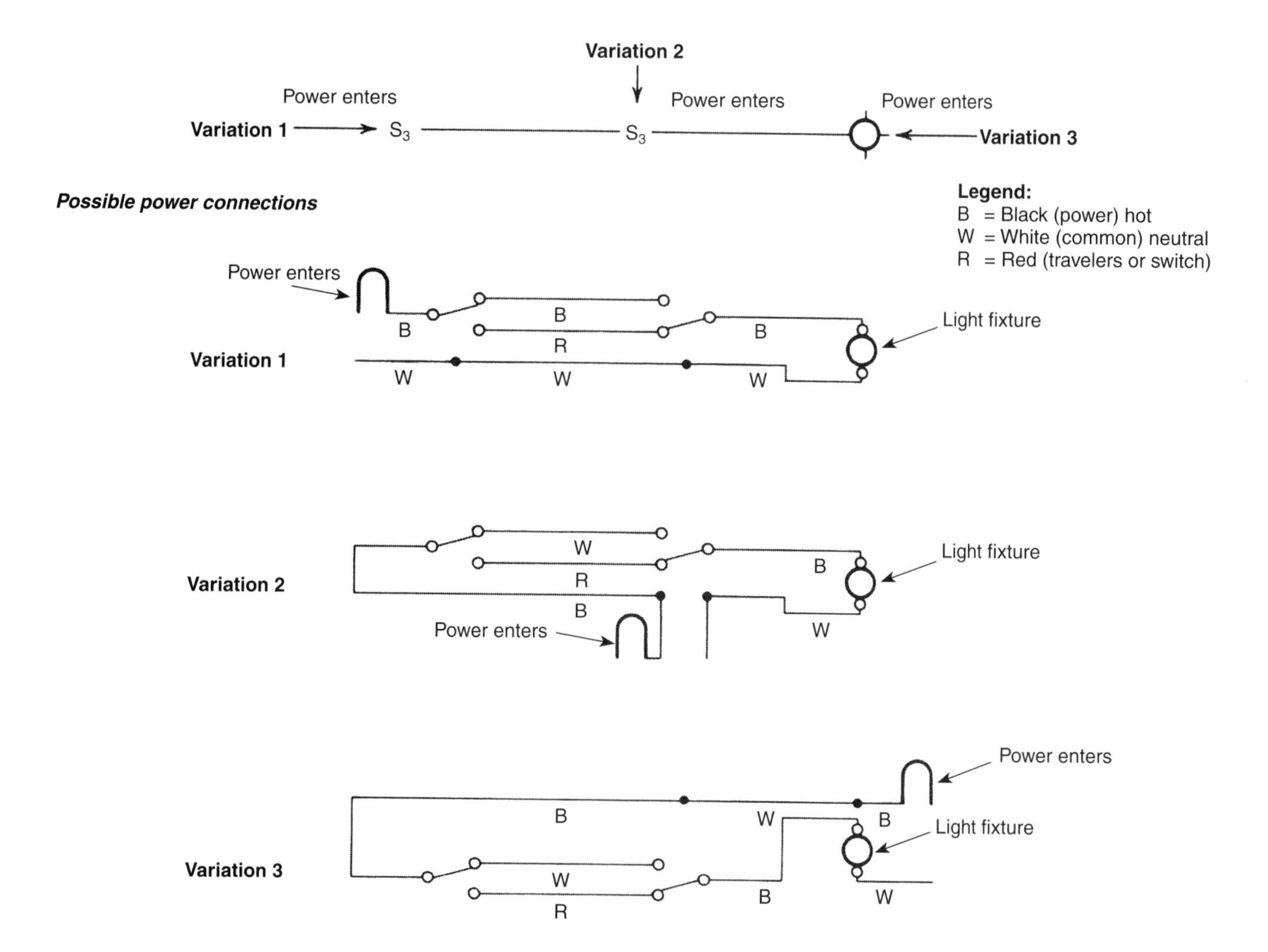

Figure 7-12
3-way switch circuits: switch, switch, load sequence

Each variation illustrated differs from the others in some respects, while in others all are the same. It'll be useful to note the similarities common to the three variations shown:

1. White common from the power source goes directly to the load and connects to nothing else.
2. Power leads going into the first switch and out of the second are both black.
3. One traveler is always red.

In variation 2, one traveler is white; and in 3 not only is one traveler white, but the incoming power wire changes from black to white and back to black again. Since all wires going to and from switches in this circuit are power wires, these applications are not code violations by *NEC* Section 200.7(C), referred to earlier. Be careful and always remember that these uses of white wires on the power side of a circuit are permitted *only in cable* and they must be reidentified wherever visible, per *NEC* Section 200.7(C).

To wire the same switching circuits in conduit, both travelers will have to be power colors: two blacks, two reds, or better yet, two blues, to help distinguish them from anything else sharing the same pipe. Remember, in conduit, don't cut and splice wires that only pass through a box without connecting to anything. Simply leave a loop and pass them on to the next box.

Figure 7-13 illustrates the only other possible sequence of parts that might occur if you use two 3-way switches and one load. When wiring this sequence in cable, white wires are travelers again. As permissible with the previous sequence, it's also permissible here.

Again, in conduit, both travelers will have to be proper power colors and will run directly from switch to switch, passing uncut through the light box.

In the event that two or more lights are to be controlled by a pair of 3-way switches, as shown in Figure 7-14, any desired number of additional lights may be added by tapping the load of any of the five diagrams in Figures 7-12 and 7-13. Sequences such as those diagrammed in Figure 7-14 (a) and (b) can't be done in 3-wire cable, but could be done in conduit. The sequences shown in Figure 7-14 (c) and (d) are no problem in 3-wire cable.

Wiring Combination 3-Way and 4-Way Switch Circuits

When it's necessary to switch a circuit from more than two locations, two 3-way switches are used and as many 4-way switches as necessary to take care of the additional locations. See Figure 7-15. Like the 3-way

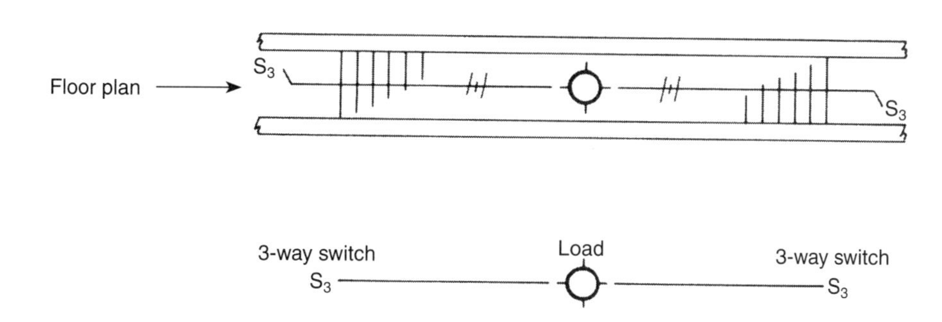

Part sequence B

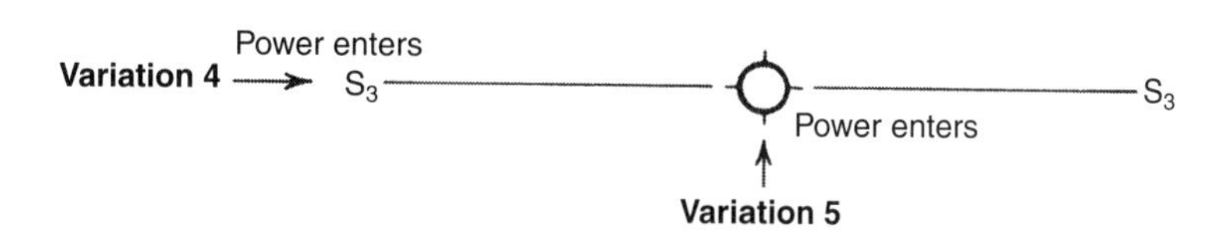

Possible power connections

Legend:
B = Black (power) hot
W = White (common) neutral
R = Red (travelers or switch)

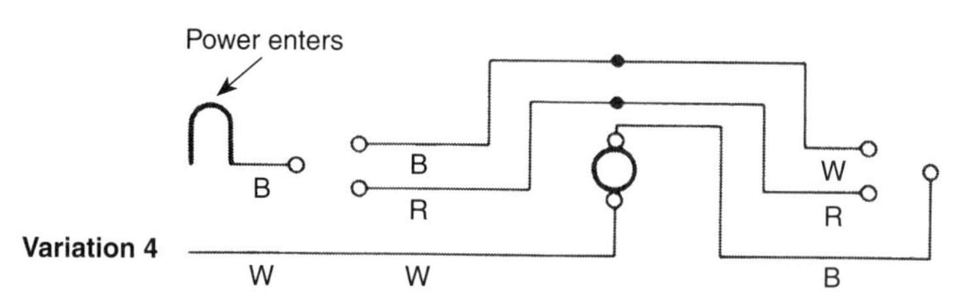

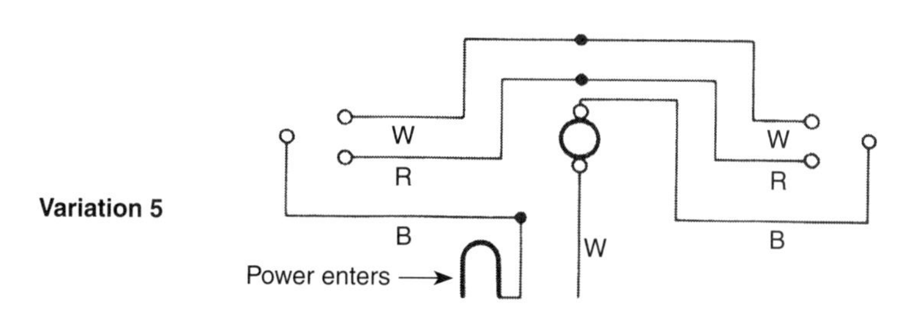

Wiring connections in cable

Figure 7-13
3-way switch circuits: switch, load, switch sequence

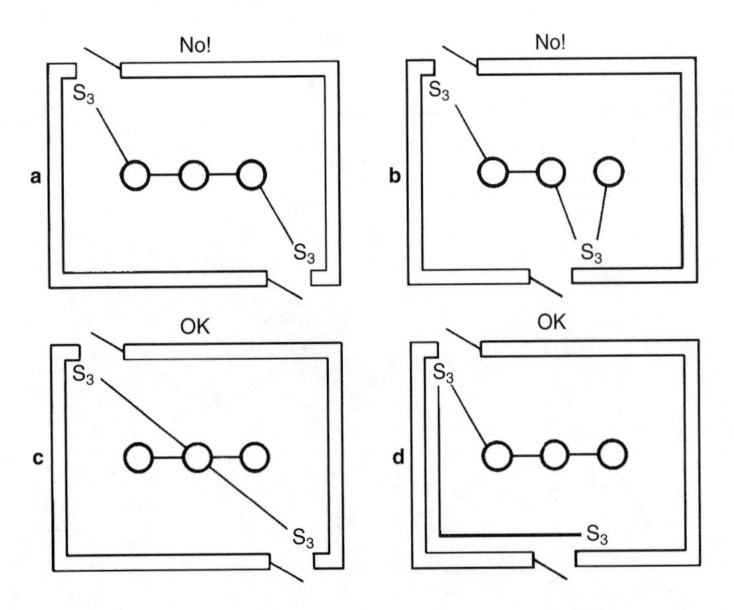

Figure 7-14
3-way switch sequences with multiple loads

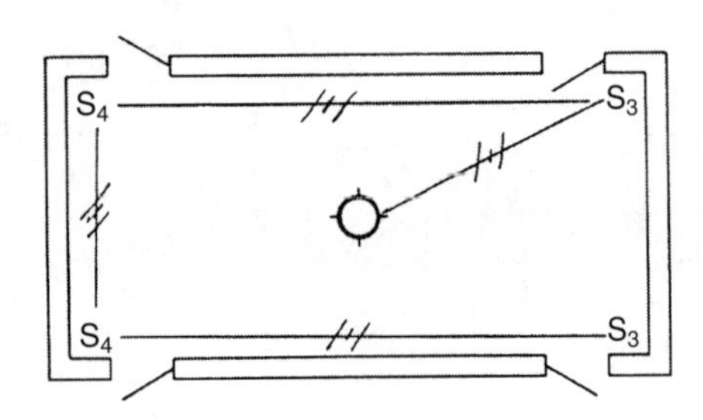

Figure 7-15
Floor plan utilizing 4-way switches

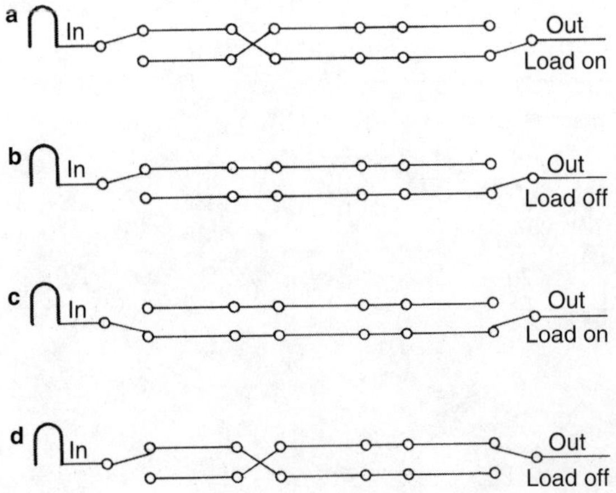

Figure 7-16
Power side of 4-way switch circuits

switch, the 4-way is an *either-or* switch, but it can only be used in a circuit that already contains two 3-ways.

Looking back at Figure 7-11, we see that 3-way switch circuits control loads from either of two locations by allowing either of the switches to connect two alternate electrical paths: one that will always connect power to the load and two that won't. As shown in Figure 7-16, by inserting 4-way switches along the two traveler lines connecting a pair of 3-way switches, the capability of switching back and forth between the two alternate electrical paths can be extended to as many switching locations as you desire.

When a circuit consists of two 3-ways, one 4-way, and a load, there are eight alternatives for wiring it in cable, depending on the part sequence and at which point along each sequence unswitched power enters the circuit. One possible sequence is to go through all three switches and end at the load. Figure 7-17 illustrates four alternative ways of wiring this sequence in cable. Figure 7-18 shows the other four alternatives you can use in cable to wire the same circuit when the sequence is changed to place the load part way along the line instead of at one end.

Once again, these diagrams of cable show many instances of white wires on the power side of the circuit. All of these are permissible under the code exception cited earlier. However, in conduit, *none* of these white wires on the power side is allowed — only power colors may be used.

Cable circuits using 3-way switches and those using 4-ways have the following similarities: white common goes only to the load; power into the first 3-way and out of the last one are both black; one traveler is always red straight through.

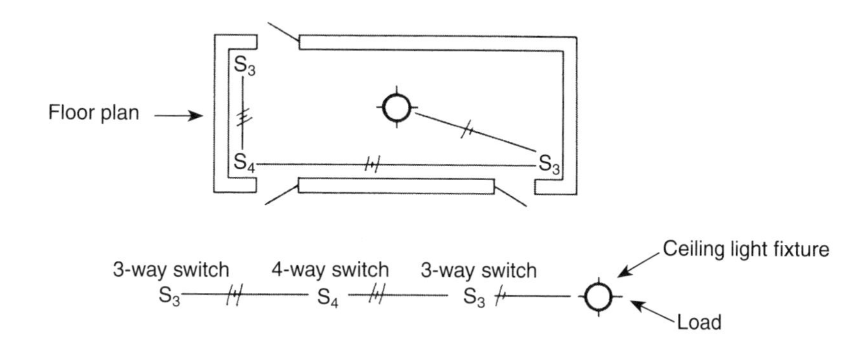

Part sequence A

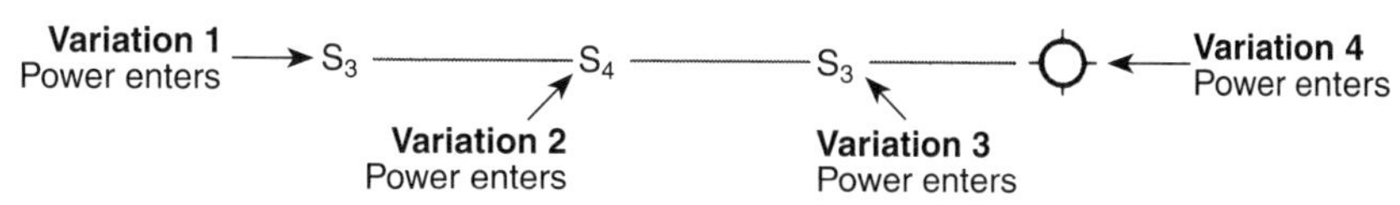

Possible power connections

Legend:
B = Black (power) hot
W = White (common) neutral
R = Red (travelers or switch)

Power enters
B B B B
R R
Variation 1
W W W W

Light fixture

W B
R R B
Variation 2
B W W
Power enters

W W
R R B
Variation 3
B B
Power enters

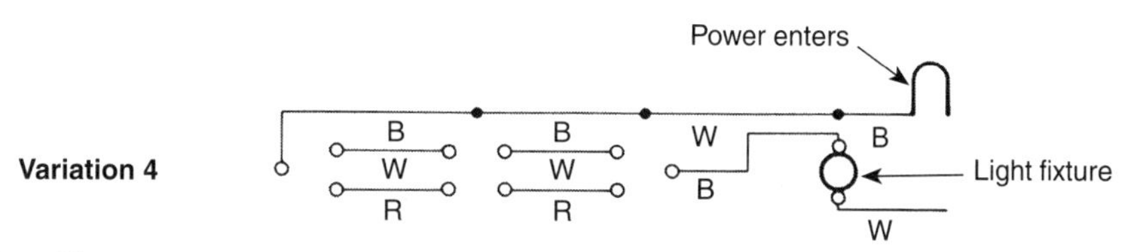

Wiring connections in cable

Figure 7-17
4-way switch circuits — switch, switch, switch, load sequence

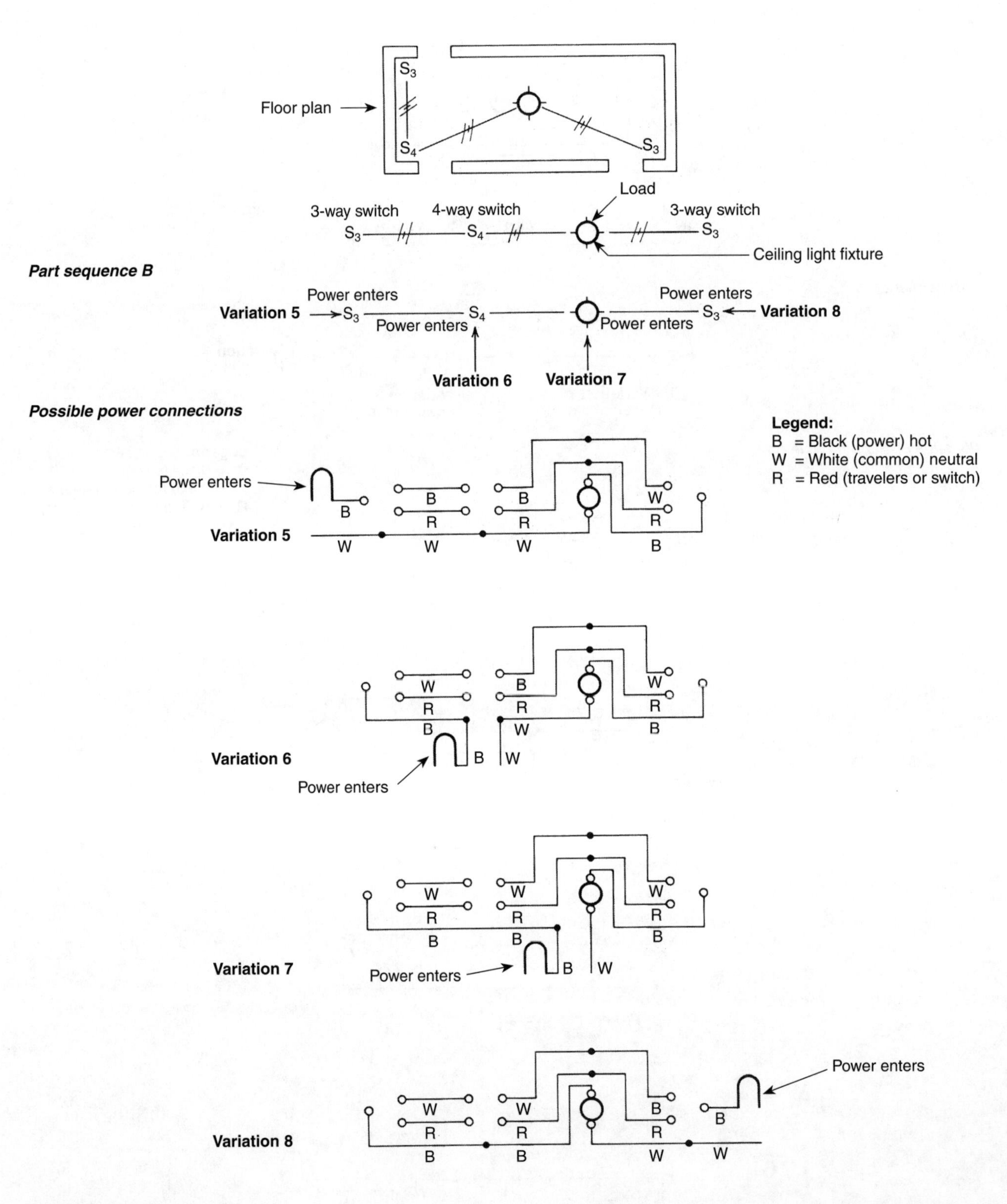

Figure 7-18
4-way switch circuits — switch, switch, load, switch sequence

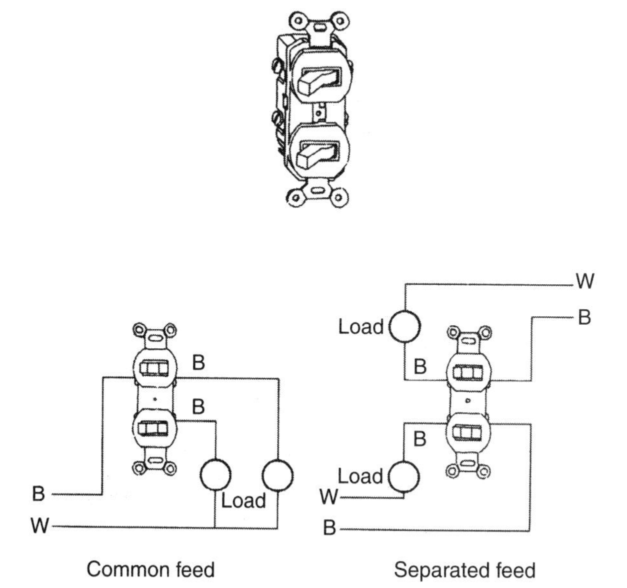

Figure 7-19
Two single-pole switches on a single yoke

Multiple loads on cable wired circuits can be tapped from the first load using any of the eight alternatives given, as long as it's done in a manner similar to what was done in Figure 7-14 (c) and (d) (the electrical sequence allows all loads to be tapped from a single output point). Again, sequences such as Figure 7-14 (a) and (b), which require multiple output locations, won't work in cable — in conduit they may, because the necessary extra wires can easily be pulled.

Wiring Dual Single-Pole Combinations

Space limitations might require the placement of two single-pole switches in an area normally occupied by only one. To meet this need, a double single-pole switch combination unit, like that in Figure 7-19, is available. While two screw terminals are found on each side of the case, the two sides *aren't* identical. Here's the difference: On one side, the two terminals are independent, and on the other, they're connected by a removable link. Because of that link, unswitched power can be supplied to both switches by attaching either of the linked terminals. The side with two independent terminals may now be used to feed and control two separate loads.

If the two loads require separate power sources originating at different breakers, the linked terminals can easily be disconnected by breaking the connecting link with a screwdriver tip or needle nose pliers, depending on the shape of the link.

Switch/Outlet Combination

Space limitations can also require a combination switch and convenience outlet that will fit in the space normally occupied by either one or the other. Figure 7-20 shows single pole switch controls and grounded power outlets. This device has two screw terminals on each side, plus a green hexagonal head ground screw at the bottom.

White common always goes to the silver terminal. The power lead to the load being switched goes on the brass terminal next to the silver common. On the other side, that leaves two brass terminals, connected to each other by a removable link. When unswitched power is connected to either of the two brass terminals, the switch will operate the load, and

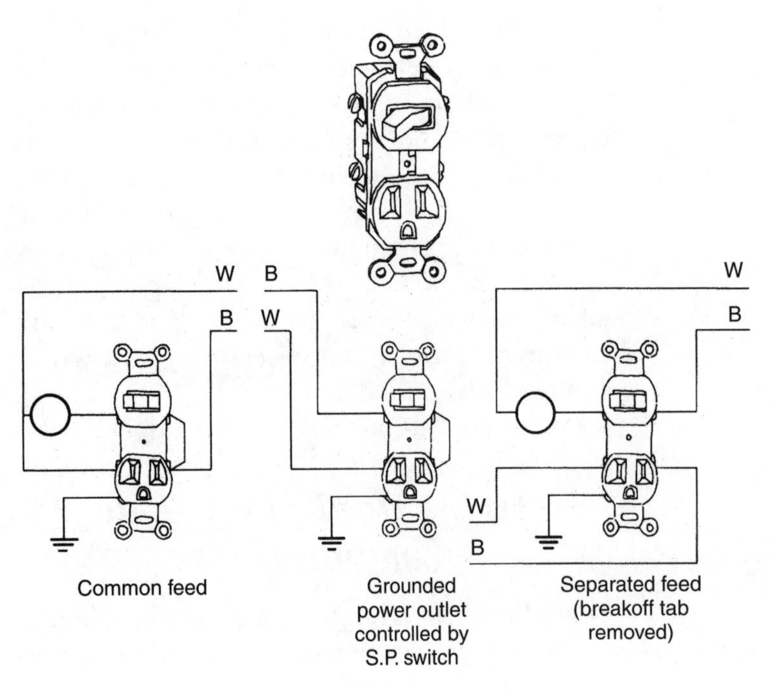

Figure 7-20
Switch and outlet combined on a single yoke

the outlet will be powered at all times — regardless of the switch position. If the switch is used to control the outlet as well as a remote load, simply reverse the power connections. Put the unswitched power on the brass terminal next to the common, and place the load on the opposite side. If switch and outlet are parts of unrelated circuits, and therefore are to be powered separately, break out the link connecting the two brass terminals and power them individually.

When cable is used, the ground wire goes on the hexagonal green terminal at the bottom. In conduit, this connection is unnecessary, as there's no ground wire.

Pilot Light Switches

Different combinations of pilot lights with single-pole switches are shown in Figure 7-21. The first, Figure 7-21 (a), contains an internal light inside the switch toggle that glows in the *ON* position. There are three terminal screws on the case: two brass heads and one silver head. White common always goes to the silver head. Power in and power out go to the two brass heads. Either one can go on either terminal.

Figure 7-21 (b) shows a pilot light that's separate from the switch, but internally it's connected. This unit has four terminal screws, two brass heads and two silver heads on each side. To wire this one, connect power in and power out to the two brass heads. It makes no difference which is which. Now put a white common on either one of the two silver heads.

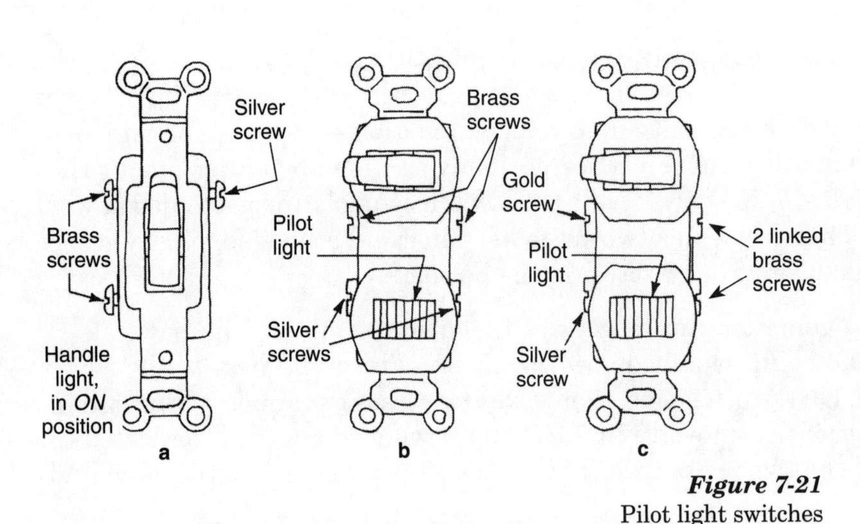

Figure 7-21
Pilot light switches

In Figure 7-21 (c), the screw terminals are the same as those on the single-pole switch/ convenience outlet combination: one brass and one silver terminal on one side and two linked brass terminals on the other. As usual, white common goes to the silver terminal.

Normally, power is connected to the brass terminal next to the silver and taken out on either of the linked brass terminals on the other side. If wired this way, the pilot lights when the switch is in the *ON* position. If the pilot doesn't shut off regardless of switch position, the power leads have inadvertently been reversed. With this switch, it does make a difference which is which.

By breaking out the connection between the linked brass terminals, the pilot light becomes completely independent of the switch and it can now be used as the indicator of some completely different activity.

Dimmer Switches

The most commonly encountered dimmer switches are the single-pole and the 3-way. The single-pole switch and dimmer combination has two black leads. The 3-way switch and dimmer has one black and two red leads. The single-pole is connected by splicing unswitched power to either one of the black pigtails and the load to the other. As usual, it makes no difference which goes where.

Either of these dimmers will fit in the space occupied by the standard switch it replaces. However, the dimmers are definitely bulkier, meaning they'll fill more of the interior space in the switch box. So if the box is full of wires, splices, and other bits and pieces, stuffing a dimmer in can be quite difficult.

As shown and mentioned earlier, those dimmers are good for loads up to 600 watts only. However, most of the time, this capacity is adequate for the loads encountered in residential and small commercial buildings. Should greater capacity be required, dimmers that fit standard electrical boxes are available with capacities as high as 2,000 watts.

> ***WITH A 3-WAY SWITCH,*** the travelers go on the two red leads, and power either *in* or *out* goes on the black. Remember, only one end of a multiple switch circuit can have a dimmer, so take care locating it in the most convenient place.

Outlets

120-Volt Duplex

While this simple device is familiar to all, it's necessary to take a closer look at several details you may not have noticed. At the angle shown in Figure 7-22, note that the vertical slots on the left of the device are longer than the ones on the right. The two screw terminals on the left have silver heads while the two on the right have brass. Both pairs are

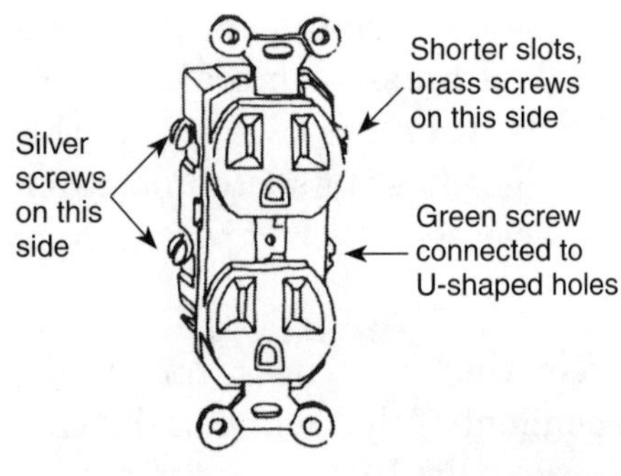

Figure 7-22
Duplex receptacle

linked. At the bottom is a single green hex-headed screw, which is internally connected to the two *U*-shaped holes. These holes are ground connections.

As always, incoming power goes to a brass terminal on the side with the shorter slots. White common goes to a silver terminal on the side with long slots. Except when you're wiring in conduit, put ground on the green hex-headed terminal. In conduit, the outlet mounting yoke makes metal-to-metal contact with the box, which is then grounded through the conduit.

Quite often, half of a duplex outlet is switched while the other half is permanently on. To accomplish this, the connecting link between the two brass terminals *only* is broken out — splitting the power side. Now, unswitched power is connected to the top brass terminal, switched power to the bottom and common to either one of the silver terminals.

In addition to splitting outlets for switching one-half while the other remains unswitched, split outlets are often *split-wired.* This means that the power sides of the two halves are connected to two completely separate circuits. The code requires kitchen countertop outlets be wired this way. When this is done, the two circuits feeding such split-wired outlets must be connected to a double-pole circuit breaker or two single breakers with linked handles [*NEC* Section 210.4(B)].

When wiring a series of outlets in sequence, instead of using all four terminals and connecting incoming power and common to one pair of terminals and outgoing power and common on the other pair, use a pigtail to splice power-to-power to one brass terminal, and common-to-common to one silver terminal. Wired in this manner, the continuity of the rest of the circuit is kept independent of the condition of any of its component outlets.

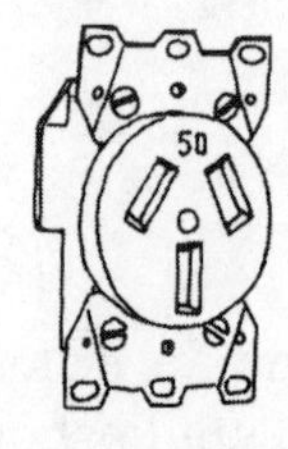

Figure 7-23
Range receptacles

240-Volt Outlet — Range and Dryer

In a residence, the electric range will be the largest single load in the entire system. It requires a 240-volt circuit with a breaker capacity of about 50 amperes. While ranges are sometimes wired directly from a junction box, more often they're plugged into an outlet. Figure 7-23 shows two outlets normally used for ranges. The two power leads of the 240-volt circuit connect to the upper slanted slots, while the common goes into the lower vertical slot. In cable, the wire size to a range will normally be #8 for both the power wires and the common. If wired in conduit, the common could be reduced to a #10, but not smaller, per *NEC* Section 210.19(A)(3), Exception No. 2.

Figure 7-24
Dryer receptacle

NOTE: Range and dryer circuits go to those appliances *only*. Nothing else may be on either of those circuits.

The normal range outlet will fit a standard 2 x 4 outlet box, but it'll have to be done with ¾-inch knockouts. Assuming the wiring is done in cable, you'll have to use an 8-3 cable with ground, which won't go in the usual ½-inch knockout. If the job is done in conduit, ¾-inch EMT will be needed, unless the insulation type is switched to either THHN or THWN. In ½-inch conduit, using TW, let alone anything thicker, two #8 wires are the maximum allowed. The #10 common can't go in as well.

The standard electric clothes dryer outlet, as illustrated in Figure 7-24, looks very similar to the range. In fact, the two slanted upper slots are identical. The difference is in the lower slot. Instead of a simple vertical, it's now half-horizontal and half-vertical, like an upside-down, backward *L*.

The dryer will be supplied by a 30 ampere, 240-volt circuit wired with a #10 wire. Again, the two power leads go to the upper slanted slots just as for the range. The common goes to the upside down, backward *L*. A dryer outlet also fits comfortably in a standard 2 x 4 box, and the ¾-inch knockouts won't be necessary.

STUDY QUESTIONS

1. Which two switch types are *either-or* rather than *on-off* switches?

A) Pilot light; photo control
B) Key; single pole, double throw
C) Double pole, reversing; double pole, single throw
D) Single pole, double throw; double pole, reversing

2. What's the function of a pilot light switch?

A) It indicates when a remote fixture is operating
B) It indicates when a remote fixture is turned on
C) It indicates when a remote photoelectric light has come on
D) It allows the user to shut an appliance off by deactivating the pilot light

3. Where is the disconnecting means located on most switches?

A) Both the power and the grounded sides
B) The grounded side of the line only
C) The common side of the line only
D) The power side of the line only

4. How many variations can there be in the cable wiring for two 3-way switches and the load they control?

A) Three
B) Four
C) Five
D) Six

5. What would you install when it's necessary to switch a circuit from more than two locations?

A) Two 3-way switches and one or more 4-way switches
B) Two 3-way switches
C) Two 4-way switches and one 3-way switch
D) Three or more 4-way switches

6. Which is *not* true of cable circuits for both 3-way and 4-way switches?

A) The white common goes only to the load
B) The power wire into the first 3-way and out of the last one are both black
C) One traveler is always red straight through
D) White wires are never allowed on the power side

7. What do you need to use for wiring sequences that require multiple output locations?

A) Conduit
B) 3-wire cable
C) 2-wire cable
D) Multiple splices

8. When are duplex outlets required to be split-wired?

A) When one outlet is switched and the other permanently on
B) When they are kitchen countertop outlets
C) When one outlet will power a small appliance
D) The *NEC* prohibits split-wired outlets except when connected in series

9. What's the largest single load in a residential electrical system?

A) Air conditioning
B) Electric clothes dryer
C) Electric range
D) Electric radiant heat system

10. What is true of the wiring required for the 240-volt outlet powering major household appliances, such as an electric range or clothes dryer?

A) An electric range should be wired in NM cable wire size #10
B) An electric range and dryer both need a 30-ampere 240-volt circuit wired with #10 wire
C) A dryer needs a 30-ampere 240-volt circuit wired with #10 wire
D) A dryer must be wired using EMT

8 PLANS

When a home or other building is planned, the work of the electrician, like that of most other trades, begins long before ground is broken and the actual construction starts. Estimates of required labor and materials must be made first in order to bid on the work to be done. After the job has been awarded, labor and materials adequate to accomplish the necessary work must be secured. This requires detailed study of the building plans. Let's examine an average set of plans for a three-bedroom residence to familiarize ourselves with the kinds of information that we'll find there.

Plot Plan

The drawings will start with a plot plan. The plot plan shows the shape and dimensions of the lot, and the size, shape and location of the building on that lot. In order to insure the plans conform with local zoning regulations, the distances between property lines and building walls will be provided. Somewhere on the plan there'll be an arrow indicating geographic north.

Foundation Plan

The foundation plan and associated details show where foundation footings, walls, piers, posts, and slabs are to be located. Also given are:

- footing depths and dimensions
- thicknesses of walls and slabs

- material specifications
- reinforcing bars required in walls and footings
- reinforcing mesh required in slabs
- size and spacing of anchor bolts

When a structure is built over a slab on grade, the electrician may want, or need, to embed some conduits in the slab in order to distribute electrical power as efficiently as possible. Or in order to do his work, he may have to go through foundation walls. In either case, he'll have to know how the foundation is to be built in order to lay out any required electrical work at the foundation level.

Typical Sections

The typical sections reveal the true structure of the building by stating the sizing and spacing of the frame parts. Sections sometimes, but not always, give interior and exterior finishes as well. Studying the typical sections in conjunction with the floor plans, the electrician can decide how he wants to route his wiring through the building frame.

Floor Plans

Separate plans are made for each floor from the basement up (if there is a basement). Floor plans show in scale the locations and sizes of walls, doors, and windows, as well as miscellaneous additional structural information. All important features are drawn to scale and specifically detailed in the floor plans. Residential plan sets usually don't include separate electrical plans. Commercial or industrial sets normally *do* include separate electrical sheets.

On residential plans, the electrical requirements that you do find on the floor plans are designated by conventional symbols, which we'll look at shortly. There are special symbols for different types of lights, outlets, switches, and other electrical devices. These symbols are located on the plan in their proper positions according to the scale of the plan, but are seldom drawn to scale because of their small size.

Exterior Elevations

A plan set will normally include four exterior views. From these, the electrician can determine how he'll handle the installation of the exterior lighting, outlets, and other exterior electrical work that may be required.

Other Sheets

Additional sheets may show interior elevations, cabinetry details, or other finish details. Some of this information will also assist the electrician in deciding how to handle parts of the electrical installation.

Types of Plans and their Uses

In the course of his work, the electrician will have to deal with three different types of plan drawings:

1. the floor plan
2. the cable plan
3. the wiring diagram

Floor Plan

Floor plans show the locations of walls, doors, windows, and other structural features, along with dimensions. As mentioned earlier, various parts of the electrical system are also located and identified on the floor plan using the conventional floor plan symbols shown in Figure 8-1. You can see some of these symbols in the partial floor plan shown in Figure 8-2. This plan shows symbols representing a ceiling outlet controlled by a single-pole switch and two duplex outlets. The symbols that appear on a floor plan indicate where the electrical supply will be made available, and where fixtures, receptacles, and other electrical devices are to be placed, as well as where switches will go, and what each switch will control. What the floor plan doesn't show is how any of this is to be accomplished.

Cable Plan

Using the same symbol system as the floor plan, the cable plan shows what parts of the electrical system go where. In addition, it indicates how the various points are to be connected together into branch circuits, and how many wires will be required for each connecting run from box to box. The cable plan in Figure 8-3 shows the 2-wire cable run for the layout in Figure 8-2.

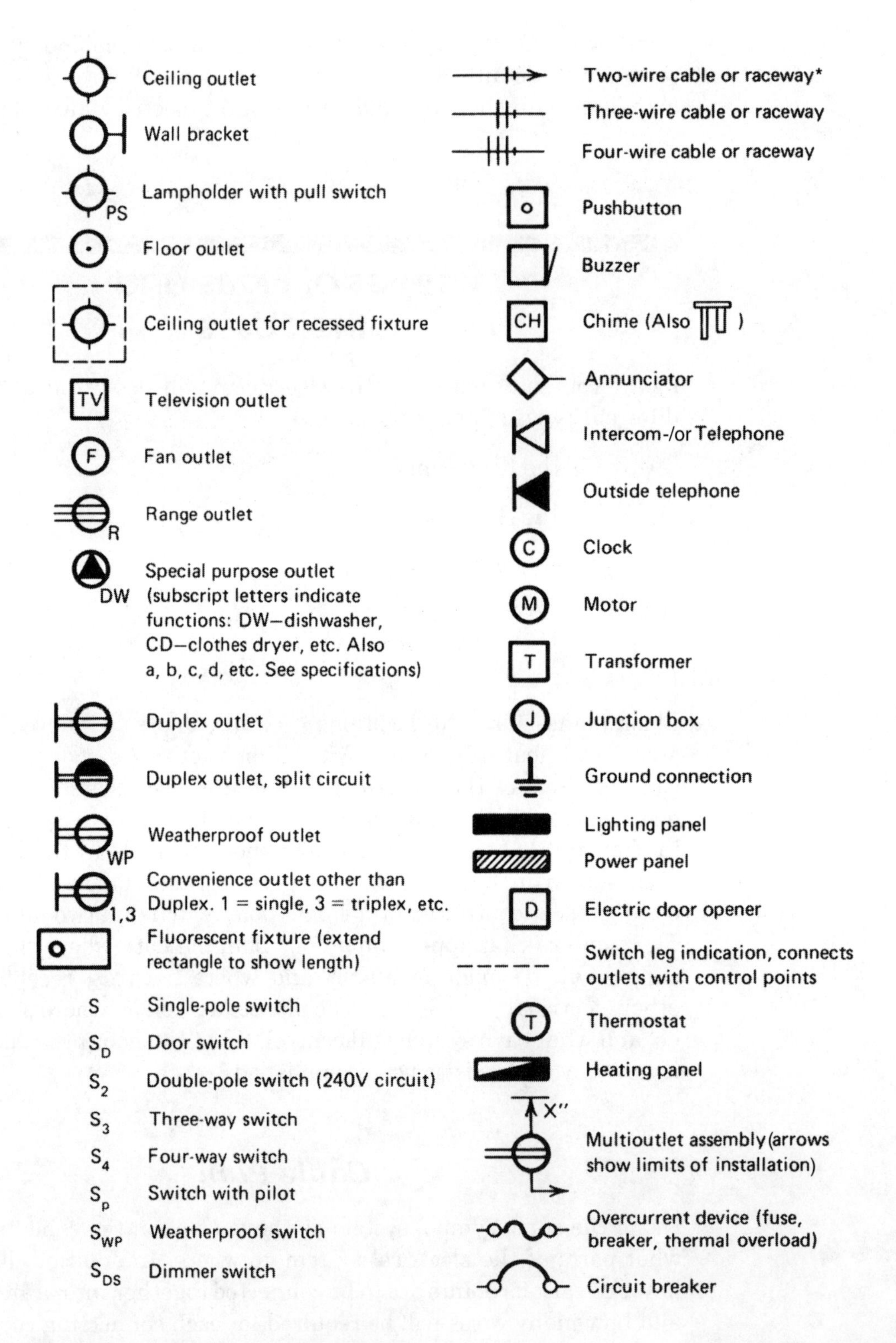

*If there is an arrow on the cable, it indicates a home run to breaker

Figure 8-1
Floor plan electrical symbols

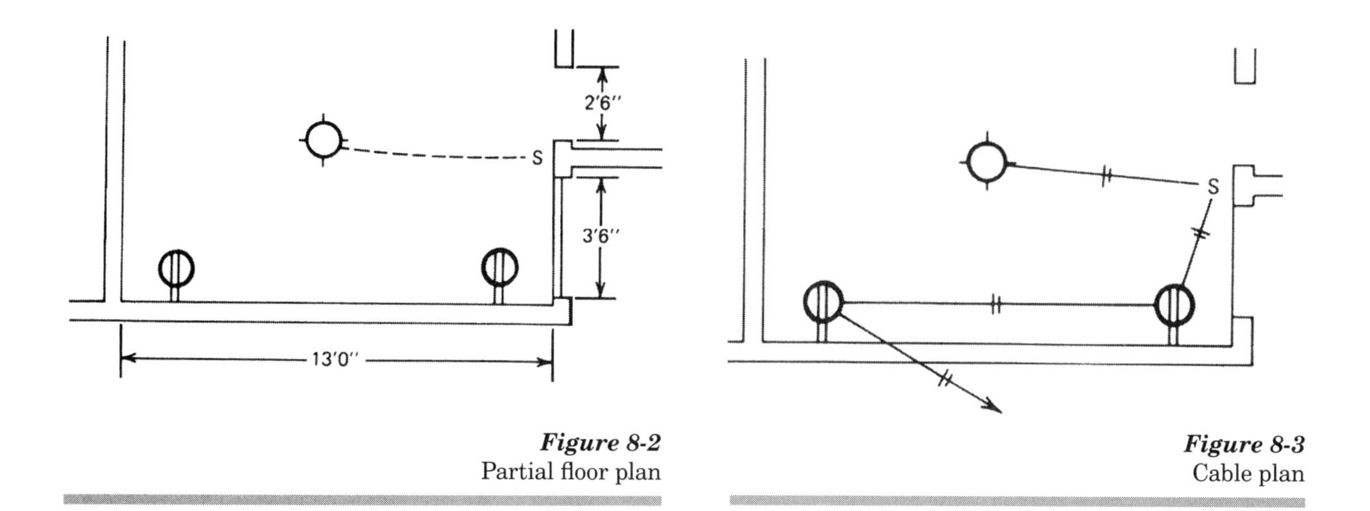

Figure 8-2
Partial floor plan

Figure 8-3
Cable plan

The number of wires entering each box, as well as where they originate, appear on the cable plan. The connections inside each box are shown on the wiring diagram. Figure 8-4 shows the wiring diagram for the cable run in Figure 8-3. It takes both the cable plan and wiring diagram to show how the electrical parts appearing on a floor plan are to be connected and made operational.

Floor Plan Symbols

The standard symbols illustrated in Figure 8-1 are used to show the location, type and function of the various parts of the electrical system. These parts are divided into two general categories: using equipment and use points, and controls.

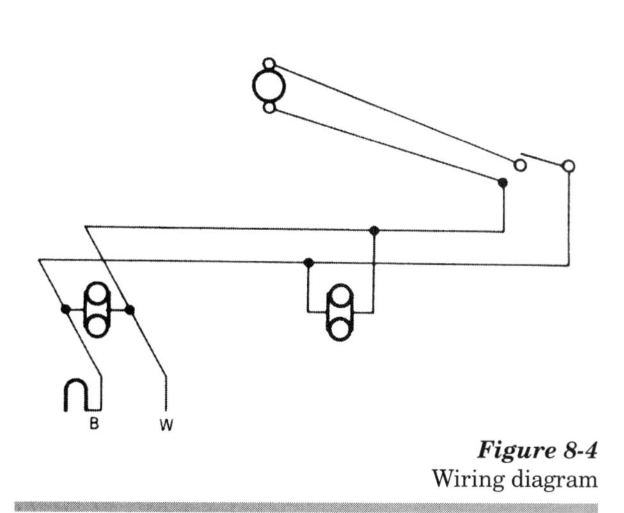

Figure 8-4
Wiring diagram

The first category includes current-using devices that are directly connected to power wiring, such as light fixtures installed on walls or ceilings, as well as receptacles into which current-using devices can be plugged. The second category includes switches, dimmers, fuses, and circuit breakers, all of which are nonconsuming devices.

The symbols most often represented on the plans are light fixtures, receptacles and switches. The remaining symbols represent parts that are required in comparatively small numbers. For example, a home might have 25 or 30 duplex outlets, but only one door chime. While some of the devices represented by these symbols consume power, they constitute minor loads. The others are either safety equipment, such as fuses or circuit breakers, or they're passive parts, such as panel boxes or junction boxes.

Design of the Electrical System

The first step in the design of a residential electrical system is to properly locate on a floor plan the minimum number of electrical outlets and controls required by code. Then add any desired electrical equipment beyond code minimums.

When you've determined all the use points and defined the power requirements at those points, divide the points into branch circuits. This must also conform to applicable code provisions. After determining the composition of the various branch circuits, prepare cable plans and wiring diagrams to show in detail exactly how to accomplish the desired results.

Minimum Residential Requirements

Since commercial and industrial requirements vary widely, the code requirements relating to them are more general in nature than those governing residential installations. Residential requirements are very detailed and specific, so we'll examine them in considerable depth.

General-Purpose Convenience Outlets

Every kitchen, dining room, den, recreation room, bedroom — essentially any room other than a hall, stairway, or bathroom — must have receptacles located so that no point along the floor line is horizontally more than 6 feet from an outlet, per *NEC*® Section 210.52(A)(1) and (2). The 6-foot distance starts at a doorway or other break in the wall. Any separate wall space (as between two doors) that's 2 feet or wider is considered individually. This measurement can go around corners and include parts of more than one wall.

Look at the floor plan in Figure 8-5. You would start measurements at the doorway in the bottom right (A), 2 feet into the corner, then around the corner 4 feet to (B). This is a total of 6 feet, at which point there must be an outlet. From this point, it's 8 feet to the corner, then 4 feet along the next wall to the outlet, a total of 12 feet. Along this 12-foot run there's no point more than 6 feet from an outlet, so it meets code. The distance from (C) to (D) is 12 feet, so again no point in between is more than 6 feet from an outlet. From (D) to (E) is only 5 feet, so no problem. The wall section at (F) is only 1½ feet wide. You don't need an

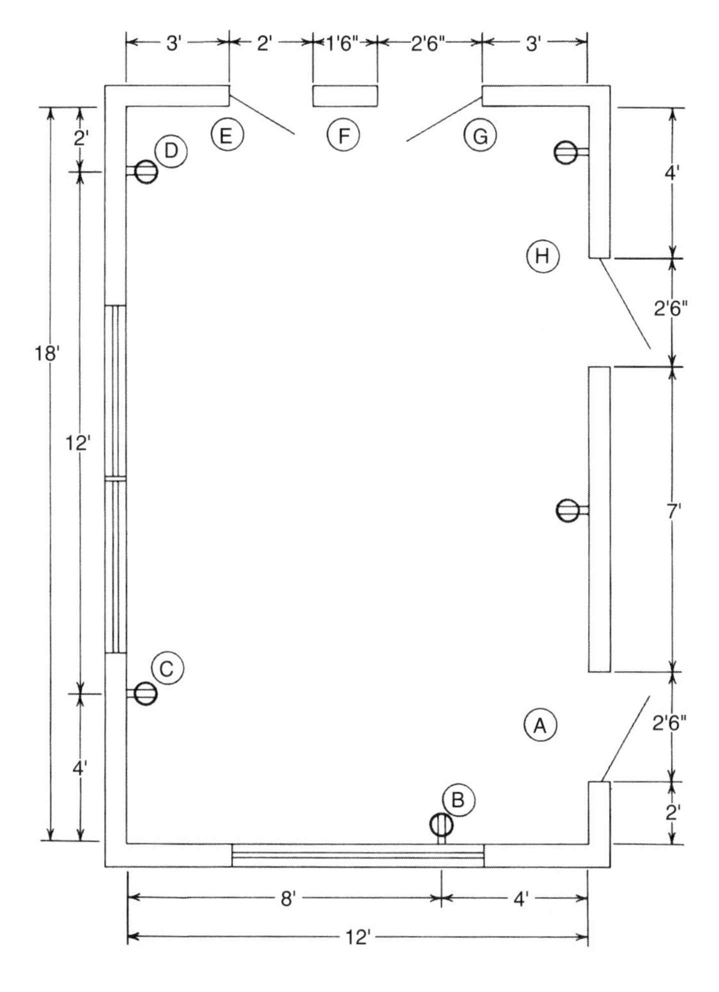

Figure 8-5
Spacing of residential wall receptacles

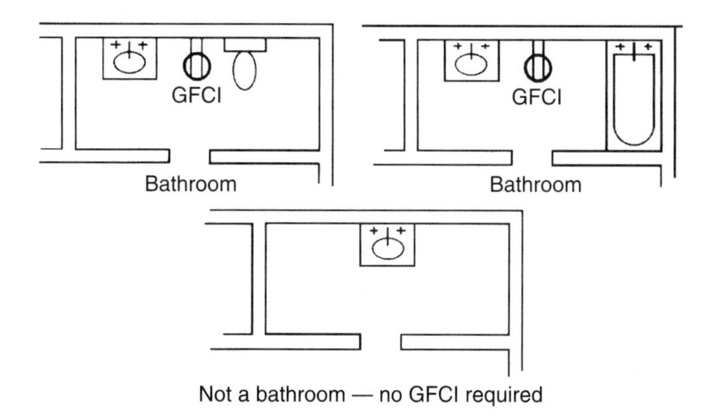

Figure 8-6
Only a bathroom must have an outlet and GFCI

outlet here, as it's less than 2 feet. From (G) to (H) around the corner is 7½ feet, so you need an outlet somewhere along here. Then, from (H) back to (A) is another 7 feet, so again you need one outlet.

A bathroom, as defined in *NEC* Article 100 — Definitions, is an area including a basin with one or more of the following: a toilet, a tub, or a shower. *NEC* Section 210.52(D) requires at least one wall receptacle adjacent to the washbasin in residential bathrooms. This outlet, and any others in a bathroom, must be in a separate 20-amp circuit, and be protected by a ground-fault circuit-interrupter (GFCI) per *NEC* Section 210.8(A)(1). Even if there's a partition between the sink and tub or toilet, a GFCI is still required. Two of the rooms shown in Figure 8-6 meet the definition of a bathroom and so require an outlet and GFCI. The third one, a lavatory with only a basin, doesn't meet the definition of a bathroom and is therefore exempt from the code requirements for an outlet with GFCI.

The GFCI required in bathrooms needs some explanation. In Chapter 3, we said that the smallest branch circuit capacity that will be found in any building is 15 amperes. We also mentioned the fact that 0.1 ampere, or 1/150th of the power in that circuit, can kill a person. If your mind is wandering, even 0.01 ampere will do a good job of getting your attention. This being true, very small defects in electrical equipment can cause serious shocks.

People often stand with bare feet on a wet floor in a bathroom. That makes resistance to ground very slight, and creates a low-resistance path to any stray current leaking from a defective appliance they might be handling. Enough serious shocks have resulted from situations of this kind that the GFCI is now required as a preventive measure in bathrooms.

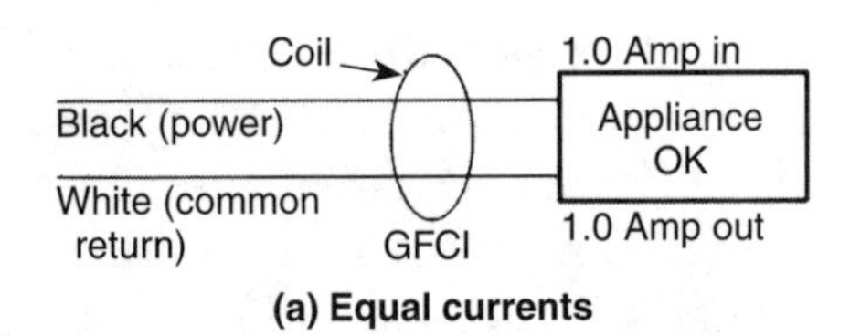

(a) Equal currents

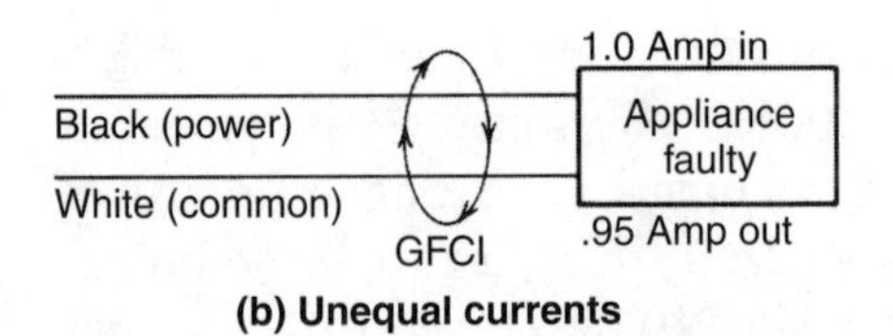

(b) Unequal currents

Figure 8-7
GFCI operation

The operation of a GFCI is very simple but extremely effective. To understand its operation, remember the elementary fact that an alternating current passing through a conductor creates a magnetic field, which fluctuates with the alternating voltage. And that fluctuating field can, in turn, induce a voltage in another conductor.

When a properly operating appliance is connected to a circuit that's protected by a GFCI, as in Figure 8-7 (a), the currents going in and out are equal and opposite. So, the magnetic fields they create are also equal and opposite, and cancel each other out. The coil in the GFCI feels no induced voltage and remains inactive.

When a defective appliance containing a small leak to ground is connected to the same circuit, as in Figure 8-7 (b), the currents in and out are no longer equal and opposite; therefore, the magnetic fields they produce no longer exactly cancel each other out. The uncanceled portion of the incoming current leaves a magnetic field, exciting the coil in the GFCI, which immediately trips it and opens the circuit.

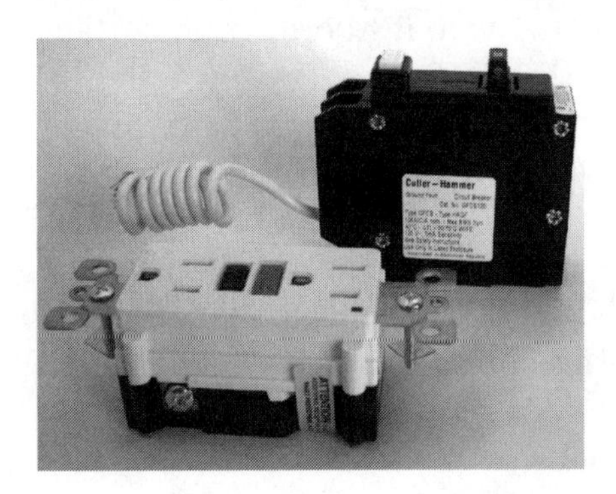

Figure 8-8
Receptacle and breaker types of GFCI

The GFCI is available in two basic models, a receptacle type and a breaker type, both shown in Figure 8-8. The receptacle type fits in the same space as a normal duplex 120-volt receptacle, and includes either one or two receptacles, plus a *Test* and a *Reset* button on its face. Some models of this type are supplied with wire pigtails, others have screw terminals, but all have both incoming and outgoing connections. This makes it possible to use them to protect their own outlets, plus any number of outlets placed beyond them in a circuit. In Figure 8-9, there are four outlets protected by one GFCI outlet.

The breaker type of GFCI is combined with a circuit breaker. The combined unit fits in the same space as a normal breaker. It looks different from the standard breaker because it has a yellow or white button on its face marked *Test*, and a long pigtail of white wire is attached to it. As we'll see later, a normal 120-volt, single-pole breaker has a single output terminal screw to which the power wire (black, red, blue or whatever color, other than white, grey or green) from the branch circuit is attached. The GFCI/ breaker combination has two screw terminals, one of which is identified by a white dot.

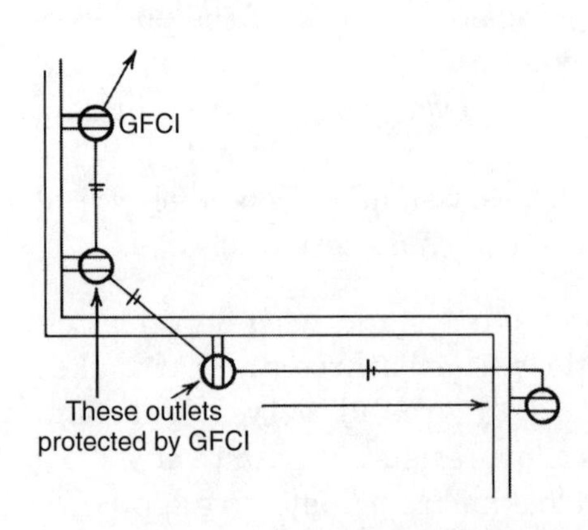

Figure 8-9
GFCI with downstream receptacles

At the breaker box, a branch circuit is normally connected with the power wire going to the output terminal on a breaker, and both the white or grey common and the ground attached to the ground buss. When the branch circuit is to be protected by a GFCI, not only does the power wire go to the breaker, but the common also goes to the breaker and attaches to the terminal identified by the white dot. The white pigtail attached to the breaker then connects to ground. While a breaker disconnects only the power side of the circuit, a GFCI disconnects both sides of the line.

The *NEC* requires, in Section 210.52(E), that at least one outdoor receptacle, accessible at grade level, be located at both the front and back of one- and two-family dwellings. For each dwelling at grade level, all outdoor receptacles must also be protected by a GFCI, per *NEC* Section 210.8(A)(3). The code defines "grade level access" as being located not more than 6 feet 6 inches above grade, and usable without entering the dwelling.

Special Outlets

In both the basement or attached garage of any one-family dwelling, there must be at least one receptacle outlet in addition to whatever has been provided for special equipment (laundry use, for example), in accordance with *NEC* Section 210.52(G). These outlets must also be covered by a GFCI, per *NEC* Section 210.8(A)(2). An unattached garage doesn't need to be supplied with electricity, but one that's attached must conform to these requirements.

NEC Section 210.52(C) also requires that a receptacle outlet be provided for small appliances along countertop areas in residential kitchens, pantries, breakfast and dining areas wherever the countertop is more than 12 inches wide. When a countertop is interrupted by a range, a sink, or a refrigerator, each section is then considered a separate space under this provision. The purpose of this requirement is to prevent small appliance cords from being stretched across a range, cooktop, or sink where either heat or water could cause problems. These countertop outlets must also be protected by a GFCI.

In a *dwelling unit*, where laundry facilities are permitted and will be installed, at least one 120-volt receptacle outlet must be provided for the laundry in addition to any 240-volt outlet provided for the dryer, per *NEC* Section 210.52(F). This outlet supplies power for the washing machine, as well as other electrical needs such as an electric iron.

A dwelling unit may be a single-family house, but it could also be a part of a *multiple dwelling* or apartment building. In the case of a multiple dwelling, where laundry facilities are provided for the use of all the occupants, a special laundry outlet isn't required in each unit. Also, in multiple dwellings where laundry facilities aren't permitted, there's no requirement for a laundry receptacle.

Lighting Requirements

In dwelling units, whether single-family or multiple-family, *NEC* Section 210.70(A) requires that at least one wall-switch-controlled lighting outlet be installed in each habitable room, plus bathrooms, hallways, stairways, attached garages, detached garages with electric power, and outdoor entrances. A habitable room, you'll recall, is a bedroom, living room, dining room, den, sun room, recreation room, etc. A switch-controlled lighting outlet in this instance means some type of light fixture. It can be fluorescent or incandescent, ceiling- or wall-mounted, flush or recessed, as long as it holds a light.

NEC Section 210.70(A) also provides a general exception to this requirement which states that in all habitable rooms other than the kitchen, the light fixture may be omitted if a switch-controlled receptacle is provided. So, in kitchens, bathrooms, hallways, stairways, attached garages, and at outdoor entrances, a switched *light fixture* must be installed, but in the other rooms a switched light fixture is optional as long as there's a switch-controlled receptacle into which a lamp can be plugged.

In an attic, crawl space, utility room, or basement, a lighting outlet (some type of light fixture) is required only if that space is used for storage or contains equipment, such as a furnace, water heater, or central air conditioning, that may require servicing. When a light is required in one of these spaces, it must be controlled by a light switch located at the point of entry to that space, per *NEC* Section 210.70(A)(3).

These receptacle and lighting requirements are code minimums. A dwelling unit electrical system providing less than these minimums won't be approved. However, the owner or the contractor might want to exceed these standards, which is certainly permissible as long as all additional electrical conforms to the applicable provisions of the code.

It's worth mentioning here that while architects and building designers are usually current with the minimum requirements of the *International Building Code®* (*IBC®*) and the *International Residential Code®* (*IRC®*), which deal with structural matters, they may at times get behind on the specific provisions of the electrical, plumbing, and mechanical codes. This means that the electrician must study the plans very carefully to make certain that all code-required lights and receptacles are included. The electrician, not the architect, is ultimately responsible for the accuracy of the electrical installation.

The electrical contract always includes a clause stating that the electrical system must be installed in accordance with drawings and specifications, *and shall be in accordance with National Electrical Code and local codes and ordinances*. If the architect omits lights, outlets, or switches that are required by code, the electrician must put them in or his job will never pass inspection.

The First Floor

With this in mind, let's study some floor plans that show the proposed electrical installation. Figure 8-10 is a first-floor plan, which includes the family room, bath 1, laundry, kitchen, dining room, living room, foyer and garage, with connecting hallways and storage closets.

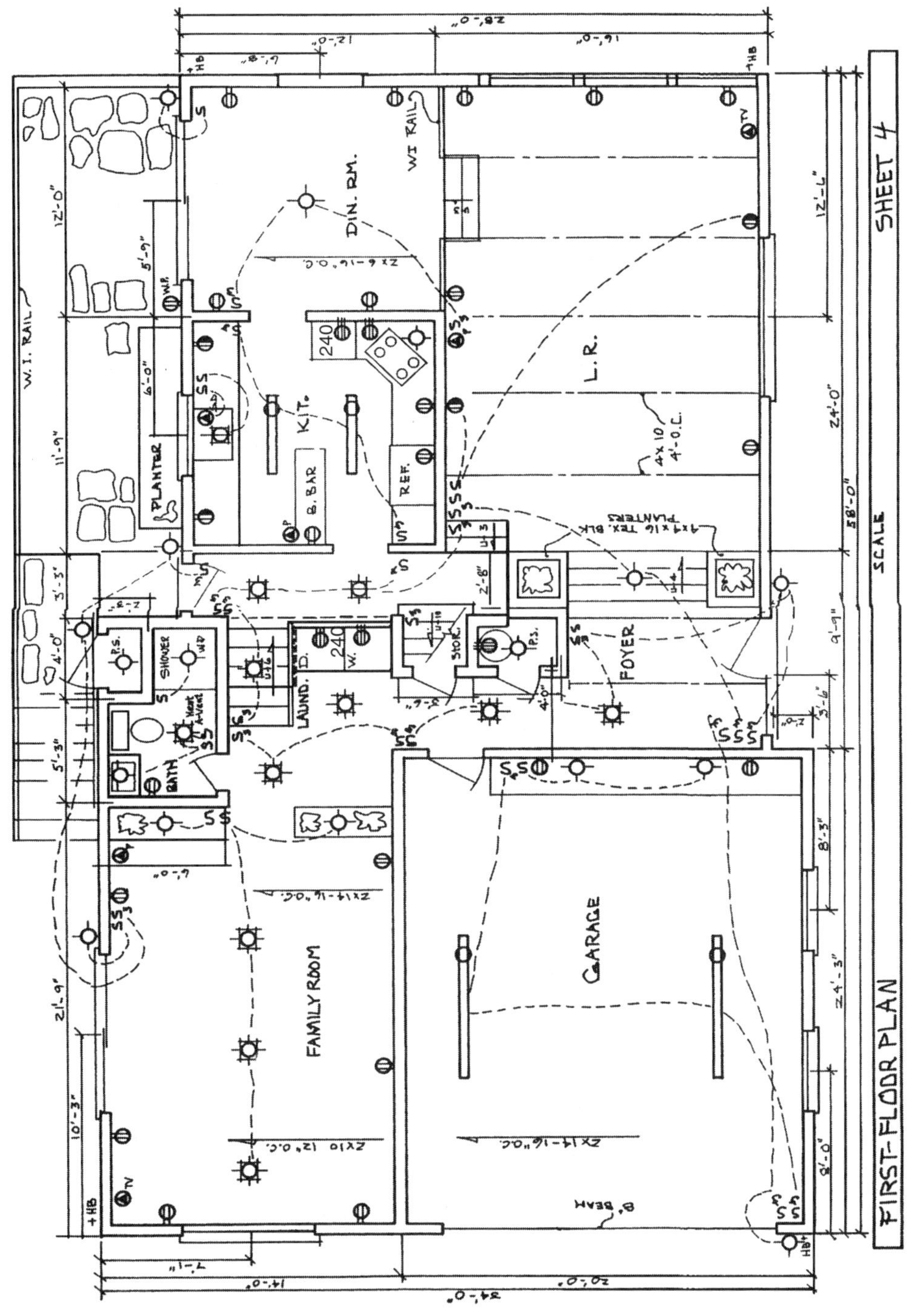

Figure 8-10
First-floor plan

Family Room

In the family room, the code requires convenience outlets spaced so that no point along the wall at the floor level is more than 6 feet from an outlet, and at least one switched lighting outlet, which can be a switched receptacle. In Figure 8-10, the lighting requirement is certainly met and exceeded by the installation of three recessed and two surface-mounted light fixtures, all switch-controlled. However, the convenience outlet requirement hasn't been met. The wall between the family room and bath 1 is given as 6 feet to the corner, then from the corner along the outside wall it's about another 2 feet to the first outlet. A convenience outlet will have to be added on the family room/bath 1 wall.

Bath 1

The bathroom must have a switched light, and an outlet adjacent to the sink protected by a GFCI. The switched light is there, as is an outlet adjacent to the sink, but GFCI isn't indicated on the plans next to the outlet. These letters must be added to the receptacle symbol on the plans to insure that the code requirement for the GFCI protection is met.

Since this bathroom doesn't have a window, the *IRC* (R303.3) requires a mechanical exhaust vent. The Heat-A-Vent with Exhaust/Heater Fan and Light will satisfy that requirement. While an electric fan in such vents isn't specifically required, it's what is generally used. In this case, the same switch that turns on the light also turns on the exhaust fan. Such combined switching is normal in a windowless bathroom. A separate switch will control the heat.

Laundry

As we discussed earlier, in the laundry you need a light, switch-controlled or not, and at least one receptacle outlet in addition to the 240-volt outlet for the dryer. There's no problem with the light requirement. However, notice that the symbol for the outlet by the washer is the same as the one for the dryer. That won't meet the requirements; the voltage is wrong. The outlet above the washer must be 120 volt, not 240.

Kitchen

Kitchens require a switch-controlled light, at least two small appliance circuits, a small appliance outlet on every section of countertop over 12 inches wide, and there should be no wall space more than 6 feet from an outlet along the floor level. The lighting requirement is met. The small appliance circuit and outlet requirements are met with the proper wiring of the outlets shown on the floor plan in Figure 8-10, with two exceptions. The receptacles on either side of the sink are within 6 feet of the sink, so they must be labeled for GFCI protection.

Dining Room

The dining room requires a switch-controlled lighting outlet, which can be a switched receptacle, and no wall space more than 6 feet from an outlet. Both requirements are met for this room on the plan.

Living Room

The living room also requires a switch-controlled lighting outlet, which can be a switched receptacle, and no wall space more than 6 feet from an outlet. Again, both requirements are met for this room on the plan.

Foyer

The foyer, hallways, and stairway all require a switch-controlled light, and the requirements are met as shown on the plan.

Garage

The attached garage must have a switch-controlled light and at least one GFCI-protected receptacle (in addition to any laundry receptacles — if the laundry were to be located in the garage). There are plenty of switch-controlled lights on this plan, plus two non-laundry receptacles, but neither receptacle shows the required GFCI protection. Both these outlets must be relabeled for GFCI protection.

Outdoor Requirements

The garage completes our survey of the first-floor interior of this single-family dwelling. Before going on to the second-floor plan, let's see what's needed to complete the outdoor requirements.

Per *NEC* Sections 210.52(E) and 210.8(A), one- and two-family dwellings must have at least one receptacle outlet in the front and one in the back of the dwelling and both must be weatherproof and GFCI-protected. There's only one outside weatherproof outlet shown on the plan. It's on the back patio, and lacks GFCI protection. Another weatherproof outlet will need to be added at the front of the home, and both outlets will need to show GFCI protection.

All the outdoor entrances to the house have switch-controlled lights, so they meet code requirements according to *NEC* Section 210.70(A).

The Second Floor

Let's look at the floor plan for the second floor, shown in Figure 8-11. There are three bedrooms: the master bedroom with its own bathroom, bedroom 1 and bedroom 2, one additional bathroom, as well as hall, stairway, utility closet, furnace room, and outside porch.

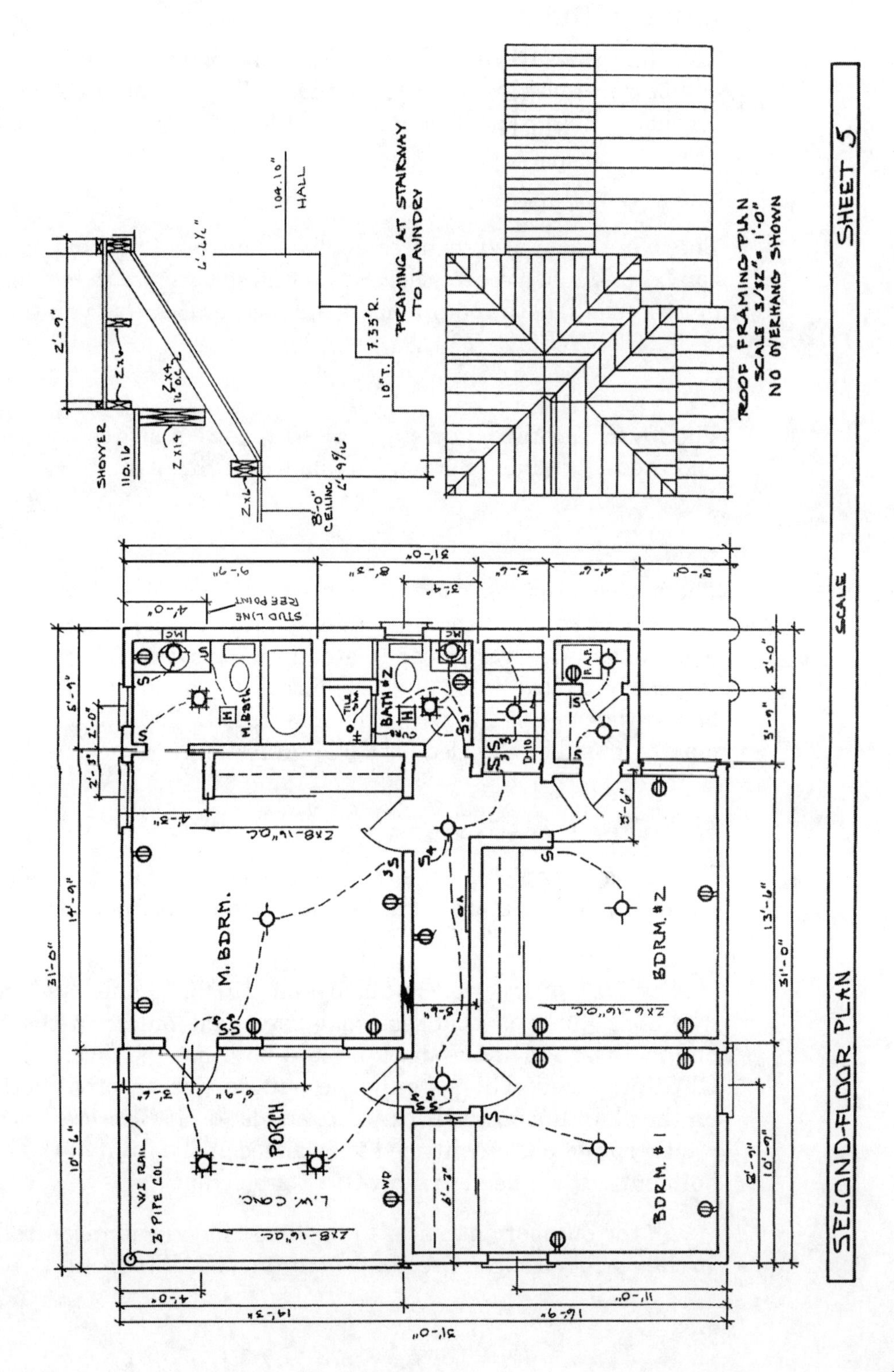

Figure 8-11
Second-floor plan

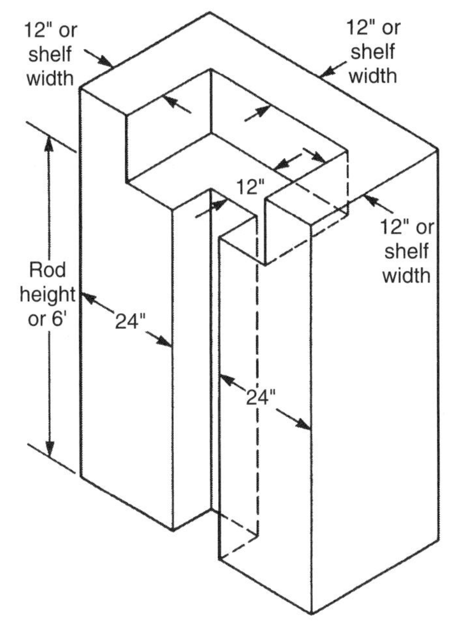

Figure 8-12
Closet light requirements

Master Bedroom, Bedroom 1, and Bedroom 2

Each room requires a switch-controlled lighting outlet (which may be a switched receptacle) and receptacles spaced so that there's no point along the wall more than 6 feet from an outlet. In all three rooms, both requirements are met. The code doesn't require lights in closets, and there aren't any lights indicated in these bedroom closets.

When closet lights are installed, they must comply with *NEC* Section 410.16(C), which says that the light fixture can be installed on the wall above the closet door or on the closet ceiling. It can be an incandescent fixture with a *completely enclosed lamp* mounted on the wall or ceiling with at least 12 inches of clearance from the storage area, as shown in Figure 8-12. You can also install surface-mounted fluorescent or recessed incandescent fixtures with completely enclosed lamps. These must have at least 6 inches of clearance from the storage area.

Bathrooms

Since both bathrooms have windows, the only requirements are the switch-controlled lights and the GFCI-protected outlets adjacent to the sinks. Both rooms are satisfactory regarding lights, and both have outlets adjacent to the sinks — but neither sink outlet indicates GFCI protection.

Hall and Stairway

The hall and stairway both have the required switch-controlled lights.

Utility Closet and Furnace Room

Both of these rooms have switch-controlled lights, and this exceeds code requirements. Each is required to have a light, but it doesn't have to be switch-controlled.

Porch

There are no requirements for a second floor porch above grade . However, since an outdoor outlet has been provided, it must be weatherproof, and a GFCI added, per *NEC* Section 210.8(A)(3).

Branch Circuits

Having corrected the original floor plans to conform to code requirements, the revised plans are as shown in Figures 8-13 through 8-16. These plans now not only meet all code requirements, but in many respects they exceed them. For example, the code doesn't require that a switch-controlled light be subject to control from more than one location — though it's very convenient, and at times a real necessity. It would

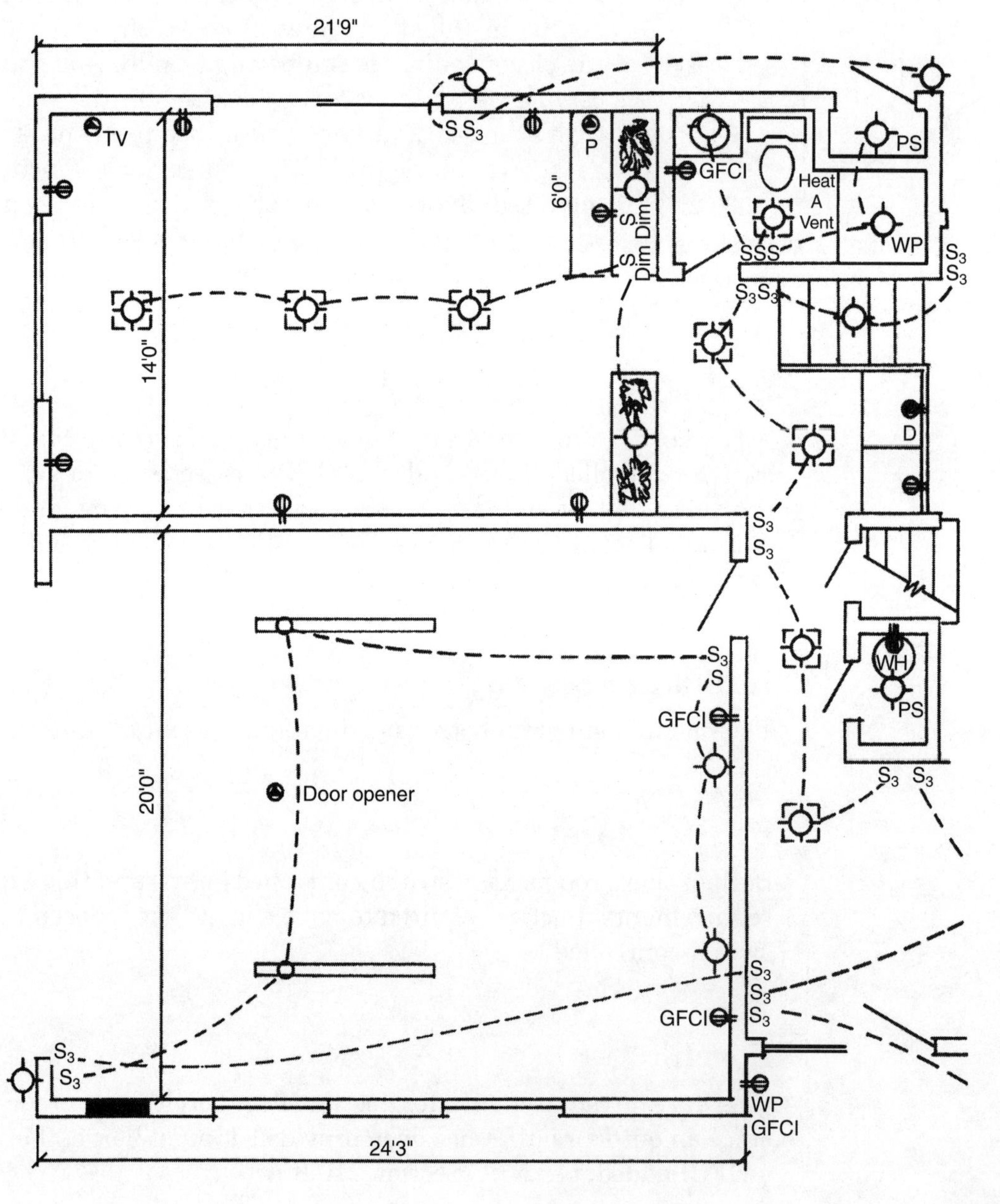

Figure 8-13
Electrical floor plan: family room, bath, garage, utility

be ridiculous to turn on a light at the bottom of a stairway in order to find your way upstairs only to have to go back downstairs to turn the light off so you can go back up in the dark. A multiple switching system is a necessity in this case.

A similar condition exists in a long hallway, or a large room with doors at both ends. On the first-floor plan, the kitchen, dining room, foyer, and garage are all places where multiple-switching circuits make good sense.

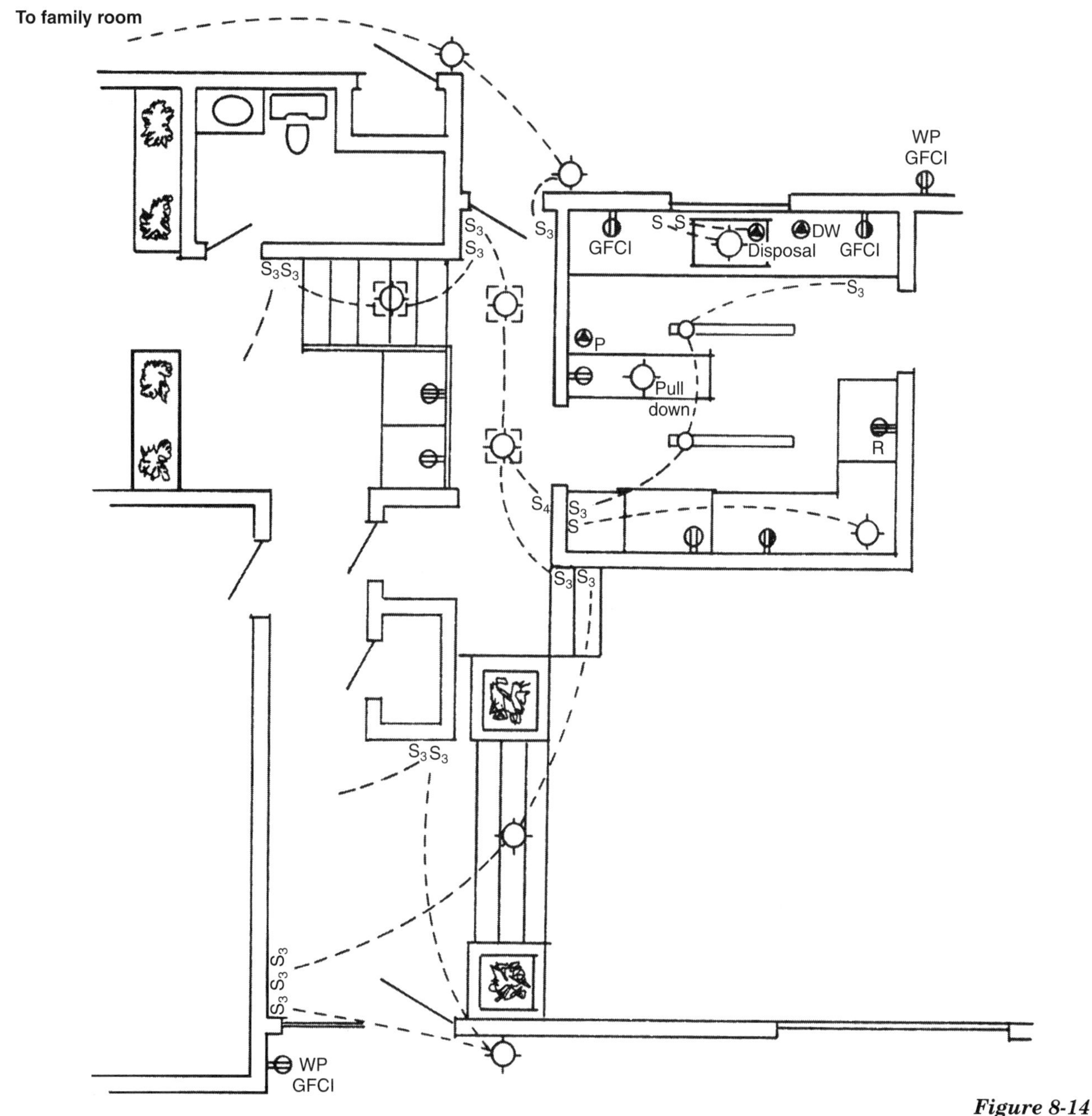

Figure 8-14
Electrical floor plan: kitchen, hall, foyer

The plans indicate where code minimums have been exceeded. For example, there are more light fixtures than required in the kitchen, family room, and garage.

Branch Circuit Loads

With the location and number of fixtures, receptacles and switches determined, it's time to divide these elements into branch circuits, plan how they'll be connected to each other, and to the incoming electrical supply. Use *NEC* Section 220.3 to compute the branch circuit loads.

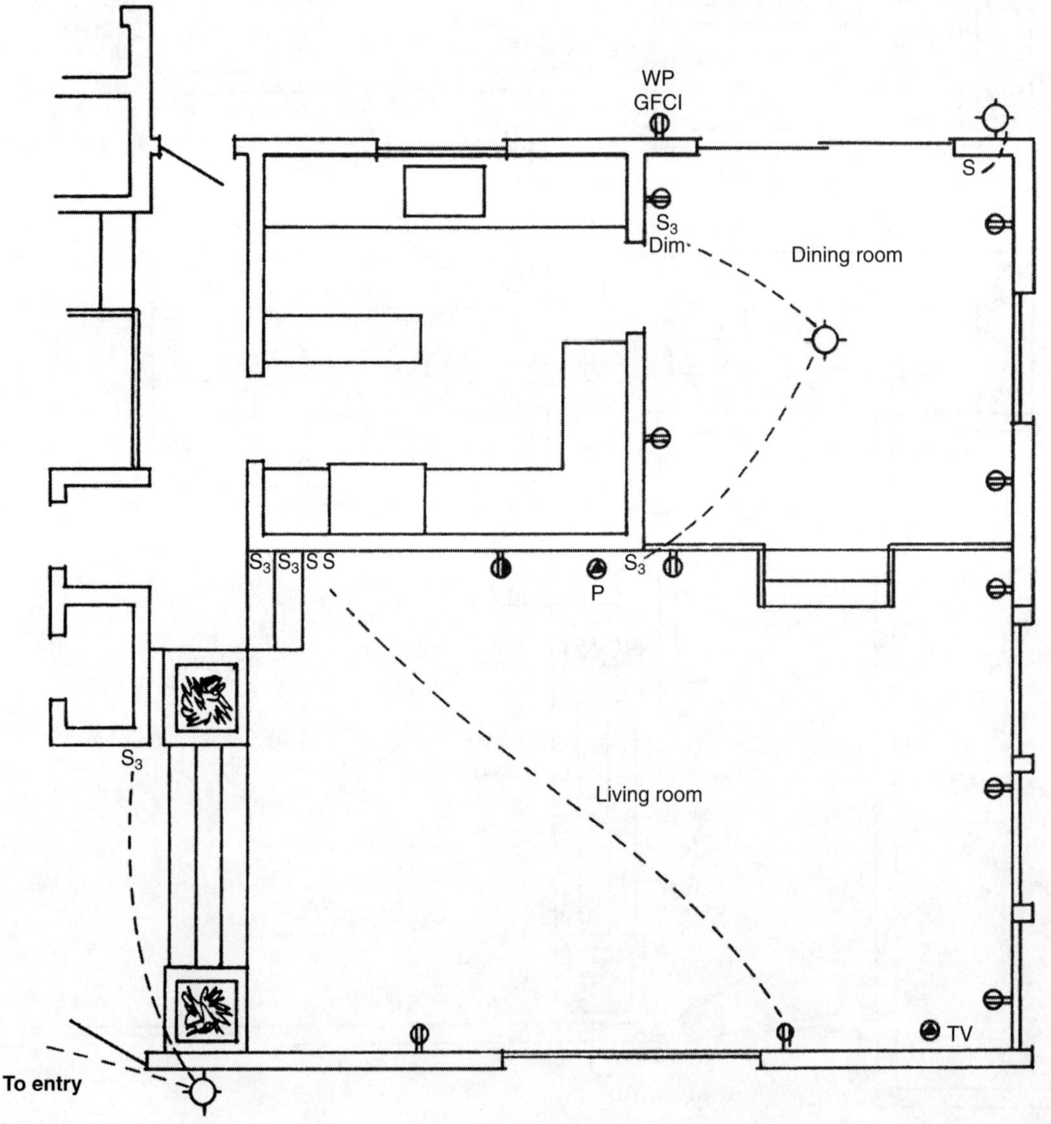

Figure 8-15
Electrical floor plan: dining room, living room

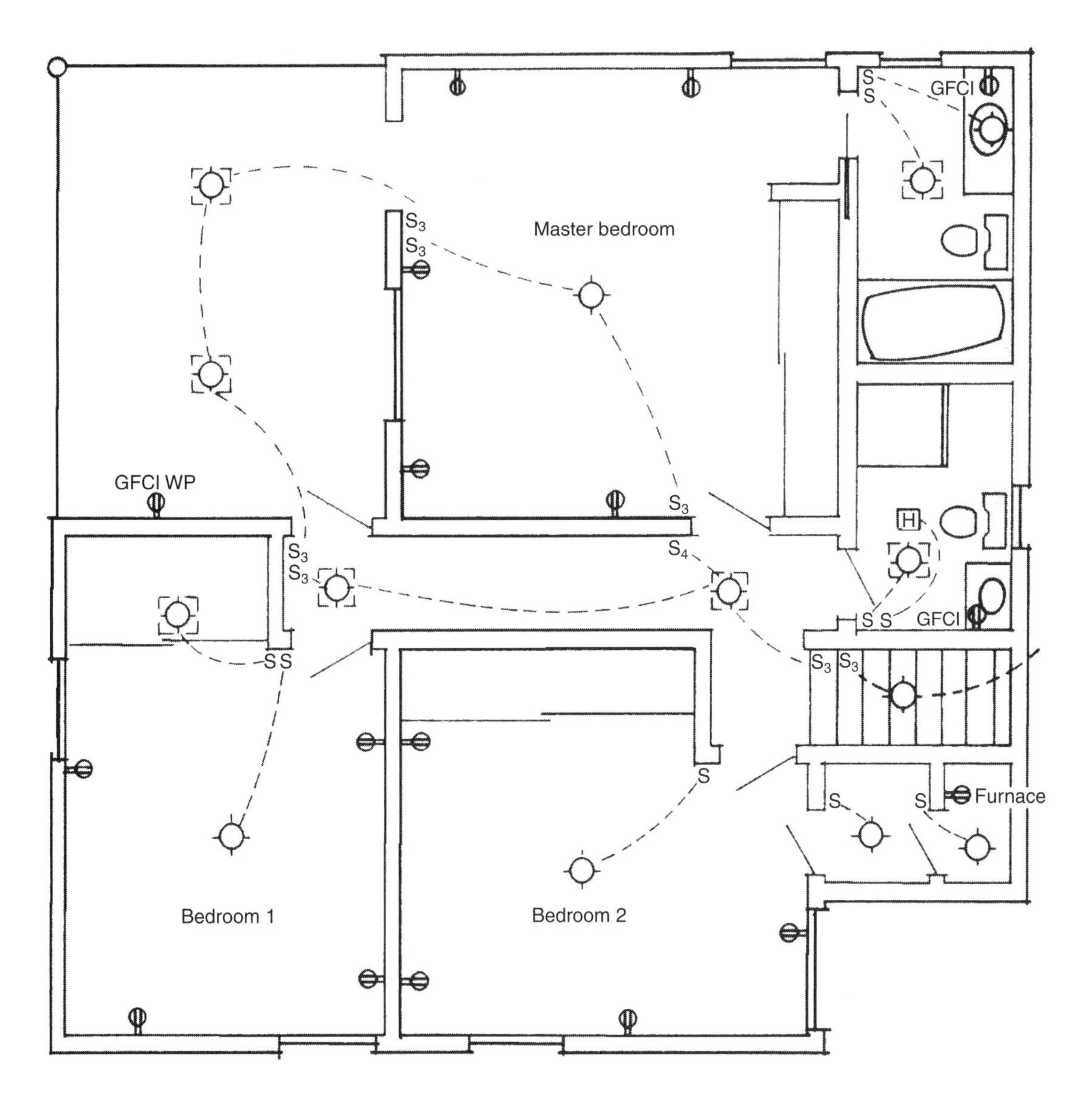

Figure 8-16
Electrical floor plan: second floor

NEC Section 210.20(A) states: *"Where a branch circuit supplies continuous loads or any combination of continuous and noncontinuous loads, the rating of the overcurrent device shall not be less than the noncontinuous load plus 125 percent of the continuous load."*

According to Watt's Law:

$$\text{Power (Volt/Amperage)} = \text{Voltage} \times \text{Amperage}$$

The most commonly used branch circuits are 15-, 20-, 30-, 40-, and 50-ampere circuits. The table, shown in Figure 8-17, lists the capacities and permissible loads of those circuits.

Amperage	*Voltage*	*Wire size*	*Capacity*	*(Volt-amperes) = 125% of permissible load or 80% of capacity*
15	120	#14	1800	1440
20	120	#12	2400	1920
30	120	#10	3600	2880
40	120	# 8	4800	3840
50	120	# 6	6000	4800

Note — Compute residential 240V circuits as two 120V circuits added together

Figure 8-17
Circuit capacities and permissible loads

Most residential lighting and convenience outlet circuits are rated at 15 amperes. *NEC* Section 240.4(D)(3) permits #14 copper wire for a 15-ampere circuit. The smallest wire size approved for permanent building wiring is #14; it's also the easiest and least expensive wire to use. The smaller the wire size, the easier it is to cut, bend, splice, and run through the building, making it also more economical in terms of labor expense.

The permissible loads in the last column of the table in Figure 8-17 are the maximum *continuous loads* the code will allow for the circuits listed. In order to conform to this requirement, it's necessary that we know just what the code means by *continuous load*, as well as how to compute it. *NEC* Article 100 defines *continuous load* as *"a load where the maximum current is expected to continue for 3 hours or more."*

A load consisting of lights that are turned on at 6:00 p.m. and off around 11:00 p.m. is a continuous load; a water heater, or any other equipment that cycles on and off automatically, is considered a continuous load, since at any given moment it may be on. But, a vacuum cleaner in use for 30 minutes at a time is not a continuous load.

The computation of individual loads for the purpose of determining total circuit loads on general purpose lighting and convenience outlet circuits is based partly on known quantities, and partly on standardized rule of thumb estimates. Let's look at a few of these.

Fluorescent Lights

Check the label pasted on the ballast stating the wattage that's drawn by that ballast and the tubes it supplies. If it's inconvenient to open the fixture to look at the label, you can estimate 10 watts per running foot of tube for each tube. Add 25 percent of the tube wattage for the ballast load. Calculated this way, the total will be a few watts above the load given on the ballast label.

Recessed Incandescent Lights

The maximum lamp wattage for which the fixture is rated must be permanently marked inside, in letters at least ¼ inch high, per *NEC* Section 410.120.

Table 220.12 General Lighting Loads by Occupancy

Type of Occupancy	Unit Load: Volt-Amperes per Square Meter	Unit Load: Volt-Amperes per Square Foot
Armories and auditoriums	11	1
Banks	39[b]	3½[b]
Barber shops and beauty parlors	33	3
Churches	11	1
Clubs	22	2
Court rooms	22	2
Dwelling units[a]	33	3
Garages — commercial (storage)	6	½
Hospitals	22	2
Hotels and motels, including apartment houses without provision for cooking by tenants[a]	22	2
Industrial commercial (loft) buildings	22	2
Lodge rooms	17	1½
Office buildings	39[b]	3½[b]
Restaurants	22	2
Schools	33	3
Stores	33	3
Warehouses (storage)	3	¼
In any of the preceding occupancies except one-family dwellings and individual dwelling units of two-family and multifamily dwellings:		
Assembly halls and auditoriums	11	1
Halls, corridors, closets, stairways	6	½
Storage spaces	3	¼

[a]See 220.14(J).
[b]See 220.14(K).

Figure 8-18
Lighting loads by occupancy

Surface-Mounted Incandescent Light with Diffuser

As a rule of thumb, allow 200 watts for one- or two-bulb fixtures.

General-Purpose Convenience Receptacles

These are considered 180 volt-amperes each. The *NEC* refers to volt-amperes, not watts, for general-purpose receptacles; see Section 220.14(I). The actual loads on general-purpose convenience receptacles vary widely. However, when the receptacles are located so that they conform to *NEC* Section 210.52(A)(l), and no point along the floor line in any wall space is more than 6 feet from an outlet, many are never used. Some are blocked by furniture and some are placed at points where the occupant has no need for electrical power. Consequently, the code writers assigned an arbitrary load of 180 volt-amperes to each one on the basis that the average of the loads on all of them will actually turn out to be less than that.

NEC Table 220.12 (Figure 8-18) lists code minimum allowances, in volt-amperes per square foot, which must be made in lighting circuits for various types of occupancies. In dwellings, the required minimum is 3 volt-amperes per square foot of floor area, excluding porches, garages, or unused or unfurnished areas. By the time the various other lighting and outlet requirements discussed earlier have been met, that 3 volt-amperes per square foot requirement has usually been exceeded.

Lighting and Convenience Outlet Branch Circuits

Assigning the lighting and convenience outlets on our sample plans to 15-ampere branch circuits that conform to code requirements could be done in several different ways. One of those ways is shown

Fixture	*Watts per fixture*	*Total watts*
7 convenience outlets	180 watts each	1260
3 recessed ceiling fixtures	100 watts each	300
2 ceiling surface fixtures	200 watts each — estimated	400
1 outdoor wall light at door	100 watts	100
Total wattage for all fixtures		*2060*

Figure 8-19
Family room loads

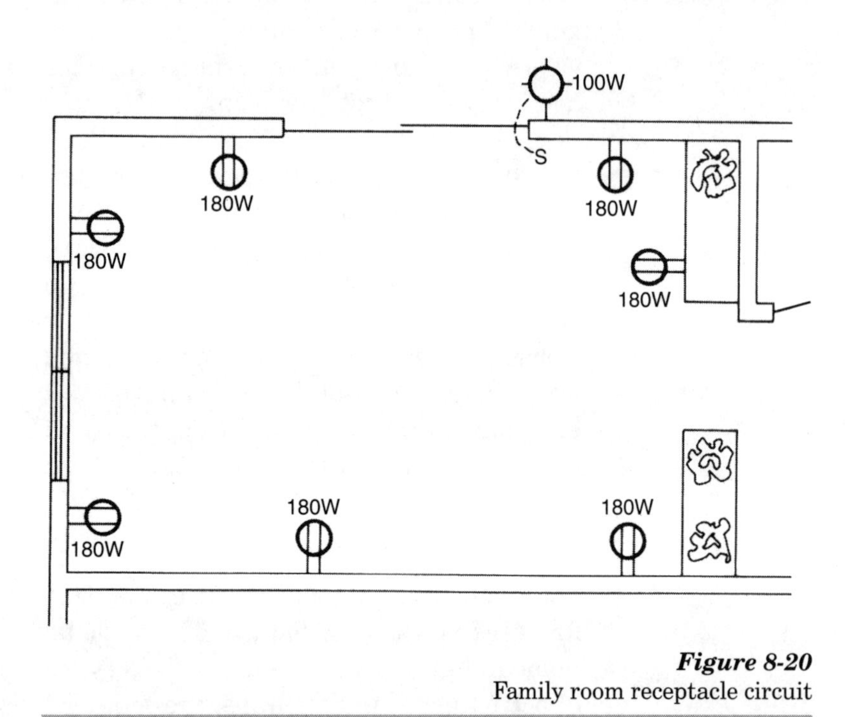

Figure 8-20
Family room receptacle circuit

in Figure 8-19. After all lights and receptacles have been divided into branch circuits, we'll give cable plans and wiring diagrams for each circuit to show in detail how the system will be assembled.

We'll start with the corrected first-floor family room plan (look back at Figure 8-13). Figure 8-19 gives the total load for this room. The special purpose outlets marked "TV" and "P" are for the cable TV and telephone and aren't loads on the electrical system, so they don't connect to branch circuits.

The total 2060-watt load for the family room can't be assigned to a single 15-ampere circuit, because 1440 watts is the load limit for the circuit, as shown in Figure 8-17. The supply to this room will have to be split over two circuits. The seven convenience outlets plus the outdoor light can go on one circuit (Figure 8-20), which will then total 1360 watts. Per code, 1360 watts is satisfactory since it's *less* than the allowable maximum.

It should be possible to combine the remaining 700 watts of lighting load with some other items since it's so far below the permitted maximum. Fortunately, next to the family room is a bathroom that calls for a total of 600 watts:

1 Heat-A-Vent	300 watts
1 medicine cabinet/vanity light	200 watts
1 shower ceiling waterproof light	100 watts

The remaining family room watts and the bathroom total 1300 watts. Next to the bathroom is a utility closet that opens to the outside and contains only a 60-watt pull chain light. We can add that amount

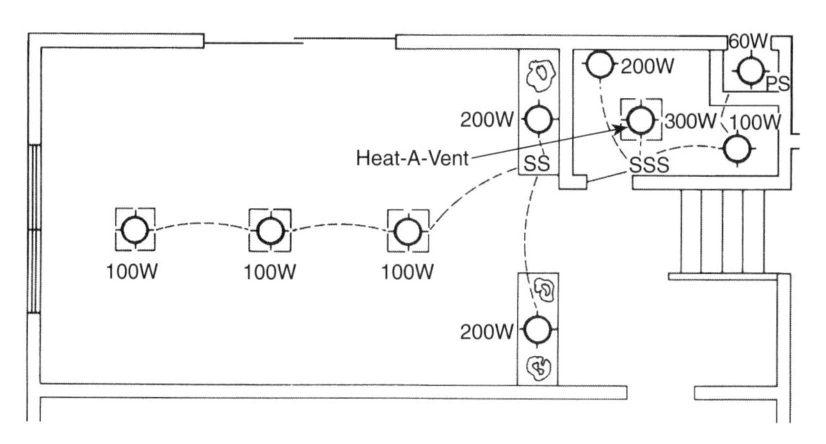

Figure 8-21
Family room, bath, utility light circuit

Fixture	*Watts per fixture*	*Total watts*
2 two-tube 4-foot fluorescents	100 watts	200
1 recessed ceiling light over sink	100 watts	100
1 surface ceiling light over counter	200 watts	200
1 pull-down light over breakfast bar	200 watts	200
Total wattage for kitchen lights		*700*

Figure 8-22
Kitchen light loads

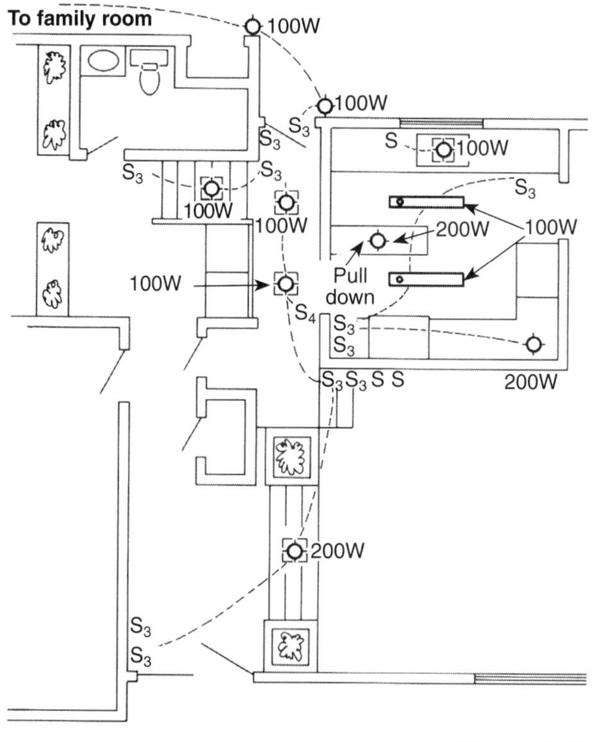

Figure 8-23
Kitchen, hall, foyer light circuit

to the other two, and have a total of 1360 watts. Figure 8-21 shows this circuit. Notice that we didn't include the GFCI-protected outlet in the bathroom. It will be covered later, along with all the other GFCI outlets in the house.

Now let's look at the kitchen in our Figure 8-14 floor plan. We have five light fixtures; see Figure 8-22. All the other receptacles and outlets in this kitchen are for appliances, and the code does *not* permit them to be included in lighting and convenience outlet circuits. They'll be covered later.

The 700 watts of kitchen lighting appears to be another likely candidate to be combined with another room circuit. If we add in the lights in the stair and hall, the two outside lights and the light in the foyer, we'll have an additional 700 watts:

Fixture	*Watts*
3 recessed ceiling lights, stair and hall	300
2 outside lights	200
1 foyer ceiling light	200

The ceiling light over the stair from the foyer to the living room certainly looks as though it belongs on a circuit with the living room. However, as we'll soon see, the living room has too many receptacles to take that light as well. So, by adding it to the kitchen-hall circuit, the total computed load becomes 1400 watts, as shown in the circuit diagram in Figure 8-23.

Fixture	Watts per fixture	Total watts
4 convenience outlets	180 watts each	720
1 ceiling light fixture	200 watts	200
1 outdoor wall light by door	100 watts	100
Total wattage for dining room		*1020*

Figure 8-24
Dining room loads

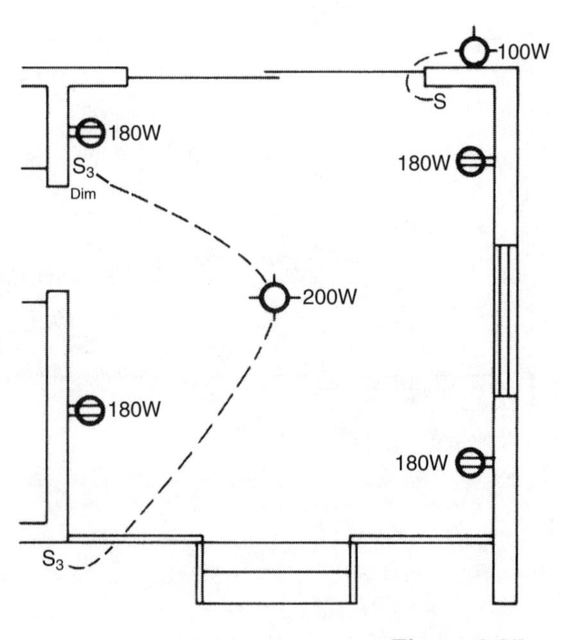

Figure 8-25
Dining room light and receptacle circuit

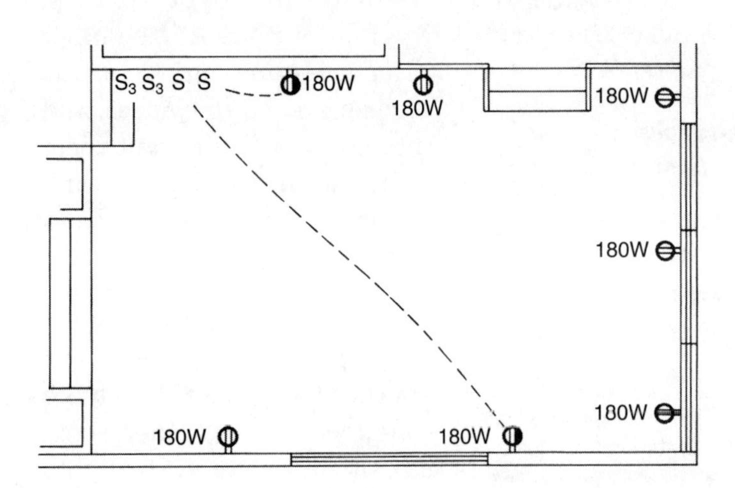

Figure 8-26
Living room receptacle circuit

Fixture	Watts per fixture	Total watts
2 two-tube 4-foot fluorescent garage lights	100 watts each	200
2 surface-mounted ceiling lights over workbench	200 watts each	400
1 outdoor light at garage door	100 watts	100
4 recessed ceiling lights in hallway	100 watts each	400
1 outdoor light at entry	100 watts	100
1 pull-switch closet light	60 watts	60
Total wattage for garage and hallway		*1260*

Figure 8-27
Garage and hallway lighting loads

The dining room and living rooms are easy, and both require their own circuits. The dining room loads, shown in Figure 8-24, total 1020 watts. The dining room light and receptacle circuit is diagrammed in Figure 8-25.

The living room has no light fixtures. It'll be illuminated by lamps plugged into the many receptacles, two of which are switched. There are seven convenience outlets in the living room at 180 watts each, for a total of 1260 watts. You can see the receptacle circuit in Figure 8-26. As mentioned earlier, the phone and cable TV outlets aren't included in the electrical power circuit.

Completing the first floor are the garage and the hall leading to it. The lighting loads are shown in Figure 8-27.

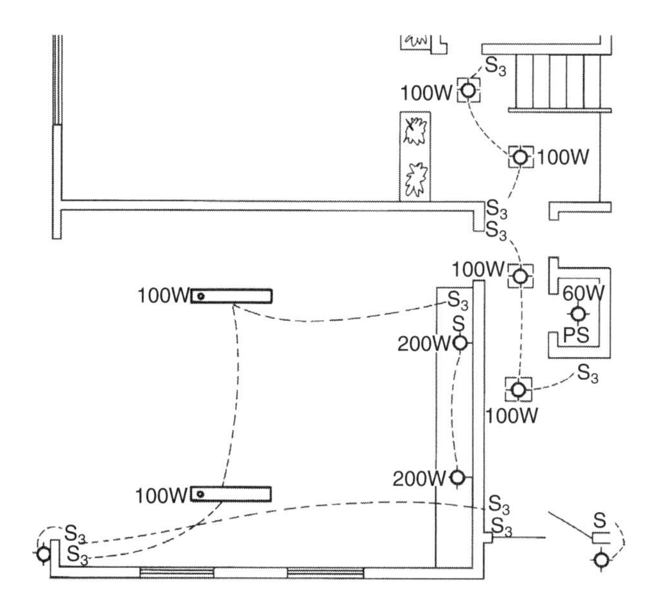

Figure 8-28
Garage and hall light circuit

The two GFCI outlets in the garage and the outside GFCI outlets aren't included in this circuit, diagrammed in Figure 8-28. They'll be dealt with later. Also the laundry outlets for the washer and dryer aren't included. As we shall see, the code specifies how they're to be handled.

Moving upstairs, we'll start with the master bedroom and porch. The lighting loads for those areas are listed in Figure 8-29, and diagrammed in Figure 8-30. The weatherproof outlet on the porch will be included with a GFCI circuit later.

Fixture	*Watts per fixture*	*Total watts*
5 convenience outlets	180 watts each	900
1 surface-mount ceiling light	200 watts	200
2 recessed ceiling lights on porch	100 watts each	200
Total wattage for master bedroom		*1300*

Figure 8-29
Master bedroom and porch loads

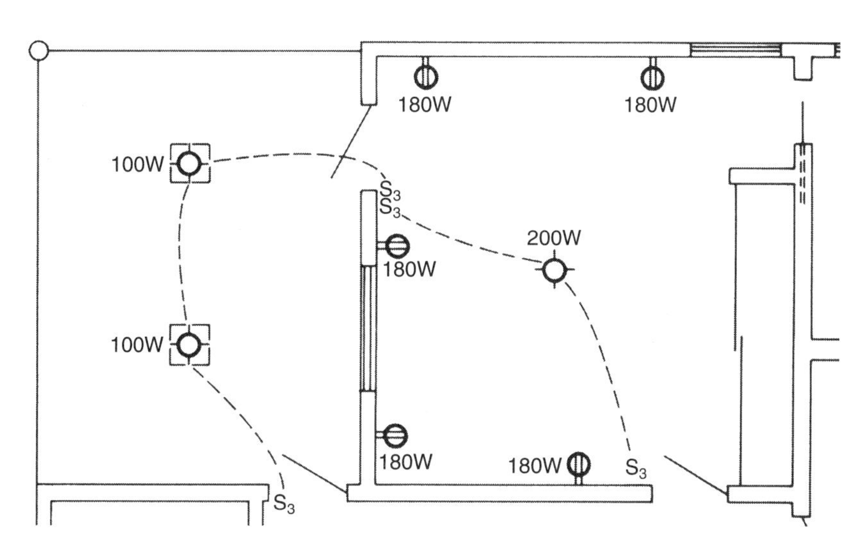

Figure 8-30
Master bedroom light and receptacle circuit

Fixture	Watts per fixture	Total watts
2 recessed ceiling lights	100 watts each	200
2 ceiling heat lamps	300 watts each	600
1 master vanity light	200 watts	200
2 two-foot fluorescent lights for bath 2 vanity	25 watts each	50
Total wattage for both bathrooms		*1050*

Figure 8-31
Upstairs bathroom loads

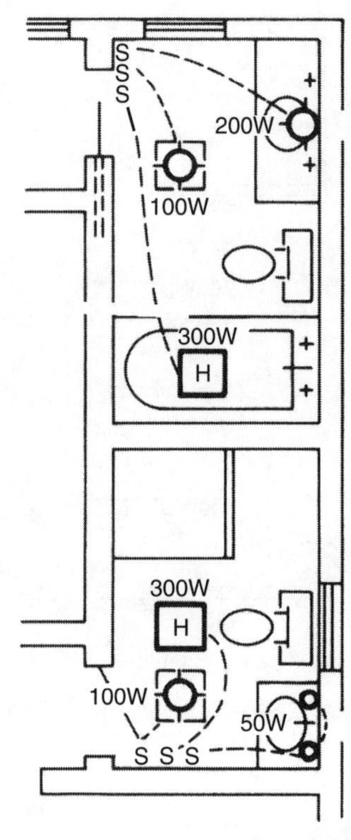

Figure 8-32
Second floor bathrooms circuit

Fixture	Watts per fixture	Total watts
4 convenience outlets	180 watts each	720
2 surface-mount ceiling lights	200 watts each	400
2 closet ceiling lights	60 watts each	120
Total wattage for bedroom 2		*1240*

Figure 8-33
Bedroom 2, closets and stairway loads

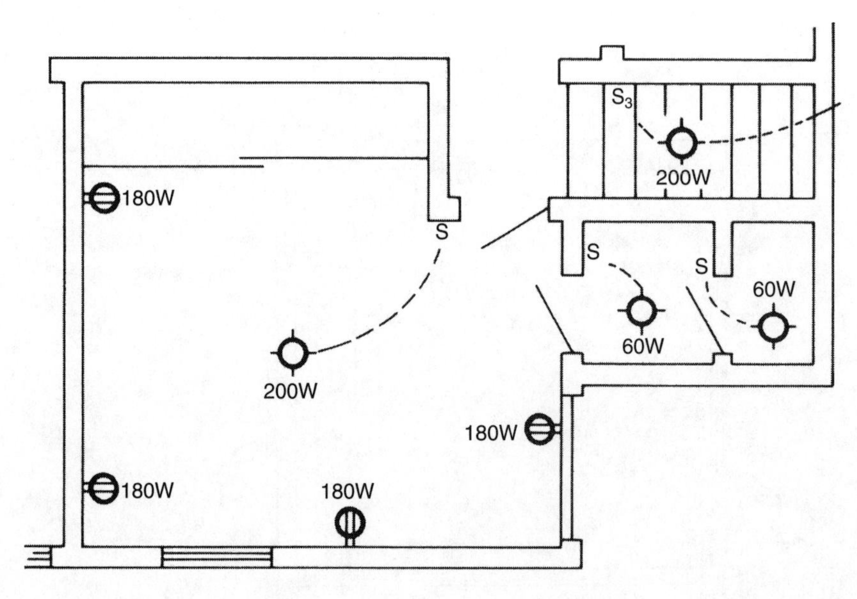

Figure 8-34
Bedroom 2, utility and stairway light and receptacle circuit

The lighting loads for the master bathroom and bath 2 can be paired on a single circuit; see Figures 8-31 and 8-32. Again, we'll deal with the two GFCI outlets in these bathrooms a little later. Bedroom 2 outlets and lights can be combined with the modest requirements of the utility and furnace closet lights, as well as the stairway light; see Figures 8-33 and 8-34.

The next circuit run includes the bedroom 1 receptacles and light and the two recessed hall ceiling lights; see Figures 8-35 and 8-36.

Fixture	Watts per fixture	Total watts
4 convenience outlets	180 watts each	720
3 recessed ceiling lights	100 watts each	300
1 surface-mount ceiling light	200 watts	200
Total wattage for bedroom 1		*1220*

Figure 8-35
Bedroom 1 and hallway loads

The lighting and convenience outlet circuits for the entire house now total 10, but in the process of setting them up, seven GFCI-protected outlets were bypassed downstairs and three upstairs. As described earlier, the GFCI is a device that can be obtained either in combination with a convenience outlet or in combination with a circuit breaker. Since it's a rather expensive device, whenever possible, all outlets in a house requiring GFCI protection are wired so that a single GFCI protects them all. Look back at Figure 8-9 to see a sequence wired in this manner. The other way of protecting a number of outlets with a single GFCI is to place them all on one circuit and put the GFCI in the breaker.

Except for kitchen countertop appliance outlets where GFCIs are required, an average house (of about 1200 to 1500 square feet) will have a minimum of four to five outlets that must be covered by GFCI per code: two outdoor outlets, one garage outlet, and one or two bathroom outlets.

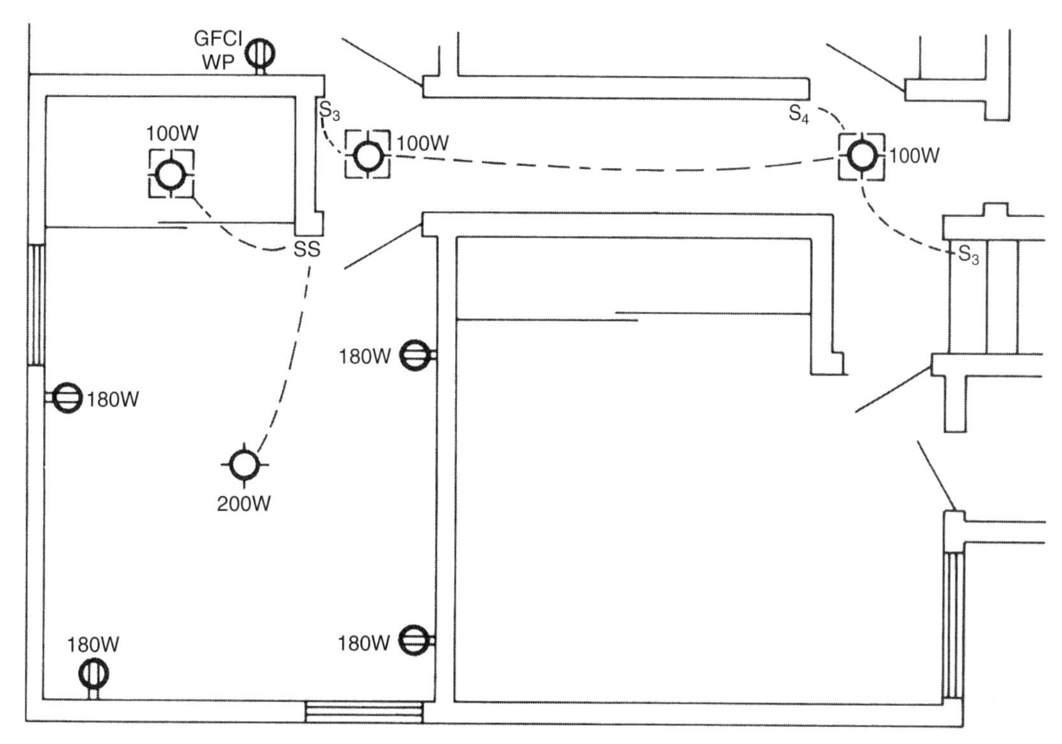

Figure 8-36
Bedroom 1 and second floor hall circuit

Area	Total watts
Family room outlets	1360
Family room, bath 1 and utility lights	1360
Kitchen, hall and foyer lights	1400
Dining room	1020
Living room	1260
Garage, hall and entry lights	1260
Master bedroom outlets and lights, porch light	1300
Second floor bathrooms	1050
Bedroom 2, utility closets, stairs	1240
Bedroom 1 and hallway	1220
GFCI-protected outlets (first floor)	1260
GFCI-protected outlets (second floor)	540
Total	*14270*

Figure 8-37
Lighting and outlet circuits with loads

If there are only four, their computed load totals 720 watts. In many instances the load of a single bathroom is 600 watts or less; so a GFCI can be placed in one bathroom and all the other required GFCI outlets can be wired to it without overloading that bathroom circuit. However, in the house we're working on, 10 outlets require protection. This constitutes a computed load of 1800 watts, which is enough to comprise another circuit as well. We'll divide them into two circuits, one for the first floor outlets and another for the second floor outlets.

Figure 8-37 shows the lighting and convenience outlet circuits, along with their computed loads, for our sample house.

NEC Section 220.12 states that a minimum lighting load must be provided of not less than the volt-amperage listed in *NEC* Table 220.12 (look back to Figure 8-18) per square foot of each structure being used for any of the purposes listed in that table. For dwelling units the requirement is 3 volt-amperes per square foot. The sample house illustrated here vastly exceeds this minimum requirement; however, the reader should be familiar with the procedure for computing a building's conformity.

The square footage is computed from the outside dimensions of the building; for dwellings, you can subtract porches, garages, and unfinished spaces from that total. The first floor on our plans is 34 feet by 58 feet = 1972 square feet, less the garage (20 feet by 24 feet = 480 square feet), giving a first floor square footage of 1492. The second floor is 31 feet by 31 feet = 961 square feet, less the porch (14 feet by 10½ feet = 147 square feet), leaving 814 square feet.

Net square footage, First Floor	1492
Net square footage, Second Floor	814
Total	*2306 square feet*
2306 square feet × 3 watts =	6918 watts

Our computed lighting circuits totaled 14270 watts, as shown in Figure 8-37, which is more than double what's required.

Other Branch Circuits

In addition to branch circuits for lighting, there are several other branch circuits required in residential structures. We've mentioned that, per *NEC* Section 210.52(C)(1), receptacle outlets must be provided for small appliances on all kitchen counter spaces over 12 inches wide. In addition, *NEC* Section 210.52(B)(3) specifies that those receptacles must be connected to at least two 20-ampere small appliance circuits and that, except for a refrigerator, those small appliance circuits can't have other outlets. The refrigerator can be on one of these circuits, or it can be on one by itself.

The laundry receptacle required in *NEC* Section 210.52(F) is also mentioned in *NEC* Section 210.11(C)(2), where it's required that a separate 20-ampere branch circuit supply that receptacle, and that circuit is to have *no other outlets.* The circuit mentioned is for a 120-volt washer. If you're installing an electric dryer, it requires a separate 240-volt, 30-ampere circuit that also has *no other outlets.* Our sample plans include an outlet for such a circuit.

An electric range calls for its own 240-volt circuit, which will usually be a 50-ampere circuit; and it too must have no *other outlets.*

An electric water heater requires another 240-volt circuit, which will be 20 amperes unless the heater is exceptionally large. This is another circuit that can have *no other outlets.*

While a refrigerator may be connected to one of the two 20-ampere small-appliance circuits required by *NEC* Section 210.52(B)(1), the dishwasher, garbage disposal, or trash compactor can't. These appliances often have individual circuits; however, sometimes the dishwasher and disposal are combined on a single 20-ampere circuit.

Electric Heating

Electric heating equipment, other than small portable heaters intended to be plugged into ordinary convenience outlet circuits, require special branch circuits. These circuits must be sized according to the rated power requirements of the units they supply. *NEC* Article 424 deals with fixed electric space-heating equipment.

Circuits for electric space-heating, according to *NEC* Section 424.3(A), can be 15, 20, 25, or 30 amperes. *NEC* Section 424.3(B) Branch-Circuit Sizing states, *"Fixed electric space-heating equipment shall be considered continuous load."* This means that it must be rated at not less than 125 percent of the total load of both motors and heaters. This restates in part the same requirement mentioned earlier in *NEC* Section 210.20(A).

Let's look at an example. A heater is rated at 1100 watts. If 1100 = 100 percent, then *NEC* Section 424.3(B) wants the circuit rated at least 125 percent of that, or at least 1375 watts. Computing in reverse, these circuits may be 15, 20, 25, or 30 amperes at 120 volts or 240 volts. Looking back at the table in Figure 8-17, a 15-ampere circuit at 120 volts can be loaded to 1440 watts, or 80 percent of its total capacity of 1800 watts. This load is only 1375 watts, which is well within code requirements for a 15-ampere circuit.

General Heating, Ventilating and Air Conditioning

Most other types of heating equipment, as well as ventilating and air conditioning devices, require some sort of electrical supply. The requirements in this area vary widely; consequently, the electrician will need to obtain specification information from the builder or owner — or whoever is in charge making this selection.

This information is needed at an early stage of plan development to insure that the necessary circuits are included in the plans, and the necessary power requirements are included in the computations for sizing the incoming service.

In our sample house, a gas-fired forced-air furnace is behind the utility closet off bedroom 2. To power its blower motor, controls, and intermittent ignition device (IID), a separate 20-ampere appliance circuit will be provided. The intermittent ignition device was developed as an electrical replacement for the gas pilot light. (It was developed in response to the emphasis on energy conservation.) It's now used on gas ranges, dryers, and water heaters, as well as furnaces.

Cable Plans

At this point, we've checked the plans in Figures 8-13 through 8-16 to insure that all outlets and switches required by code have been included. And we've grouped the switches and outlets to make up the lighting and convenience outlet circuits so that the computed loads on those circuits all conform to code loading requirements. See Figures 8-19 through 8-37. The next step is to prepare cable plans and wiring diagrams to show exactly how every circuit will be put together.

The switch, fixture, outlet, and other symbols used on cable plans are the same as those used on floor plans. However, the plans themselves are quite simple. They're drawn to scale just as floor plans are, but have few (if any) dimensions or other building construction information.

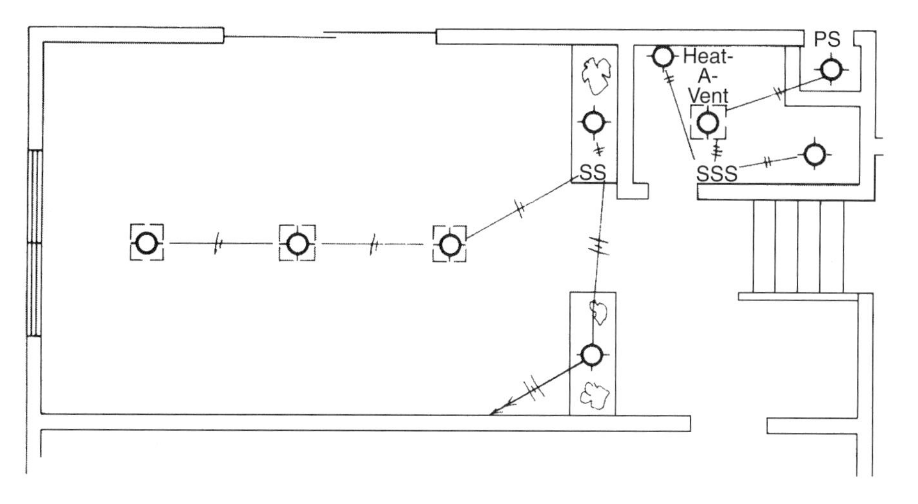

Figure 8-38
Cable plan for family room, bath, utility light circuit

All that's given are walls, doors, windows, etc., and electrical symbols. Their purpose is simply to show the electrician how many wires are to be used in each of the runs from one electrical box to another. Any unnecessary information is omitted.

Figure 8-38 is the cable plan for the lights assigned to the family room, bath, and utility light circuit that was laid out in Figure 8-21. Since the load on the circuit has already been found satisfactory, wattages are omitted. The various electrical symbols are connected by solid lines, each of which is crossed by two or more hatch lines indicating the number of circuit conductors to be used in that run. (Look back at Figure 8-1 to see electrical symbols.) Remember, the circuit conductors are white or grey common, and power is black, red, blue, or other color. The ground wire isn't a circuit conductor and doesn't appear on either cable plans or wiring diagrams because it's normally an inactive part of the system. This house is being wired in NM cable, so the wire runs are cross-hatched by either two or three lines because only 2- or 3-wire cables will be used. In any of the conduits — rigid, EMT, or flexible — many additional wires are commonly used. When cross-hatching a line to show how many wires are in a conduit or cable, the longer hatch indicates a power wire, the shorter one a common. So, if a conduit line is crossed by four long hatches and two short, it contains four power wires and two commons.

In Figure 8-38, three 3-wire runs are shown; the remaining runs are 2-wire. Note that while the cable plan is intended to indicate how many wires are required in every run from one box to another, they're *not* to be regarded as indicating the actual route of that wire run through the building.

Wiring Diagrams

Cable plans show how many wires make up each box-to-box run, and how many cables or conduits enter each box. The wiring diagrams tell us what to do with those wires after they're in the boxes. For these

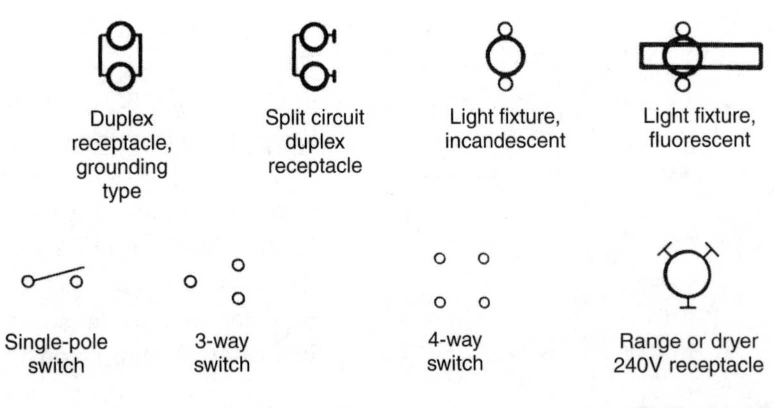

Figure 8-39
Wiring diagram symbols

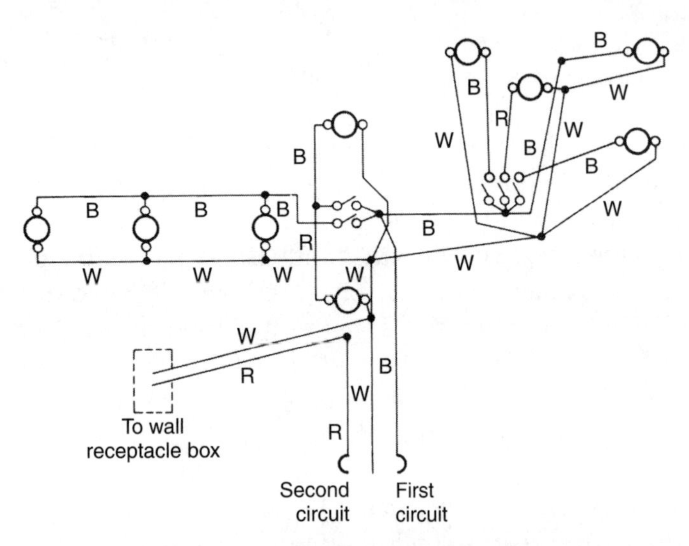

Figure 8-40
Family room, bath, utility light circuit

diagrams, you use a different set of symbols to indicate the terminal points where you make connections to the devices in each box. The wiring diagram symbols are shown in Figure 8-39.

Family Room Light Circuit

Using the cable plan in Figure 8-28 as the guide to how many wires enter each box, and the symbols in Figure 8-39 to show the details of how they're connected, you can create a wiring diagram for the family room light circuit. It should look like the diagram in Figure 8-40. Each splice is indicated by a black dot. The number of lines that meet at each splice dot is the number of wires in that splice. The short lines from dots to lights or switches are pigtails.

For example, the three switches in the bathroom are fed by three pigtails spliced to the incoming unswitched power wire coming from the switch box in the family room. The same splice contains the unswitched black power wire that goes to the Heat-A-Vent and connects there to the cable supplying the pull-switched light in the utility closet.

The home run cable going back to the breaker box contains three wires instead of the two you might expect. Connecting a circuit to the breaker box requires two wires: a power wire protected by the breaker and a common connected to ground. The safety grounding wire is in the same cable, but doesn't show on cable plans or wiring diagrams.

The third wire on this cable is the power wire for a second circuit that'll supply the wall outlets in the family room. That circuit will share the same common and ground wires in this home run cable. Two circuits can be fed through a 3-wire cable sharing one common and one ground *only* when the two power wires are connected to two breakers on the opposite phases of the incoming service.

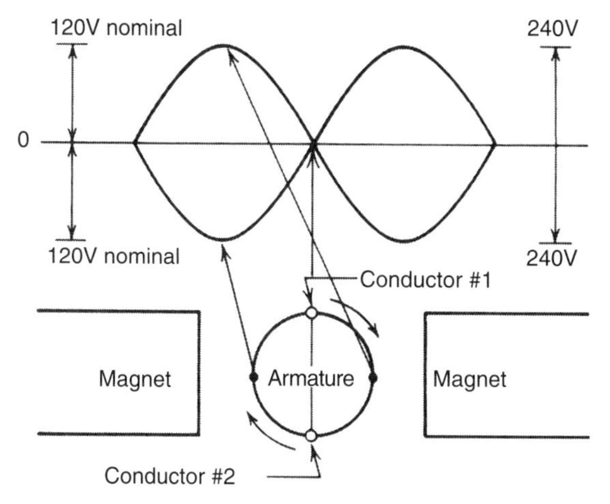

Figure 8-41
240V single-phase current

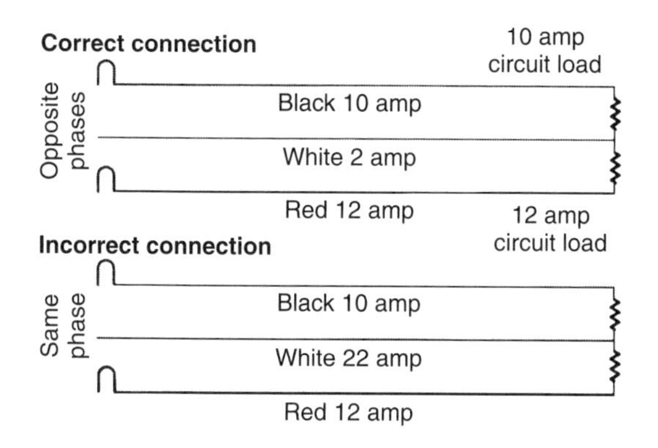

Figure 8-42
3-wire feed to two circuits

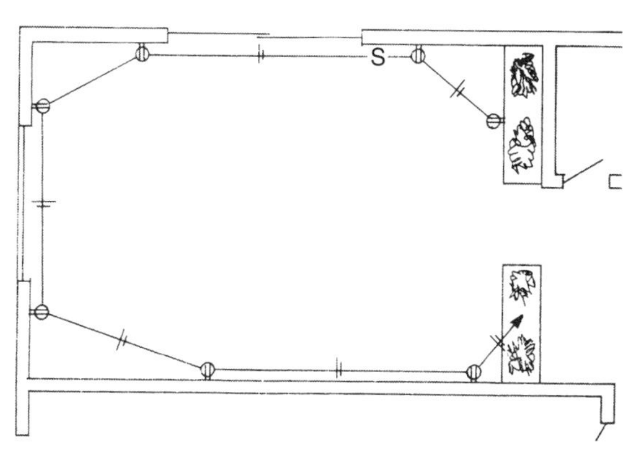

Figure 8-43
Cable plan: family room receptacle circuit

Looking back to Figures 2-10 and 2-11 in Chapter 2, the voltage changes illustrated are those that occur in the conductors on one side of a generator armature rotating in a magnetic field. That set of conductors is always matched by another set directly opposite them on the same armature. The cycle of voltage changes in the second conductor is exactly opposite to what occurs in the first. See Figure 8-41. When conductor #1 is at +120 volts, conductor #2 is at −120 volts. The difference across the two is 240 volts.

This is the source of the 240 volts across the two power wires in the service drop coming into the building, and the source of the 240-volt power used for ranges, dryers, water heaters, air conditioners, space heaters, and other 240-volt equipment.

When two circuits are properly connected to a 3-wire cable, as shown in Figure 8-42, the neutral common carries only the current imbalance between the two power lines. If the two power lines are connected in error to the same side of the incoming service, the amperages drawn will add to each other rather than subtract. In most modern breaker boxes, any two breakers *next* to each other on the same side of the box will be on opposite phases of the supply. Any two breakers *opposite* each other will be on the *same* phase.

Carrying the feed for two circuits via a 3-wire cable as far as possible through the building is a common practice and will be done repeatedly in this house. This obviously reduces the cost. The feed for a single circuit is a cable containing three wires: power, common and ground; so two circuits will require six wires: two power, two commons, and two grounds. When, as here, two circuits are fed with a 3-wire cable, the job is being done with only four wires instead of six. There are two separate power leads, but only one common and ground.

The second circuit in this 3-wire supply cable drops down from the light to a wall receptacle box (see Figure 8-40) in the family room, where it becomes the incoming feed to the receptacle circuit in Figure 8-43. With the exception of the

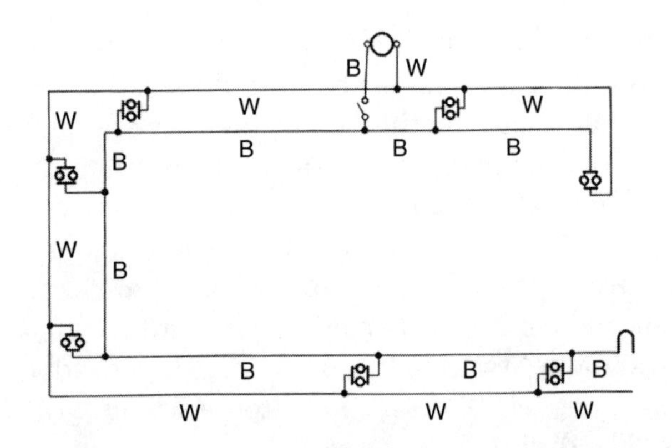

Figure 8-44
Circuit wiring: family room receptacles

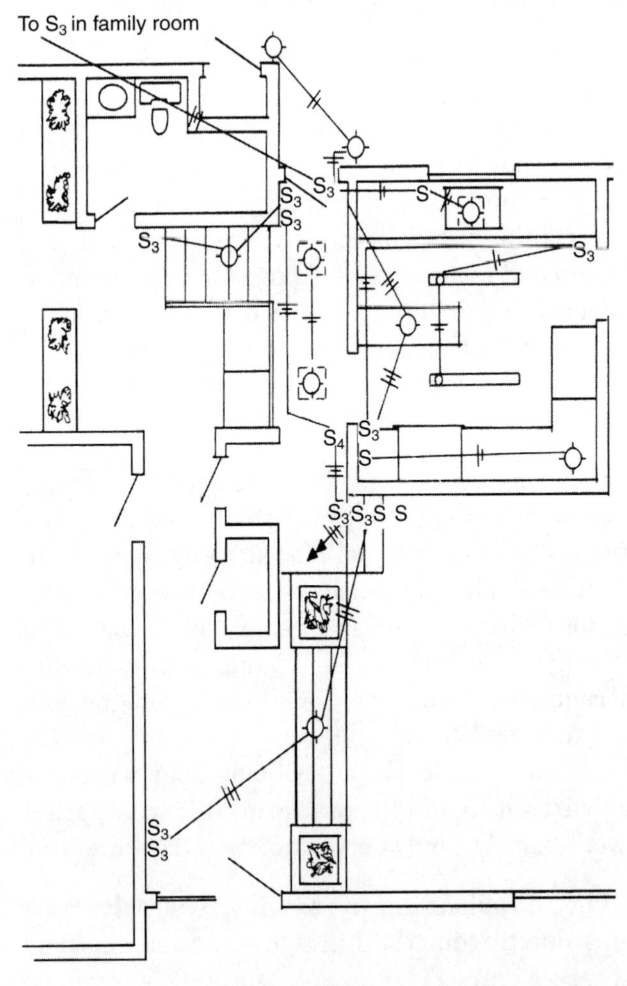

Figure 8-45
Cable plan: kitchen, hall, foyer light circuit

short branch to the outside light, the wiring of this circuit is a simple repetition of splices and pigtails to outlets as the cables circle the room. See Figure 8-44.

Kitchen-Hall-Foyer Light Circuit

The cable plan for the kitchen, hall, foyer light circuit, shown back in Figure 8-23, appears in Figure 8-45. The feeder from the panel box in the garage comes in at a four-gang switch box on the wall between the kitchen and the living room. Again, it contains two circuits in a 3-wire cable like we saw in the family room. This time the second circuit will take care of the living room, which we'll discuss later.

The switch circuits that are indicated in Figure 8-45 aren't complicated. They're all just like the circuit examples we saw back in Chapter 7. However, there are so many in this section of the house, it's difficult to keep track of them. All together, there are five multiple switch circuits here. Three of them control two lights each, the other two control single lights. An area with this many switch circuits is more easily understood by analyzing each of them individually, then dealing with the interconnections afterwards.

The wiring diagram for this section is given in Figure 8-46. The upper part of this diagram shows a pair of 3-way switches controlling two outside lights. This is the same as circuit Variation 2 in Figure 7-12 of Chapter 7, except that a second light has been added. Just below this is a light over the stairs controlled by two 3-way switches, a similar circuit to circuit Variation 4 in Figure 7-13 of Chapter 7. The circuit at the bottom controls the light over the steps into the living room in exactly the same way.

The 3-way, 4-way, 3-way circuit leading to the two recessed hall lights is similar to circuit Variation 3 in Figure 7-17 of Chapter 7, with a second load added. The kitchen fluores-

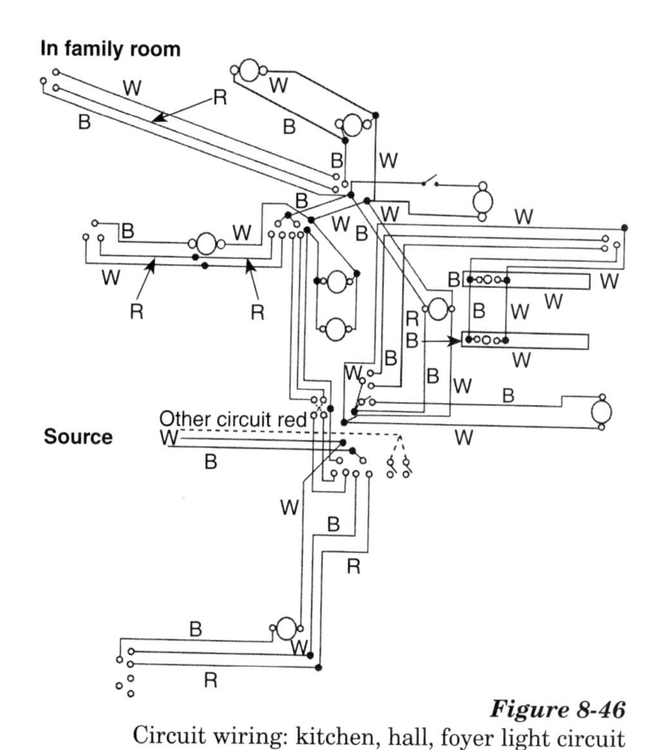

Figure 8-46
Circuit wiring: kitchen, hall, foyer light circuit

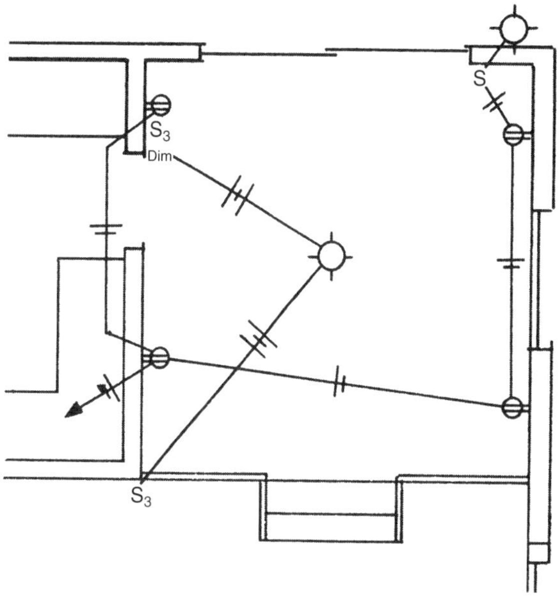

Figure 8-47
Cable plan: dining room receptacle and light circuit

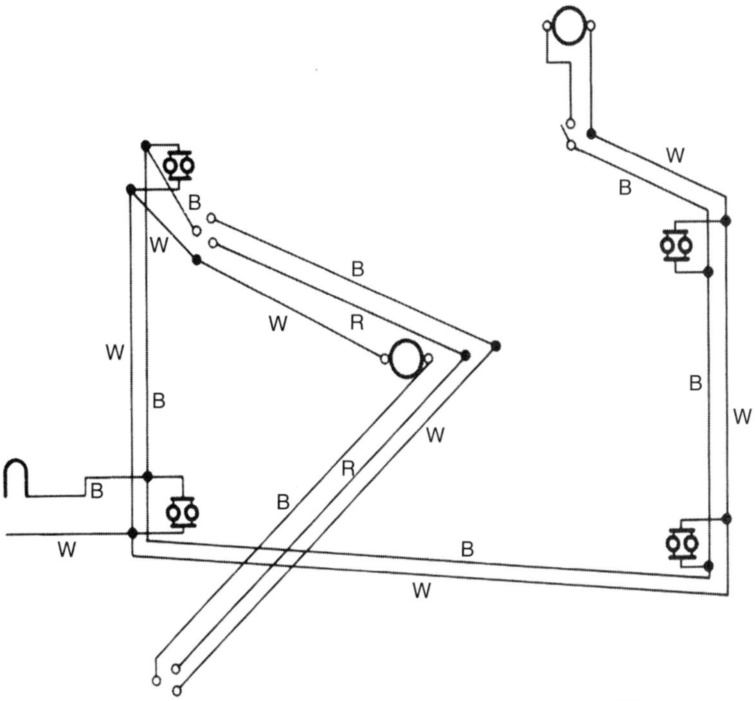

Figure 8-48
Circuit wiring: dining room receptacle and light circuit

cents are fed by a circuit based on circuit Variation 1 in Figure 7-12 of Chapter 7.

Dining Room Lights and Outlet Circuit

Compared to the wiring diagram we just looked at, the cable plan (Figure 8-47) and circuit wiring (Figure 8-48) for the dining room is extremely simple. The one 3-way switch circuit is a repetition of Figure 7-13 Variation 4 in Chapter 7, except that one of the 3-ways is a dimmer switch. Remember that the 3-way dimmer switch has no screw- or push-in terminals, but has three wire pigtails attached. Incoming unswitched power goes to the black one, while the two travelers

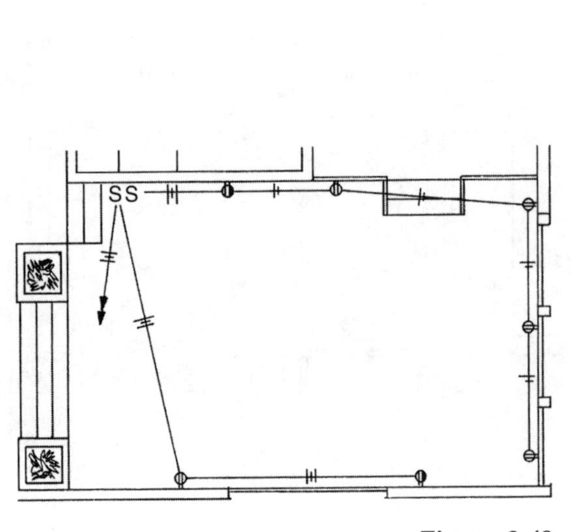

Figure 8-49
Cable plan: living room receptacle circuit

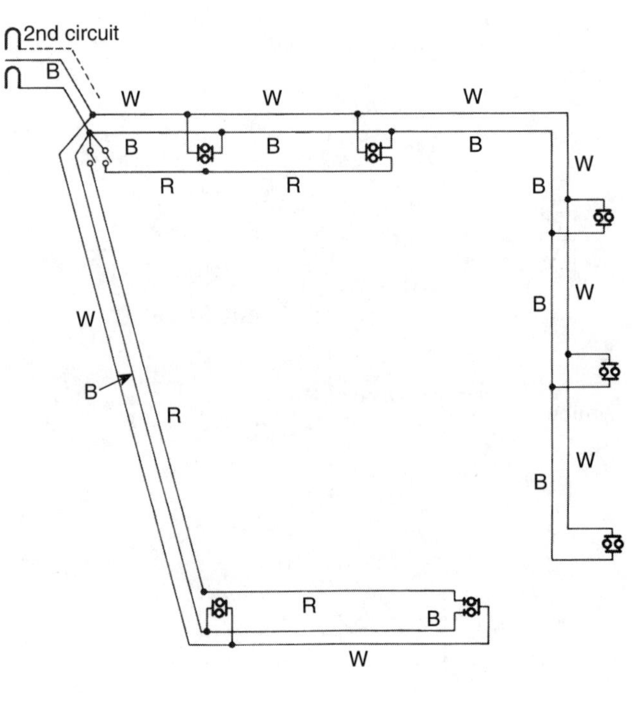

Figure 8-50
Wiring: living room receptacle circuit

— black and red — go to the two red pigtails. The rest of this circuit consists of routine duplex outlet connections with a single-pole switch to an outdoor light at the end of the line.

Living Room Wall Outlet Circuit

The power for the living room wall outlet circuit comes to the living room switch box in a 3-wire cable along with the power for the kitchen, hall and foyer light circuit. See the cable plan in Figure 8-49 and the detailed wiring plan in Figure 8-50. This is a very simple circuit to wire, consisting only of seven duplex receptacles. Two of these are split, with switch-operated receptacles on the lower half.

Garage and Hall Light Circuit

The garage and hall comprise another circuit with many multiple switches. Figure 8-51 shows five pairs of 3-ways. Power enters this circuit, shown in Figure 8-52, at a box containing two 3-way switches. Those two 3-way circuits are powered at that point. However, to power the remainder of the lines, an unswitched power wire is carried along

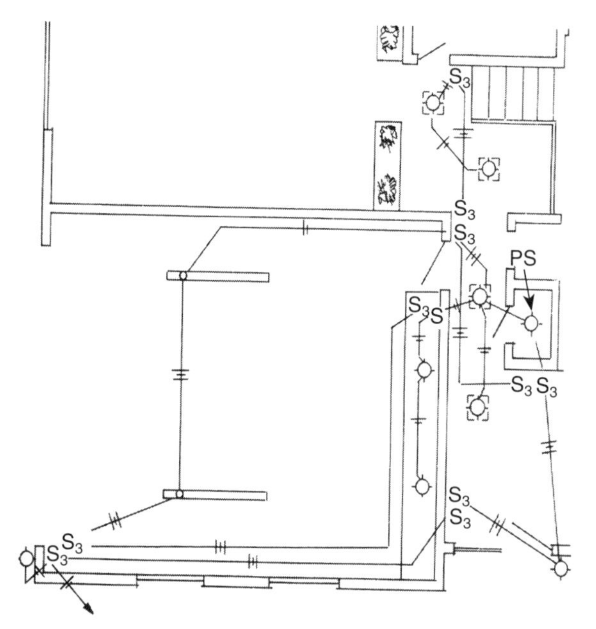

Figure 8-51
Cable plan: garage and hall light circuit

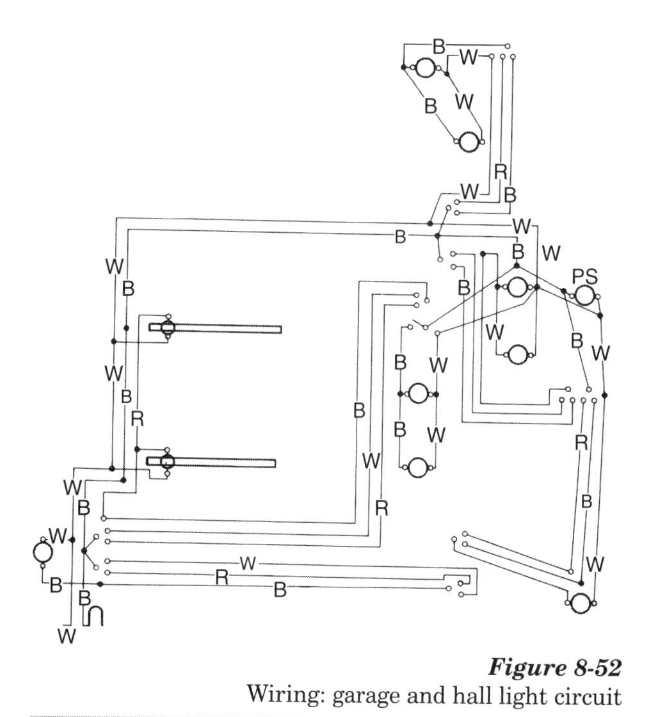

Figure 8-52
Wiring: garage and hall light circuit

with the switched power up to the ceiling fluorescents. This unswitched line, along with the common, comes down on the other side of the garage to go through the wall and power the hall lights. It also returns to the garage from the hall to power the workbench lights.

The use of 3-wire cable to carry both switched and unswitched power is common practice, as shown with the split receptacles in the living room, and in the family room light circuit (Figure 8-40). The hall circuit uses 3-wire cable as well.

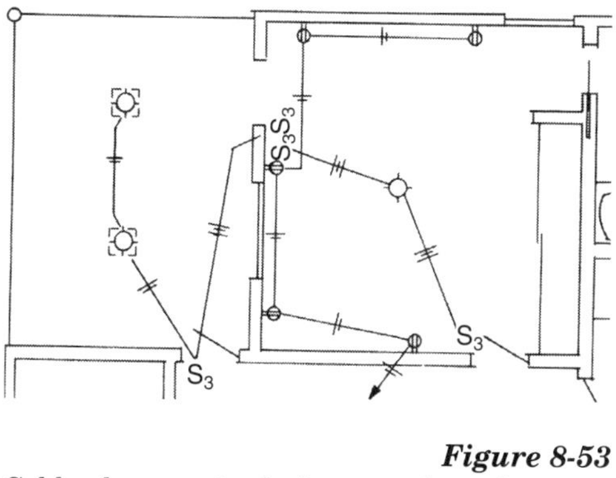

Figure 8-53
Cable plan: master bedroom and porch circuit

Master Bedroom and Porch Lights and Receptacles

The cable plan in Figure 8-53 shows two 3-way switch circuits plus five duplex receptacles, with the power entering at one of the wall receptacles. Both 3-way switch circuits are familiar arrangements by now; see Figure 8-54. Power to the light switches comes from the receptacle below them. To avoid the labor of cabling around the porch door and a framed corner as well, the power feed to the two receptacles on the far wall runs under the floor from the same box that feeds the switches.

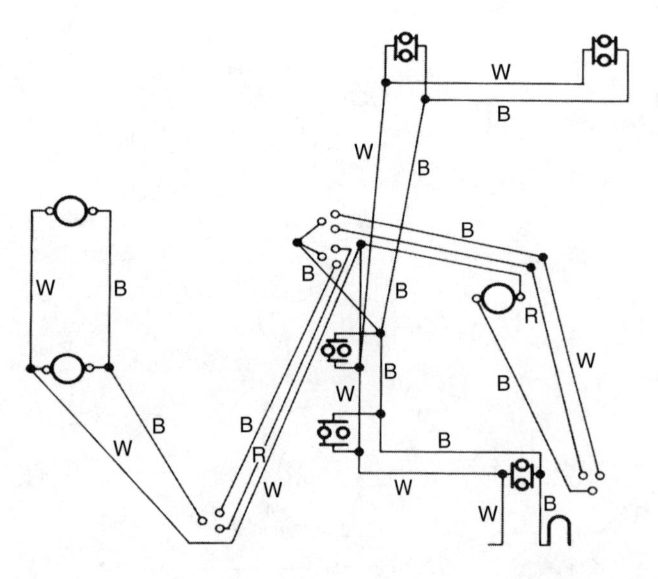

Figure 8-54
Wiring: master bedroom and porch circuit

Second Floor Bathrooms

Notice that the cable plan or the second floor bathrooms in Figure 8-55 does not show the required GFCI outlets that were very clearly indicated on the final electrical floor plan of this area (Figure 8-16). While they will be installed, they don't appear in either Figure 8-55 or 8-56 because they're part of a separate GFCI circuit.

Bedroom 2, Utility and Stairs

Figure 8-57 shows the power entering through the wall from bedroom 1. Notice the wiring of the 2-wire cable from the switches to the lights in each utility closet (Figure 8-58). These are standard *switch loops*, per *NEC* Section 200.7(C). White is *not* common, but is power into the switches, and must be reidentified as such. Black is properly power back from switches to loads.

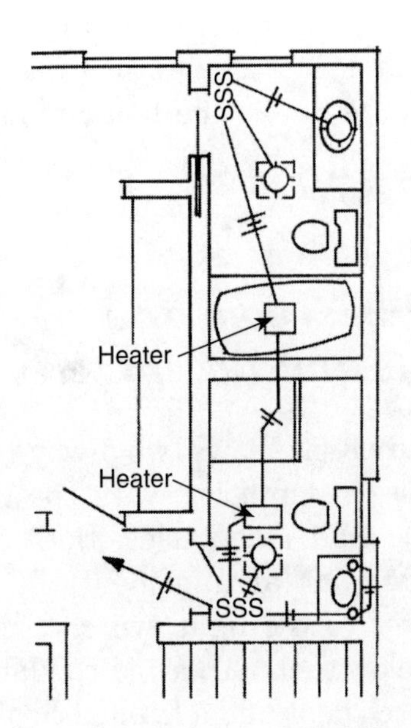

Figure 8-55
Cable plan: second floor bathrooms circuit

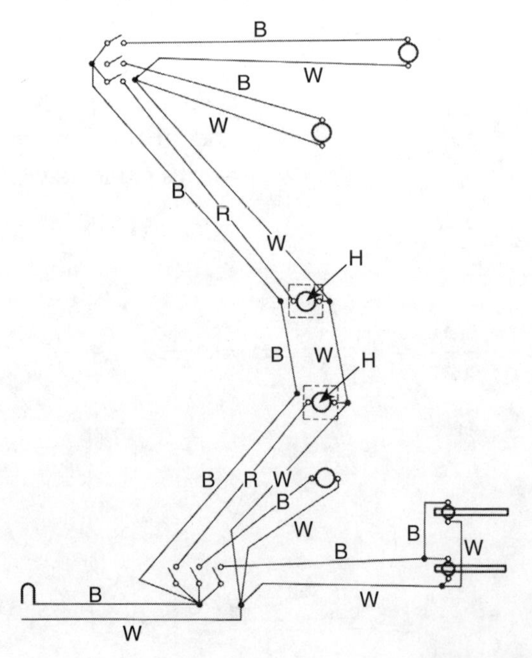

Figure 8-56
Wiring: second floor bathrooms circuit

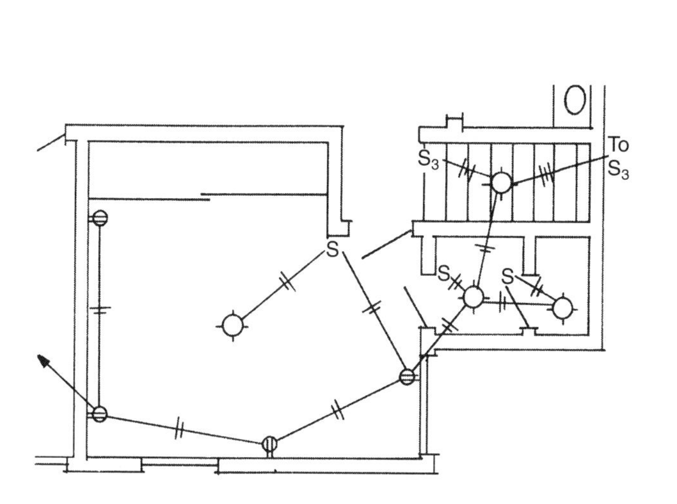

Figure 8-57
Cable plan: bedroom 2, utility, stairway circuit

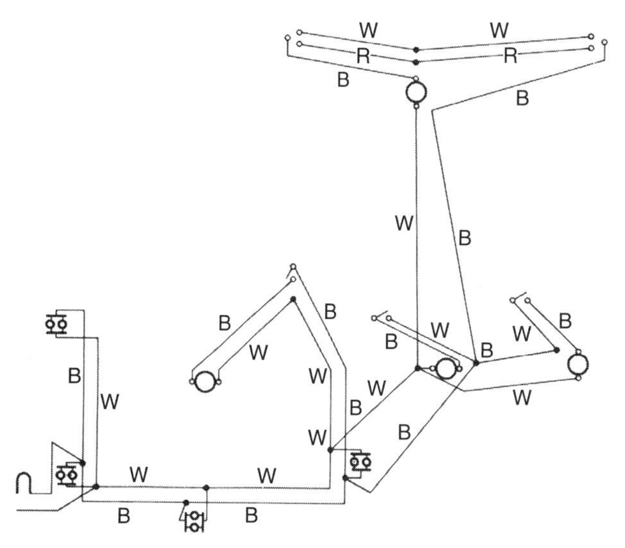

Figure 8-58
Wiring: bedroom 2, utility, stairway circuit

Bedroom 1 and Hall

The incoming power supply to bedroom 1 is another case of two power circuits carried by a 3-wire cable. In Figure 8-59, the 3-wire power cable enters through the lower left hand wall receptacle. One power line feeds this room and the hall lights, the second goes to the lower right receptacle box and from there through the wall to feed the bedroom 2 circuit. There's nothing unusual or exotic in the wiring of this circuit, as you'll see by examining Figure 8-60.

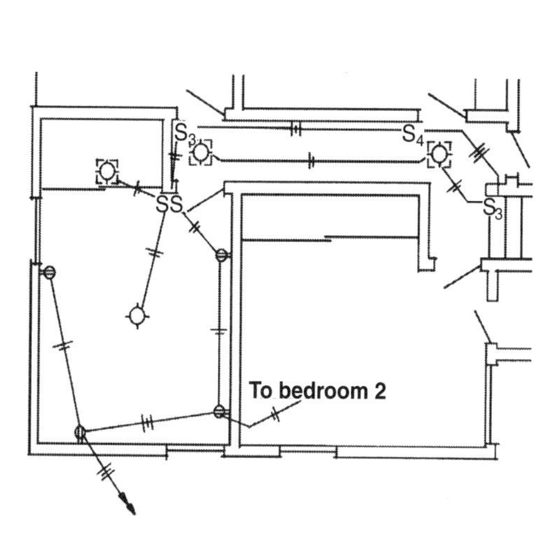

Figure 8-59
Cable plan: bedroom 1, hall circuit

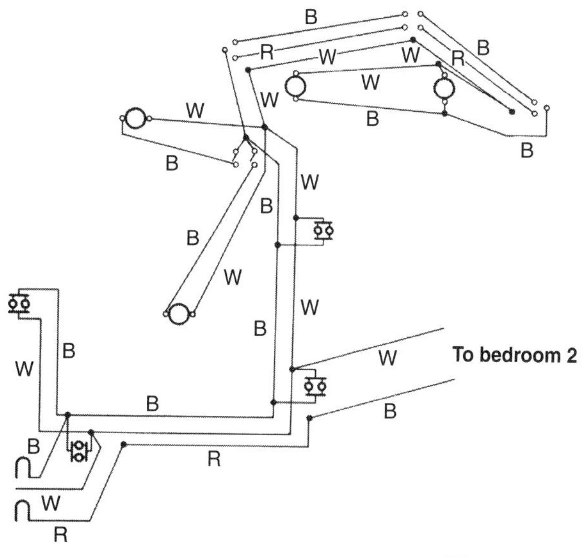

Figure 8-60
Wiring: bedroom 1, hall circuit

GFCI Circuits

As mentioned earlier, GFCIs can be installed in any receptacle by merely replacing a normal duplex receptacle with a receptacle-type GFCI. The procedure is to then use one GFCI receptacle, and wire the other receptacles requiring GFCI protection to it. However, that first GFCI may be part of a circuit that already has a partial load on it. If that load is small enough that all the remaining GFCIs can be added without overloading the circuit, the receptacles may be wired together.

In this case, GFCI protection is required for two receptacles in the garage, one in the first floor bathroom, one weatherproof receptacle outside the dining room at the back of the house and another one outside the entry at the front, two receptacles in the two upstairs bathrooms, plus one more weatherproof receptacle on the upstairs porch. This makes a total of eight, and constitutes a computed load of 1440 watts, an amount appropriate for a separate GFCI circuit. Cabling for this circuit is shown in Figure 8-61 and wiring in Figure 8-62.

Appliances

In addition to the GFCI receptacles on their own circuit just mentioned, there are two GFCI-type receptacles in the kitchen, which are included on the small appliance circuits. The separate circuits for fixed appli-

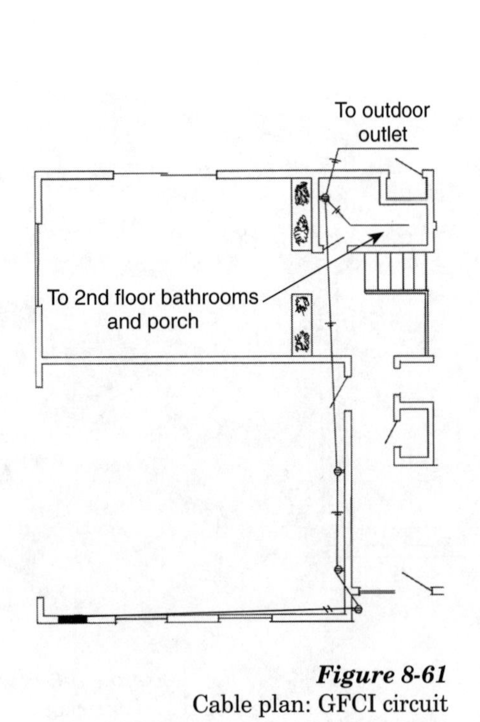

Figure 8-61
Cable plan: GFCI circuit

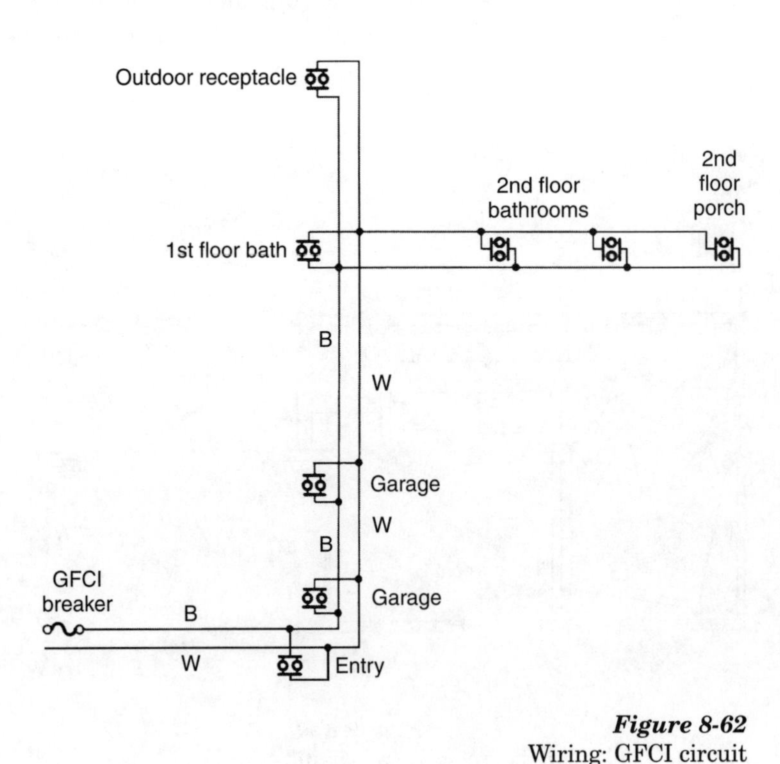

Figure 8-62
Wiring: GFCI circuit

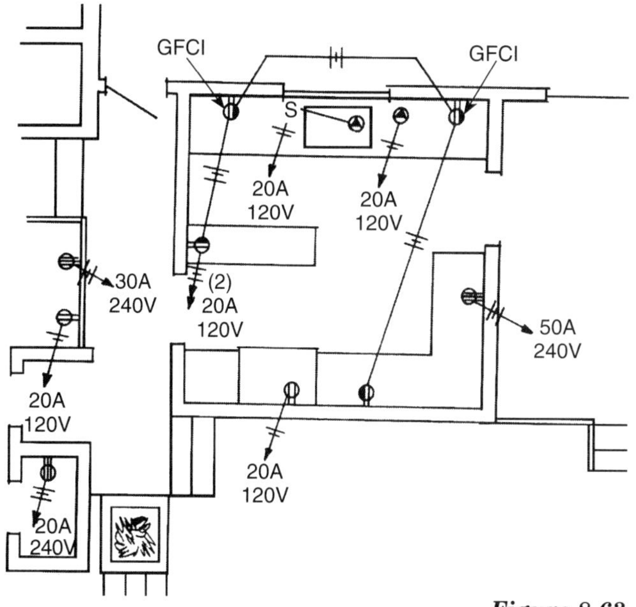

Figure 8-63
Cable plan: appliance circuits

ances are also shown on the appliance circuit cable and wiring plans in Figures 8-63 and 8-64. With the exception of the forced-air furnace on the second floor, they are all physically close to each other.

According to *NEC* Section 210.52(B), the two code-required small appliance circuits feed four split receptacles on the kitchen countertops. This is another case of taking two circuits from the breaker box via a 3-wire cable. When the two halves of a duplex receptacle are powered by two separate circuits fed through two different breakers, then, per *NEC* Section 210.4(B), those two breakers must be linked so that any defect that trips either one of them trips them both simultaneously. What this means, in simple terms, is that those two circuits must originate at

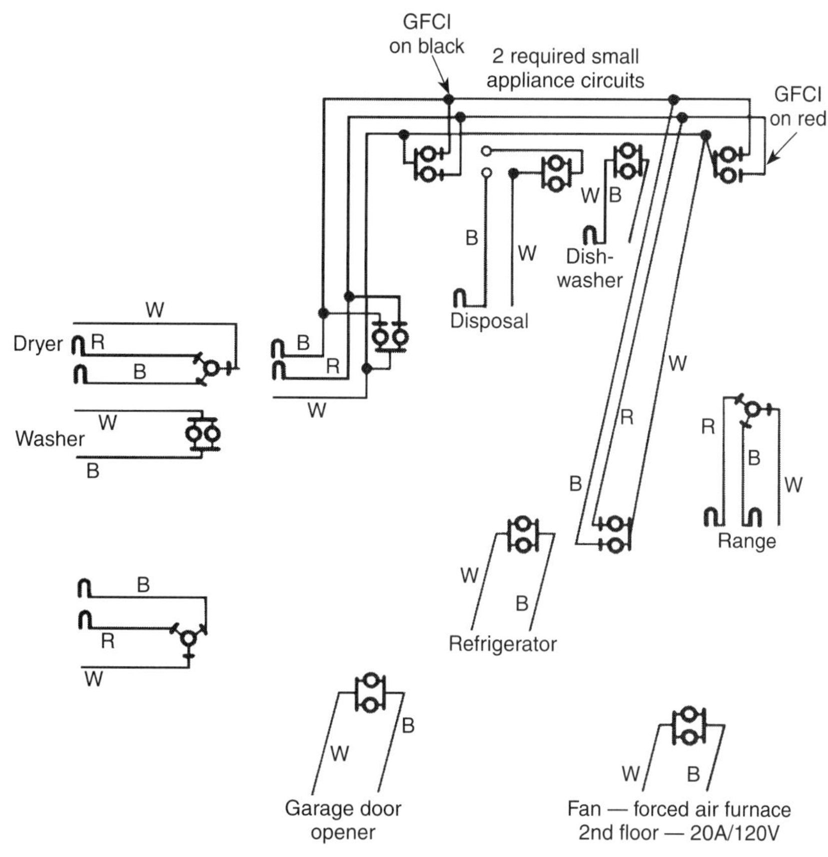

Figure 8-64
Wiring: appliance circuits

the breaker box from a linked double breaker. These circuits will be 20 amperes and will be wired with #12 copper. Since two of those outlets are within 6 feet of the sink, they must be GFCI-protected, per *NEC* Section 210.8(A)(7). We can use GFCI-type receptacles for both of these outlets, one on the black circuit and one on the red circuit, and these will protect the two required outlets plus the other two not requiring protection.

The refrigerator here has its own circuit at 20 amperes, wired in #12 copper. While, according to *NEC* Section 210.52(B) the refrigera tor can be attached to one of the required small appliance circuits just discussed, quite often, as in this case, it's separated from everything else.

The code specifically permits refrigeration to be attached to a small appliance circuit, but the dishwasher, disposal, and trash compactor aren't included in that code mention. The dishwasher and disposal are sometimes placed together on a single circuit, and sometimes separated, as in this case. This home has no trash compactor, but if there were one, it would be on another circuit terminating under the kitchen counter.

The range is the largest single load in the building. It's provided with a 50-ampere, 240-volt circuit. This circuit should be wired with #6 copper, per *NEC* Table 310.16 (look back to Figure 4-6 in Chapter 4). However, according to *NEC* Table 220.55 in Figure 8-65, it's permissible to compute the branch circuit load of a single range at 80 percent of its rating. If the nameplate rating of a range were the full 50 amperes, then 80 percent of that would be 40 amperes — a load that can be handled by #8 copper. That means you can wire for a range from a 50-ampere, 240-volt breaker using #8 copper. *Don't wire any other 50-ampere circuits anywhere else using less than #6 copper.* The range is a special case. Only the range outlet may be on this circuit.

The dryer requires a 30-ampere, 240-volt circuit. It's wired in #10 copper from a 30-ampere, 240-volt breaker. No other outlets may be on this circuit.

Ranges or dryers may plug into receptacles as described in Chapter 7, or they may be directly wired. Water heaters are normally directly wired. Depending on heater size and requirements, the circuit will usually be 20 ampere, 240 volt, wired with #12 copper. A larger than average water heater might require a 30-ampere, 240-volt circuit, in which case it's wired with #10 copper. Again, no other outlets may be on the same circuit with the water heater.

In the laundry, the 20-ampere circuit, required by *NEC* Section 210.52(B), which supplies the second receptacle, required by *NEC* Section 210.52(F), is separate from the dryer. Again, as we've seen so many times before, here's a single item referred to in two different places in the code, each reference supplying only partial information.

Table 220.55 Demand Factors and Loads for Household Electric Ranges, Wall-Mounted Ovens, Counter-Mounted Cooking Units, and Other Household Cooking Appliances over 1¾ kW Rating (Column C to be used in all cases except as otherwise permitted in Note 3.)

	Demand Factor (%) (See Notes)		
Number of Appliances	**Column A (Less than 3½ kW Rating)**	**Column B (3½ kW through 8¾ kW Rating)**	**Column C Maximum Demand (kW) (See Notes) (Not over 12 kW Rating)**
1	80	80	8
2	75	65	11
3	70	55	14
4	66	50	17
5	62	45	20
6	59	43	21
7	56	40	22
8	53	36	23
9	51	35	24
10	49	34	25
11	47	32	26
12	45	32	27
13	43	32	28
14	41	32	29
15	40	32	30
16	39	28	31
17	38	28	32
18	37	28	33
19	36	28	34
20	35	28	35
21	34	26	36
22	33	26	37
23	32	26	38
24	31	26	39
25	30	26	40
26–30	30	24	15 kW + 1 kW for each range
31–40	30	22	
41–50	30	20	25 kW + ¾ kW for each range
51–60	30	18	
61 and over	30	16	

Notes:

1. Over 12 kW through 27 kW ranges all of same rating. For ranges individually rated more than 12 kW but not more than 27 kW, the maximum demand in Column C shall be increased 5 percent for each additional kilowatt of rating or major fraction thereof by which the rating of individual ranges exceeds 12 kW.
2. Over 8¾ kW through 27 kW ranges of unequal ratings. For ranges individually rated more than 8¾ kW and of different ratings, but none exceeding 27 kW, an average value of rating shall be calculated by adding together the ratings of all ranges to obtain the total connected load (using 12 kW for any range rated less than 12 kW) and dividing by the total number of ranges. Then the maximum demand in Column C shall be increased 5 percent for each kilowatt or major fraction thereof by which this average value exceeds 12 kW.
3. Over 1¾ kW through 8¾ kW. In lieu of the method provided in Column C, it shall be permissible to add the nameplate ratings of all household cooking appliances rated more than 1¾ kW but not more than 8¾ kW and multiply the sum by the demand factors specified in Column A or Column B for the given number of appliances. Where the rating of cooking appliances falls under both Column A and Column B, the demand factors for each column shall be applied to the appliances for that column, and the results added together.
4. Branch-Circuit Load. It shall be permissible to calculate the branch-circuit load for one range in accordance with Table 220.55. The branch-circuit load for one wall-mounted oven or one counter-mounted cooking unit shall be the nameplate rating of the appliance. The branch-circuit load for a counter-mounted cooking unit and not more than two wall-mounted ovens, all supplied from a single branch circuit and located in the same room, shall be calculated by adding the nameplate rating of the individual appliances and treating this total as equivalent to one range.
5. This table shall also apply to household cooking appliances rated over 1¾ kW and used in instructional programs.

Figure 8-65
Cooking appliance load demand factors

Service Load Computation

The design of this electrical system is very nearly complete. We've studied the architect's floor plans and revised them as necessary to conform to code requirements for lights, switches, and receptacles. Then we designed circuits to supply these various use points, also in keeping with code requirements. Next we prepared cable plans and wiring diagrams, detailing exactly how power is to reach each of those locations. All that remains is to determine how much power must be brought into the building to adequately supply all of the various uses.

This total is *not* calculated by adding the capacities of the various circuits. That total would be much too large. For example, the computed load on a branch circuit can't exceed 80 percent of the circuit rating, so the total of the circuit capacities is going to be higher than the computed loads, which are already purposely higher than the actual loads.

The writers of the code are also aware that, in a dwelling, all the *actual* loads are never on at once. Try to picture such an absurd situation: every light is on; so are the TVs and the stereo; every burner on the range and oven are at starting load, as well as the dryer; all the countertop appliances are on; both the water heater and refrigerator just cycled on; the disposal is running; the dishwasher is on the drying cycle; someone is ironing; someone else is running an electric lawnmower on an outdoor outlet; all heating units, plus any air conditioning equipment, are on; and last, but not least, there's a hair dryer and an electric toothbrush being used. Obviously, the total of the actual connected loads is going to be greatly in excess of the real power requirements at any given time. As a result, the code provides several different *demand factors* to be applied to different types of loads. The size of the service required is then determined by totaling all loads after applying these demand factors.

As we'll see in the next chapter, the present requirements say that no new residential service can be installed with a capacity less than 100 amperes at 240 volts, per *NEC* Section 230.79(C). The reason we compute the load on this system in accordance with code instructions is to insure that the capacity is adequate, and if not, to know what capacity we need.

Lighting Load Demand Factors

For computing feeder demand, the lighting load consists of all of the various 15-ampere lighting and convenience outlet circuits. There are 10 in our sample house. To this total we add the small appliance circuits, the laundry circuit, and the GFCI circuit, per *NEC* Section 220.16(A) and (B).

We'll compute our lighting load using the totals from Figure 8-66:

Lighting circuits	*Watts*
Family room receptacles	1360
Family room, bath 1 and utility lights	1360
Kitchen, hall and foyer lights	1400
Dining room	1020
Living room	1260
Garage, hall and entry lights	1260
Master bedroom outlets and lights, porch light	1300
Second floor bathrooms	1050
Bedroom 2, utility closets, stairs	1240
Bedroom 1 and hallway	1220
Subtotal	*12470*

Other circuits	
2 small appliance (1500 watts each)	3000
Laundry	1500
GFCI (see Note below)	1500
Subtotal	*6000*
Total	*18470*

Note — In this instance, the GFCI-protected receptacles aren't part of another circuit, but are completely separate. Since the reason for the GFCI protection of these receptacles is that they're intended to power appliances that may be defective, some jurisdictions would classify this as a small appliance circuit. There are eight receptacles counted on the one GFCI circuit. If rated as general-purpose receptacles at 180 watts each, they would total 1440 watts. Considered together as a small appliance circuit, they're rated at 1500 watts. The difference isn't great. In addition, each of the small appliance circuits listed includes one GFCI-type kitchen receptacle, which also protects the other receptacle on the same circuit.

Figure 8-66
Lighting and other circuit loads

NEC Table 220.42, in Figure 8-67, shows that the first 3000 watts of lighting load for dwelling units is to be taken at 100 percent. From 3001 to 120,000 watts, the load is figured at 35 percent. In our example, the total load is 18470 watts.

3000 @ 100%	3000 watts
15470 @ 35%	5415 watts
Lighting load demand	*8415 watts*

Table 220.42 Lighting Load Demand Factors

Type of Occupancy	Portion of Lighting Load to Which Demand Factor Applies (Volt-Amperes)	Demand Factor (%)
Dwelling units	First 3000 or less at	100
	From 3001 to 120,000 at	35
	Remainder over 120,000 at	25
Hospitals*	First 50,000 or less at	40
	Remainder over 50,000 at	20
Hotels and motels, including apartment houses without provision for cooking by tenants*	First 20,000 or less at	50
	From 20,001 to 100,000 at	40
	Remainder over 100,000 at	30
Warehouses (storage)	First 12,500 or less at	100
	Remainder over 12,500 at	50
All others	Total volt-amperes	100

*The demand factors of this table shall not apply to the calculated load of feeders or services supplying areas in hospitals, hotels, and motels where the entire lighting is likely to be used at one time, as in operating rooms, ballrooms, or dining rooms.

From the *National Electrical Code* © 2007, NFPA

Figure 8-67
Lighting load demand factors

Table 220.54 Demand Factors for Household Electric Clothes Dryers

Number of Dryers	Demand Factor (%)
1–4	100
5	85
6	75
7	65
8	60
9	55
10	50
11	47
12–23	47% minus 1% for each dryer exceeding 11
24–42	35% minus 0.5% for each dryer exceeding 23
43 and over	25%

From the *National Electrical Code* © 2007, NFPA

Figure 8-68
Dryer load demand factors

Dryer Demand

According to *NEC* Section 220.54, the load for a dryer is 5000 watts (volt-amperes), or the nameplate rating, whichever is larger. Demand factors for dryers are listed in *NEC* Table 220.54 (Figure 8-68). A single dryer is computed at 100 percent, which in our example will be 5000 watts.

Range, Oven, Cooktop and Cooking Units Rated Over 1750 Watts Demands

Dealing with load demands for cooking appliances can be a bit confusing. However, an average residential oven-range combination is rated at approximately 12 kilowatts (12000 watts), making the demand for that unit 8 kilowatts (8000 watts), as given in *NEC* Table 220.55, Column A; look again at Figure 8-65. In our sample house, we'll assume our oven-range combination is rated at just under 12 kilowatts, producing a demand of 8 kilowatts or 8000 watts.

If the nameplate rating on a cooking appliance is over 12 kilowatts, then you must increase the demand given in Column A by 5 percent for each kilowatt, or major fraction thereof, over 12 kilowatts. Suppose a large oven-range combination carries a nameplate rating of 14750 watts, or 14.75 kilowatts. The demand for this unit would be figured as 8000 watts for the first 12000 watts (12 kilowatts) of nameplate rating, plus 15 percent of that 8000 for the additional 2750 watts of nameplate rating above 12000:

Column A demand for 12 kilowatts	8 kilowatts or 8000 watts
Plus 15 percent of 8 kilowatts	1.2 kilowatts or 1200 watts
Total demand	*9.2 kilowatts or 9200 watts*

When, as often happens, a counter-mounted cooktop and one or two wall-mounted ovens are located in the same room and supplied by the same branch circuit, the demand must be computed by adding the nameplate ratings of the individual appliances together, and then treating the total as though it were a single unit.

Fixed Appliances

Appliance	*Watts*
Water heater	3200
Disposal	764
Dishwasher	1000
Garage door opener	720
Total	*5684*
75 percent of 5684 watts =	*4263*

Figure 8-69
Appliance demands

Most homes contain several other electrically powered appliances in addition to the range and dryer. These appliances are served by small appliance circuits. Other appliances you must consider are a water heater, disposal, dishwasher, trash compactor, attic fan, garage door opener, or exhaust fan. For all of these items, the demand is determined by adding the nameplate ratings together and taking 75 percent of the total, according to *NEC* Section 220.53. This applies to four or more fixed appliances. This house has four, which comes to a demand of 4263 watts, as shown in Figure 8-69.

Fixed Electric Space Heating and Air Conditioning

Fixed space heating means permanently installed electric baseboards, radiant ceilings, or other non-portable heating elements. These are all to be computed at 100 percent of their rated connected loads, per *NEC* Section 220.51.

NEC Section 220.60 states that when two dissimilar loads exist that won't be in use at the same time, the smaller of the two can be omitted from the load calculations. So, when a building contains both electric heating and air conditioning, omit the smaller of the two. If there's air conditioning and oil, gas, or coal heating, then the air conditioning is figured at 100 percent and the smaller electrical requirement of these heating systems is omitted.

Our sample house doesn't include either electrical heating or air conditioning. The only electrical requirement for the heating system is the blower for the forced-air gas furnace. This requires 1100 watts, which we'll figure at 100 percent.

Load	*Watts*
Lighting	8415
Dryer	5000
Range	8000
Fixed appliances	4263
Furnace	1100
Total system load	*26778*

Figure 8-70
Sizing the service loads

Table 310.15(B)(6) Conductor Types and Sizes for 120/240-Volt, 3-Wire, Single-Phase Dwelling Services and Feeders. Conductor Types RHH, RHW, RHW-2, THHN, THHW, THW, THW-2, THWN, THWN-2, XHHW, XHHW-2, SE, USE, USE-2

	Conductor (AWG or kcmil)	
Service or Feeder Rating (Amperes)	**Copper**	**Aluminum or Copper-Clad Aluminum**
100	4	2
110	3	1
125	2	1/0
150	1	2/0
175	1/0	3/0
200	2/0	4/0
225	3/0	250
250	4/0	300
300	250	350
350	350	500
400	400	600

Figure 8-71
Conductor types and sizes for dwelling services

Sizing the Service Entrance

We've now calculated all the various individual demands on the service entrance. Let's add them up to determine the amperage that our service will be required to deliver. The total system load is shown in Figure 8-70.

Now calculate the service entrance load using the following formula:

$$\text{Watts} \div \text{Volts} = \text{Amperage}$$

$$26{,}778 \div 240 = 111.58 \text{ amperes}$$

The required minimum 100-ampere service won't work for this house after all. The service will have to be increased to 125 amperes, and the service entrance wires sized at #2 copper to meet the requirements of *NEC* Table 310.15(B)(6), shown in Figure 8-71.

Now that we know how to calculate the size of the service, we'll next discuss the service entrance that brings the electrical supply into the building.

STUDY QUESTIONS

1. Which three of the following plans do electricians primarily work with?

A) Plot plan, floor plan, wiring plan
B) Floor plan, cable plan, plot plan
C) Cable plan, wiring diagram, floor plan
D) Exterior elevations, floor plan, wiring plan

2. Which parts of the electrical system are shown on a floor plan?

A) Fixture, receptacle, and switch locations
B) Wiring runs, receptacles and fixtures
C) Receptacles, switches and supply wiring
D) Fixtures, switches and cabling runs

3. Dwelling units *must* have a switch-controlled light fixture in which areas?

A) Each habitable room, plus bathrooms, kitchen, stairways and garage
B) Each habitable room, plus kitchen, garage, hallways and entrances
C) Each habitable room, plus bathrooms, kitchen, stairways and hallways
D) Kitchen, bathrooms, hallways, stairways, outdoor entrances and garage

4. Who is ultimately responsible for the accuracy of the electrical installation?

A) The architect
B) The general contractor
C) The electrician
D) The building designer

5. Which does the code consider a *continuous* load?

A) A water heater that cycles on and off automatically
B) House lights on from 9:00pm to 11:30pm daily
C) Having both the cooktop and the oven on high for 2 hours
D) Operating a vacuum cleaner continuously for 30 minutes

6. What is the permissible load for a 15-ampere branch circuit?

A) 1260 watts
B) 1440 watts
C) 1620 watts
D) 1800 watts

7. What is the minimum lighting load required for dwelling units?

A) 100 watts per room
B) 10 watts per square foot
C) 10 volt-amperes per square foot
D) 3 volt-amperes per square foot

8. Which appliance is *not* required to have a separate branch circuit?

A) Range
B) Refrigerator
C) Dryer
D) Water heater

9. If the nameplate rating for a range were 50 amperes on a 240-volt breaker, what is the smallest size wire you could use?

A) #6
B) #8
C) #10
D) #12

10. How do you determine the size of the service that must be brought into a house to supply power to all the various uses?

A) Add up the capacities of the various circuits
B) Add up the total load of every electrical device
C) Calculate it at 80 percent of the total computed load
D) Total all loads after applying appropriate demand factors

9

THE SERVICE ENTRANCE

The service entrance connects the electrical supply from the local utility to the building distribution network inside the home. Normally, only one service entrance is allowed per building (*NEC*® Section 230.2). There are specific exceptions to this one-service rule, but you won't often find those situations in the residential work this book covers.

In addition to the service entrance limitation, there are also restrictions, per *NEC* Section 230.3, on the use of one service to supply two or more additional buildings. It's permissible to run wiring from the service entrance on building 1 through the interior of that building, then out and across to building 2 when both buildings are occupied or managed by the same people. For example, if building 1 is a single-family residence and building 2 is its detached garage, they both belong on the same supply, as would a permitted workshop or granny flat.

What would *not* be acceptable is for the supply to go through residence 1, pass through its garage (building 2), and be used by one or more neighboring, nonconnected buildings (3 and/or 4) not belonging to the owner of building 1. Each separately owned building must have a separate supply.

Overhead Service Entrance

When the incoming wires bringing the electrical supply to a building pass above ground from a power pole to the building, it's called an *overhead service*; see Figure 9-1. The wires coming to the building are called the

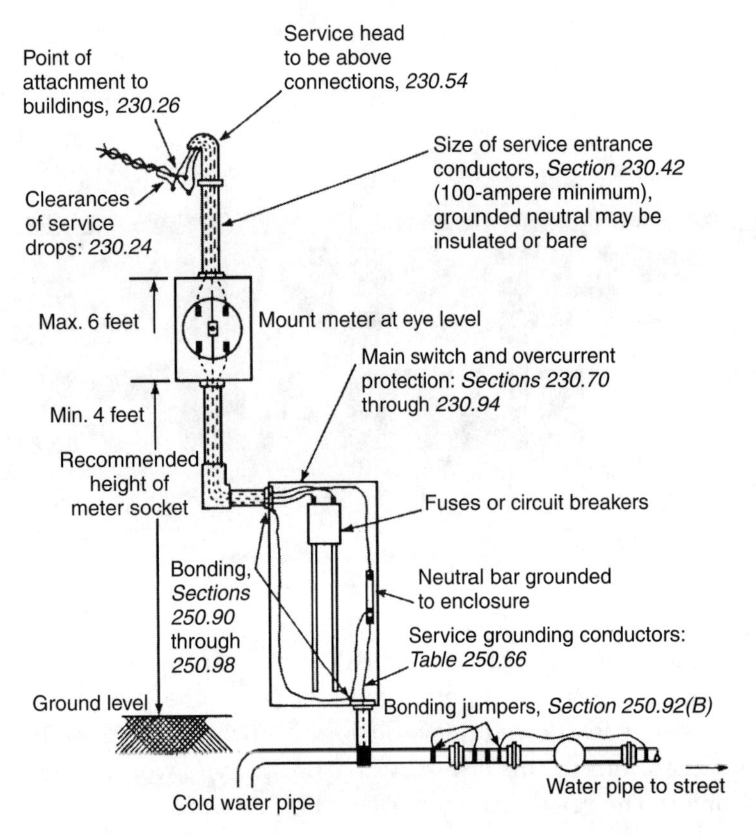

Figure 9-1
Typical overhead service installation with Code references

service drop. These wires terminate in connections to another set of wires that hang down from the service head (also called the weatherhead). The wires from the service head are the *service entrance wires*, and must be marked as approved for this purpose, per *NEC* Section 310.11(B)(1). The service head itself must be a raintight device, extending above the point at which the service drop wires attach to the service entrance wires, per *NEC* Section 230.54(A) and (F). The service entrance wires *must*, therefore, extend down from the head. This prevents rain or water from any source from getting into the service head.

The service drop attachment can never be less than 10 feet above grade (*NEC* Section 230.26). However, it can be more, if needed, to conform to the additional clearances required by *NEC* Section 230.24, illustrated in Figure 9-2.

When the service drop terminates at a pipe mast that extends above the roof of the building, the *NEC* requirements can become just a bit confusing. *NEC* Section 230.24 begins by stating that a service drop must have a vertical clearance of at least 8 feet from all points of a roof over which it passes. That sounds very simple and definite, but it's quickly followed by Exceptions. *Exception No. 2* allows a reduction in clearance to 3 feet if the voltage doesn't exceed 300 volts and the roof has a slope of at least 4 inches in 12 inches. That, too, sounds simple, but read on. *Exception No. 3* permits a further reduction in clearance to 18 inches when the voltage isn't over 300 and the mast is located so that no more than 4 feet of service drop conductors pass over the roof. This time, nothing is mentioned about the roof slope. These clearances are shown in Figure 9-3.

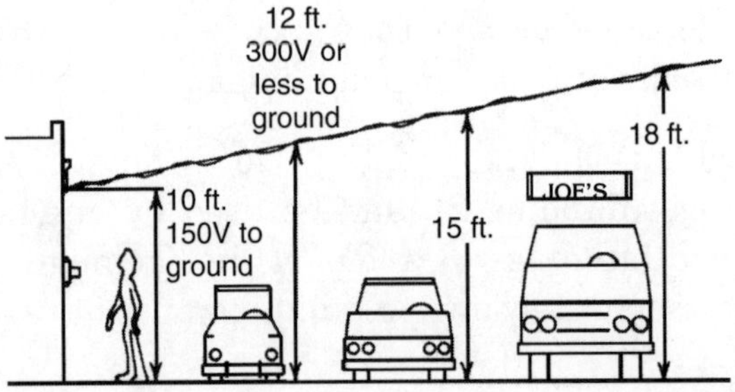

12 ft. 300V or less to ground
18 ft.
10 ft. 150V to ground
15 ft.
JOE'S

Figure 9-2
Minimum clearances under service drop

If you think making sense of these seemingly contradictory provisions is somewhere between difficult and impossible, don't give up. Most overhead service drops that terminate on a roof are within the 4-foot limit covered by *Exception No. 3*. For this reason, most rooftop masts

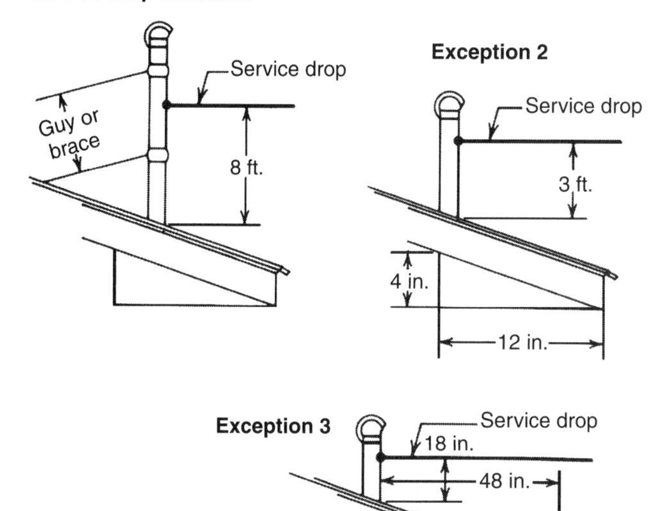

Figure 9-3
Service drop clearances required above roof, per *NEC* 230.24

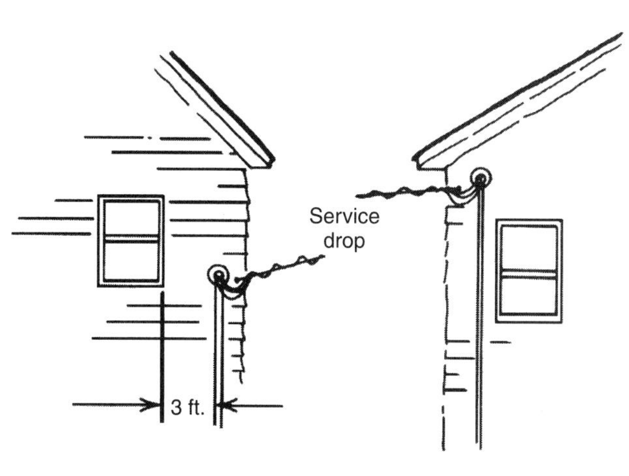

Figure 9-4
Weatherhead on the side of building, per *NEC* 230.24

only need to provide an 18-inch clearance for the service drop wires. Note that the 18 inches mentioned is the required clearance for the *service drop wires*. The mast and weatherhead need to be higher than that to allow the service entrance wires to come down from the weatherhead and connect to the drop wires at a point where they maintain that 18-inch distance.

When an overhead service drop terminates on the side of a building instead of on the roof, the attachment point still has to be at least 10 feet above grade, but it must also be at least 3 feet horizontally from any windows, porches, or fire escapes, as shown in Figure 9-4. If it comes in above the top of a window, the 3-foot horizontal clearance isn't necessary. These clearances are designed to insure that service drop and service entrance wires are safely out of reach of any people.

From the point where the service drop is attached, power is transmitted through the service entrance wires to the electric meter. The meter is in a meter box that's sealed by the utility company. The meter box is supplied and owned by the building owner, but the utility company owns the meter itself. It's normally mounted on an outside wall of the building, 4 to 6 feet above grade, where it can be easily read by the meter reader without requiring him to have access to the interior of the building. Refer back to Figure 9-1.

The service entrance conductors leading from the service head down to the meter must be sized correctly to meet the power requirements of the service supply. Determine the size of the service by calculating the anticipated loads, as we discussed in the last chapter. In the case of a small one-family dwelling, the minimum supply required by the *NEC* is 100 amperes; see *NEC* Section 230.79(C).

In wiring the service entrance for a dwelling, the entrance conductors can be one size smaller than the rated amperage of the service according to the ampacity ratings shown in *NEC* Table 310.15(B)(6). See Figure 9-5. So, if you're wiring a house requiring 100 amperes, one size

Table 310.15(B)(6) Conductor Types and Sizes for 120/240-Volt, 3-Wire, Single-Phase Dwelling Services and Feeders. Conductor Types RHH, RHW, RHW-2, THHN, THHW, THW, THW-2, THWN, THWN-2, XHHW, XHHW-2, SE, USE, USE-2

	Conductor (AWG or kcmil)	
Service or Feeder Rating (Amperes)	Copper	Aluminum or Copper-Clad Aluminum
100	4	2
110	3	1
125	2	1/0
150	1	2/0
175	1/0	3/0
200	2/0	4/0
225	3/0	250
250	4/0	300
300	250	350
350	350	500
400	400	600

From the *National Electrical Code* © 2007, NFPA

Figure 9-5
Ampacity ratings

Copper	*Al & Cu clad Al*	*Service Amperage*
4	2	100
3	1	110
2	1/0	125
1	2/0	150
1/0	3/0	175
2/0	4/0	200

Figure 9-6
Conductor sizes permitted for residential occupancies

smaller than the #4 copper wire indicated in *NEC* Table 310.15(B)(6) is permitted. Figure 9-6 shows the sizes permitted for residential occupancies only.

From the meter, the power goes to the *service disconnecting means*, required per *NEC* Section 230.70. Here again we encounter one of those vague parts of the code. Section 230.70 clearly requires that:

> *"Means shall be provided to disconnect all conductors in a building or other structure from the service-entrance conductors."*

As that Section continues, it seems to be calling for a single main switch or main breaker to be that disconnect means. However, *NEC* Section 230.71 very clearly and specifically permits the *service disconnecting means* to consist of up to six switches or breakers. Although six are permitted, in normal field practice, the majority of small building electrical services contain a single main disconnect, which most commonly is a breaker. When it's not a breaker, it'll be a fused knife switch or perhaps a pull-out twin fuse holder.

The service disconnecting means may or may not be incorporated in the same box in which the electric meter is mounted. In the warmer and dryer areas of the country, you'll commonly find a single combination outdoor box housing the meter, the main breaker/disconnect, and many, if not all, of the branch circuit breakers. In the colder and damper areas, a more practical arrangement is an outdoor meter box and a separate indoor main breaker and distribution box. The separate indoor distribution box has the advantage of being protected from both the elements and meddlesome or mischievous tampering. However, locking an outdoor box to prevent tampering isn't a good idea. A breaker box contains a group of safety devices that should be accessible at all times. In an emergency, should it be necessary to get into the box quickly, the time spent finding the key and unlocking it could be critical.

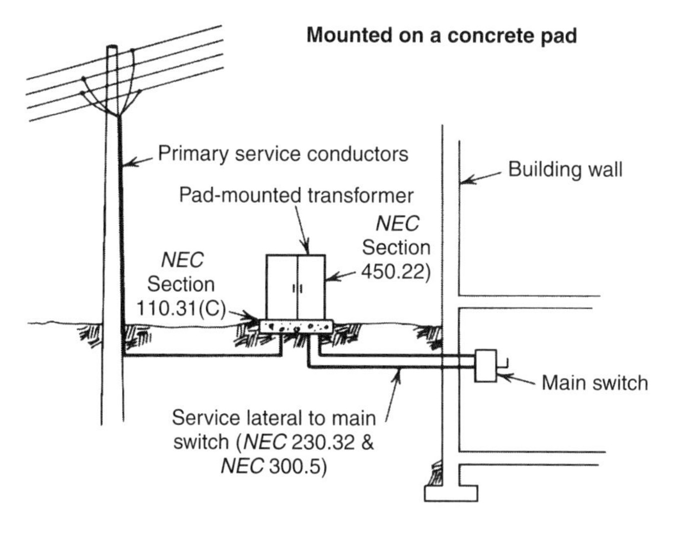

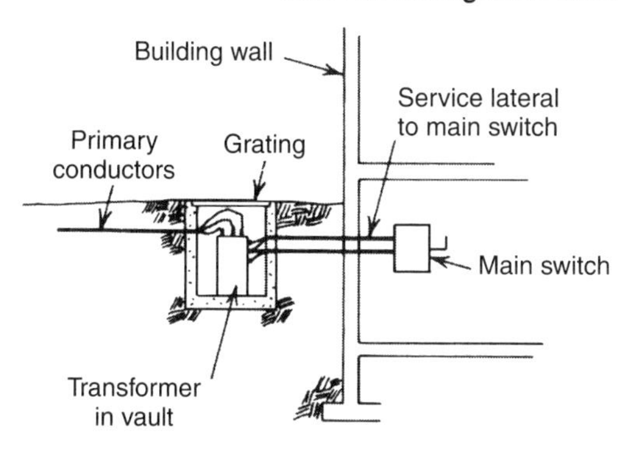

Figure 9-7
Underground service transformers

Underground Services

The installation of underground services has become more common. In fact, in some areas, all new services are installed underground. Typical underground service entrances are illustrated in Figure 9-7. The underground method involves a somewhat different distribution system on the part of the utility company. When the overhead method is used, the utility's final step-down transformer is mounted on a power pole and the power reaches the building via a service drop. The underground method permits the power company to mount the final step-down transformer on or under the ground. Power is delivered to the building through an underground *service lateral* that runs from the transformer to the building, then up and into the meter box.

The transformer is either mounted on a concrete slab at ground level and protected with a heavy steel casing, or it's placed underground in a concrete vault. Primary service conductors run in conduit down the power pole and underground to the pad-mounted enclosure or the underground vault. Service lateral conductors then run underground from the transformer to the main switch. Both installations are shown in Figure 9-7. The advantage of this method is that in the event stormy weather endangers power poles, a transformer on or under the ground is in a far safer place.

The service laterals from the transformer can terminate either inside or outside the building wall. See Figure 9-8. In either case, the building must have a meter box on the exterior wall where the utility can mount its meter. Again, the meter box should be mounted 4 to 6 feet above grade and at a location that will be visible to the meter reader. From the meter box and on through the remainder of the building electrical system, there's no difference between overhead and underground services.

Another advantage of the underground service entrance is that it's less conspicuous and looks better. The overhead installation, with dangling wires, the service head and the mast, doesn't exactly constitute an

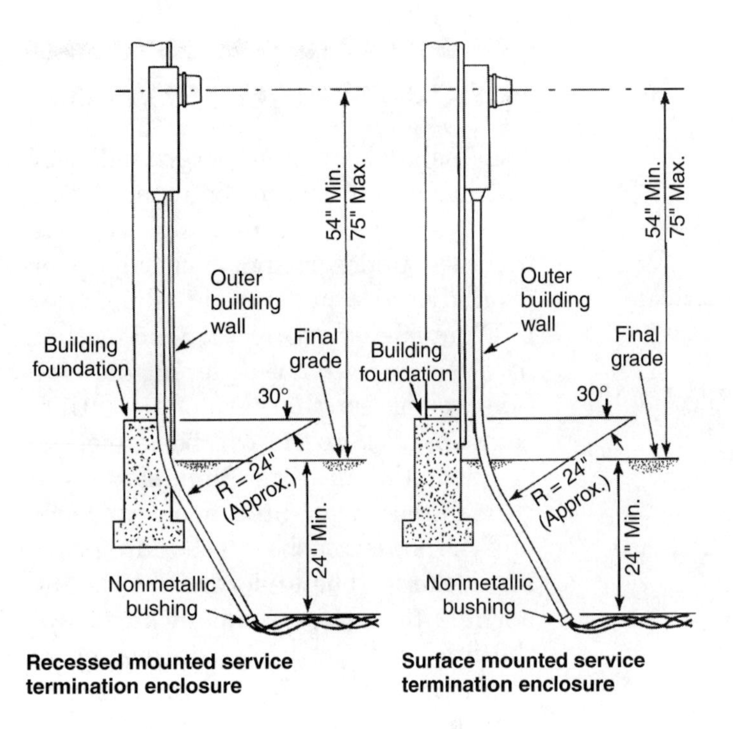

Figure 9-8
Underground service entrances terminating on outside wall

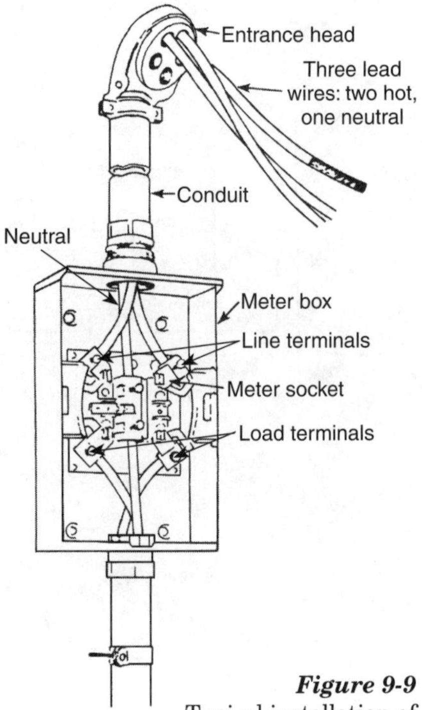

Figure 9-9
Typical installation of outdoor meter box

aesthetic triumph. With underground service, not only are the transformers safer from bad weather, so are the wires. Its major disadvantage is that, when you want to increase the capacity of the building's electrical system, it's much more difficult to dig up and replace a buried service lateral than to replace an overhead service drop.

Meter Box

Regardless of which way the electrical supply reaches a building — overhead or underground — it'll be metered when it gets there. The service entrance wires will terminate at a meter socket, which should be mounted in an outdoor weatherproof metal box, as shown in Figure 9-9.

For a 120/240-volt single-phase service, the service drop or service lateral consists of three wires. Two of these are technically termed *ungrounded*, or in lay terms, *hot*. These will each read 120 volts above ground, and 240 volts apart from each other. The third wire is the grounded common return. When these are attached to the service entrance wires, the one used as the common will have white insulation or, more often, be identified with white tape at both ends.

Figure 9-10
Typical combination outdoor meter box and distribution box

When the service entrance wires reach the meter box, the two hot leads go to the *line* terminals (top terminals) on the meter socket. The common, marked by white insulation or tape, goes to the grounded terminal. From the *load* terminals (bottom terminals) on the meter socket, two more leads connect the now metered supply to the required service disconnecting means discussed previously. Most of the time, this is going to be a single main breaker (occasionally a fused knife switch) that cuts power from everything on the load side of the meter.

When you use a combination meter box containing the meter, main breaker and a number of branch circuit breakers, the meter is internally wired to the main breaker. The main breaker, in turn, connects directly to two-phase buss bars from which the individual branch circuit breakers draw their power. A combination meter box is shown in Figure 9-10. If the meter box is separate from the main and branch circuit breakers, it'll be connected to the distribution panel, usually by means of a metal conduit containing the two hot leads and the common. See Figure 9-11. Notice the ground wire attachment to the water pipe in the figure.

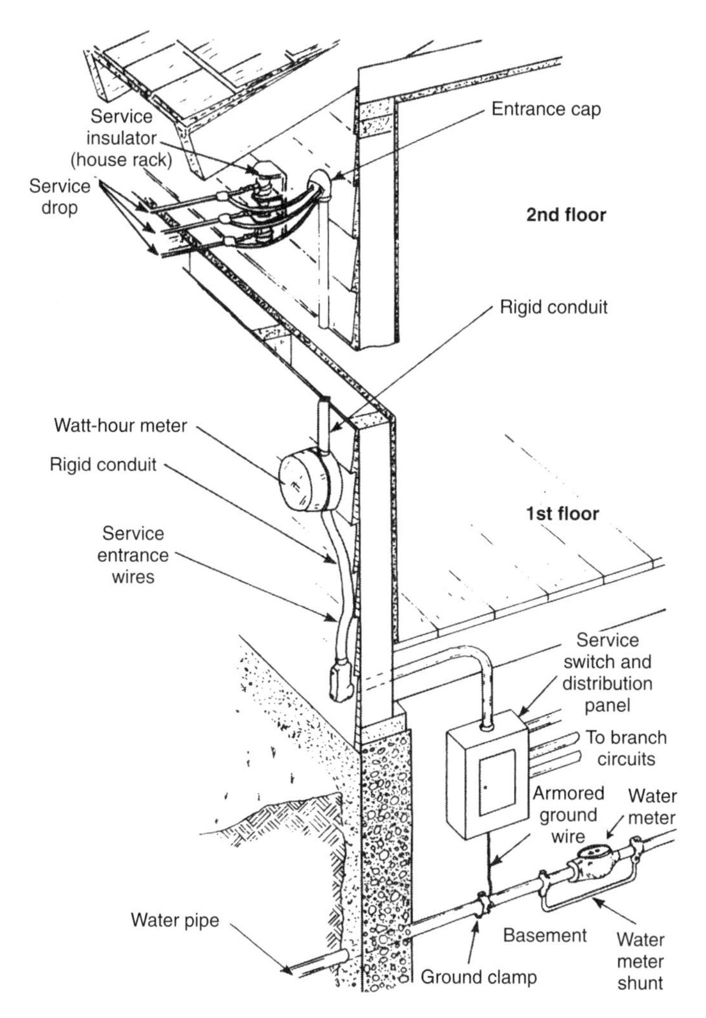

Figure 9-11
Overhead service with outdoor meter box and indoor distribution box

Breaker Box

The size of the breaker box you install depends on the number of branch circuit breakers you need to enclose. That's determined during the planning of the system, which we covered in the last chapter. There are, however, a couple of general code limitations that you need to remember in connection with sizing the breaker box.

Single-family residences with six or more 2-wire branch circuits must be supplied by at least a 100-ampere, 3-wire service, per *NEC* Section 230.79(C). The standard 100-ampere service panel contains 12 breaker spaces in addition to the main breaker. So, whether you need that many breakers or not, 12 spaces is the minimum number in a new residential circuit breaker panel.

According to *NEC* Section 240.5(A), all wiring installed in buildings must be protected against overcurrent by breakers or fuses appropriate to the amperage rating of that wiring. (Refer to *NEC* Table 310.16, in Figure 9-12, for the allowable ampacities of various wire types.) However, nowhere is there any requirement stating that all of the overcurrent devices in a single building must be in a single panelbox. Nor is there any requirement that all the panelboxes containing overcurrent devices be located together. This allows you to place as many subpanels as you find useful or economical in areas around the building where they will be convenient to the devices they protect.

Grounding

Virtually all alternating-current building wiring circuits and systems must be grounded. (*NEC* Section 250.20(B) gives the specifics.) And each branch circuit must contain a continuous ground return path throughout. That branch circuit path, and the common (white circuit wire), all terminate at the neutral bar in the upper left of the main panel box. See Figure 9-13. From there, the building electrical system is tied to true ground by means of an appropriately-sized grounding conductor.

The grounding conductor must be made of copper, aluminum, or copper-clad aluminum, per *NEC* Section 250.62. It has to be resistant to any corrosive conditions around it, or else be suitably protected from any such conditions. It doesn't matter whether it's solid or stranded wire, insulated or bare, as long as it's in *one continuous length*. There can't be any splices between the connection to the ground buss in the panel box and the point where it attaches to the grounding electrode system. You must attach the conductor to the grounding electrode system using a pressure connection, clamp, lug or other listed means, per *NEC* Section 250.70. It can't be soldered.

NEC Section 250.64(B) says that you must protect grounding electrode conductors smaller than #6 against physical damage using rigid metal conduit, intermediate metal conduit, rigid nonmetallic conduit, EMT, or cable armor. Flexible cable armor is the material most commonly used for this purpose. You must also protect a grounding electrode conductor larger than #4 if it will be exposed to physical damage. Secure the grounding electrode conductor or cable to the building structure according to the standard requirements for the material.

Table 310.16 Allowable Ampacities of Insulated Conductors Rated 0 Through 2000 Volts, 60°C Through 90°C (140°F Through 194°F), Not More Than Three Current-Carrying Conductors in Raceway, Cable, or Earth (Directly Buried), Based on Ambient Temperature of 30°C (86°F)

	Temperature Rating of Conductor [See Table 310.13(A).]						
	60°C (140°F)	75°C (167°F)	90°C (194°F)	60°C (140°F)	75°C (167°F)	90°C (194°F)	
Size AWG or kcmil	Types TW, UF	Types RHW, THHW, THW, THWN, XHHW, USE, ZW	Types TBS, SA, SIS, FEP, FEPB, MI, RHH, RHW-2, THHN, THHW, THW-2, THWN-2, USE-2, XHH, XHHW, XHHW-2, ZW-2	Types TW, UF	Types RHW, THHW, THW, THWN, XHHW, USE	Types TBS, SA, SIS, THHN, THHW, THW-2, THWN-2, RHH, RHW-2, USE-2, XHH, XHHW, XHHW-2, ZW-2	
	COPPER			ALUMINUM OR COPPER-CLAD ALUMINUM			Size AWG or kcmil
18	—	—	14	—	—	—	—
16	—	—	18	—	—	—	—
14*	20	20	25	—	—	—	—
12*	25	25	30	20	20	25	12*
10*	30	35	40	25	30	35	10*
8	40	50	55	30	40	45	8
6	55	65	75	40	50	60	6
4	70	85	95	55	65	75	4
3	85	100	110	65	75	85	3
2	95	115	130	75	90	100	2
1	110	130	150	85	100	115	1
1/0	125	150	170	100	120	135	1/0
2/0	145	175	195	115	135	150	2/0
3/0	165	200	225	130	155	175	3/0
4/0	195	230	260	150	180	205	4/0
250	215	255	290	170	205	230	250
300	240	285	320	190	230	255	300
350	260	310	350	210	250	280	350
400	280	335	380	225	270	305	400
500	320	380	430	260	310	350	500
600	355	420	475	285	340	385	600
700	385	460	520	310	375	420	700
750	400	475	535	320	385	435	750
800	410	490	555	330	395	450	800
900	435	520	585	355	425	480	900
1000	455	545	615	375	445	500	1000
1250	495	590	665	405	485	545	1250
1500	520	625	705	435	520	585	1500
1750	545	650	735	455	545	615	1750
2000	560	665	750	470	560	630	2000

CORRECTION FACTORS							
Ambient Temp. (°C)	For ambient temperatures other than 30°C (86°F), multiply the allowable ampacities shown above by the appropriate factor shown below.						Ambient Temp. (°F)
21–25	1.08	1.05	1.04	1.08	1.05	1.04	70–77
26–30	1.00	1.00	1.00	1.00	1.00	1.00	78–86
31–35	0.91	0.94	0.96	0.91	0.94	0.96	87–95
36–40	0.82	0.88	0.91	0.82	0.88	0.91	96–104
41–45	0.71	0.82	0.87	0.71	0.82	0.87	105–113
46–50	0.58	0.75	0.82	0.58	0.75	0.82	114–122
51–55	0.41	0.67	0.76	0.41	0.67	0.76	123–131
56–60	—	0.58	0.71	—	0.58	0.71	132–140
61–70	—	0.33	0.58	—	0.33	0.58	141–158
71–80	—	—	0.41	—	—	0.41	159–176

* See 240.4(D).

Figure 9-12
Ampacities of insulated conductors

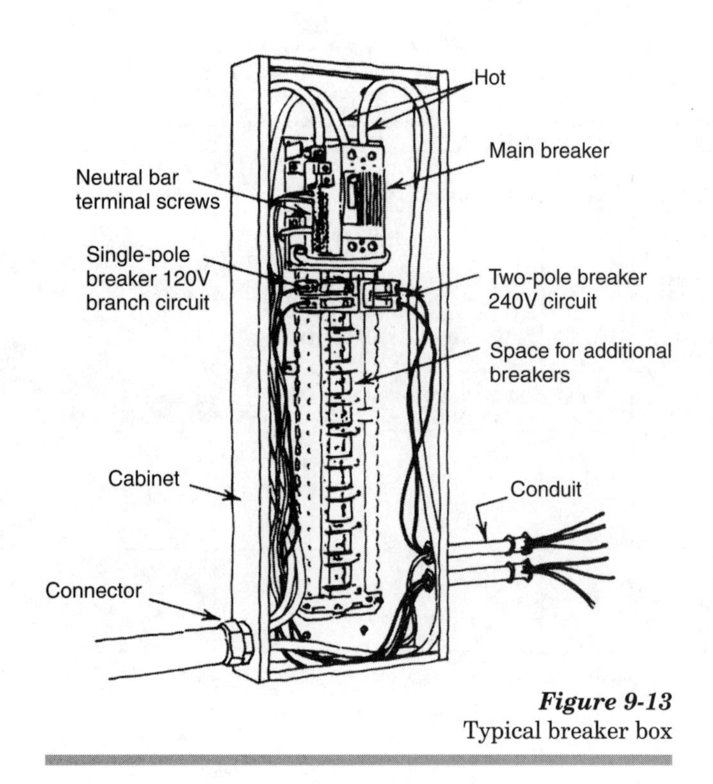

Figure 9-13
Typical breaker box

Sizing

Size the grounding electrode conductor to conform to requirements in *NEC* Table 250.66, shown in Figure 9-14. Regardless of the size of the service, the grounding conductor can never be smaller than #8 AWG copper, or #6 aluminum.

Grounding Electrode System

The code requirement for a grounding electrode system, which first appeared in the 1978 *NEC* Section 250.50, replaced the single ground system used up until that time. That single ground was commonly the underground metal water pipe coming into the building.

The present requirements found in *NEC* Section 250.52 call for the following items, if available, to be bonded together to form the grounding electrode system:

- A metal underground water pipe in direct contact with earth for a distance of at least 10 feet;
- The metal frame of the building, with at least 10 feet or more of a single structural member in direct contact with the earth;
- A concrete-encased electrode, which can be either a minimum of 20 feet of at least ½-inch steel foundation reinforcing bar or at least 20 feet of bare copper wire no smaller than #4 AWG, embedded in the foundation concrete;
- A ground ring consisting of at least 20 feet of bare wire, no smaller than #2 AWG, buried in the ground no less than 2½ feet deep.

If the second and third items listed above aren't available at the job, dig a trench to install a ground ring. Sometimes, however, digging a trench just isn't possible. If there's nothing but the underground metal water pipe to use for grounding the system, supplement the ground with one of the following other electrodes, per *NEC* Section 250.52.

Table 250.66 Grounding Electrode Conductor for Alternating-Current Systems

Size of Largest Ungrounded Service-Entrance Conductor or Equivalent Area for Parallel Conductors[a] (AWG/kcmil)		Size of Grounding Electrode Conductor (AWG/kcmil)	
Copper	Aluminum or Copper-Clad Aluminum	Copper	Aluminum or Copper-Clad Aluminum[b]
2 or smaller	1/0 or smaller	8	6
1 or 1/0	2/0 or 3/0	6	4
2/0 or 3/0	4/0 or 250	4	2
Over 3/0 through 350	Over 250 through 500	2	1/0
Over 350 through 600	Over 500 through 900	1/0	3/0
Over 600 through 1100	Over 900 through 1750	2/0	4/0
Over 1100	Over 1750	3/0	250

Notes:
1. Where multiple sets of service-entrance conductors are used as permitted in 230.40, Exception No. 2, the equivalent size of the largest service-entrance conductor shall be determined by the largest sum of the areas of the corresponding conductors of each set.
2. Where there are no service-entrance conductors, the grounding electrode conductor size shall be determined by the equivalent size of the largest service-entrance conductor required for the load to be served.
[a]This table also applies to the derived conductors of separately derived ac systems.
[b]See installation restrictions in 250.64(A).

From the *National Electrical Code* © 2007, NFPA

Figure 9-14
Sizing grounding electrode conductors

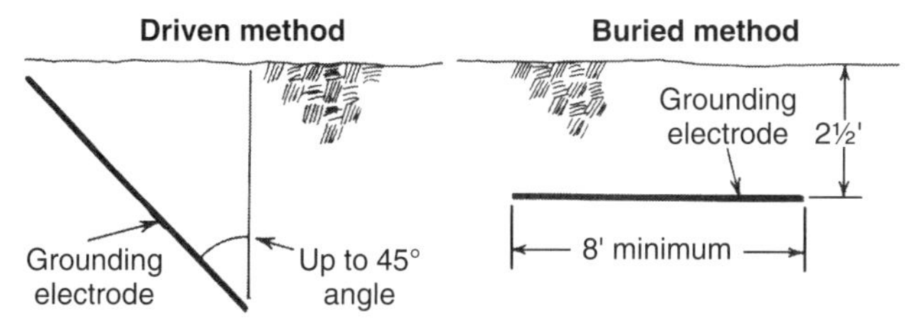

Figure 9-15
Supplementary grounding electrodes

- A rod or pipe — either one must be a minimum of 8 feet long. If using a pipe, it has to be at least ¾-inch trade size. Steel pipe must be galvanized or coated with other metallic protection. If using a rod, ferrous metals must be at least ⅝-inch in diameter, while nonferrous rods need only be ½-inch diameter. Drive the pipe or rod into the earth anywhere in the range from straight down to 45 degrees off vertical. If you can't do this, bury the entire 8-foot length a minimum of 2½ inches deep. Figure 9-15 illustrates both methods.
- A plate electrode can also be used as an additional alternative. It should be at least 2 square feet in area. If a ferrous metal is used, it needs to be ¼ inch thick; nonferrous metals only need to be 0.06 inch thick
- Other local metal underground systems or structures such as piping or tanks can also be used. Be sure that the underground metal you use isn't protected by paint or a nonmetallic material that isolates it from the earth. If a material isolated, it can't function as a grounding electrode.

The use of a metal underground gas piping system is now specifically forbidden. Under the Uniform Plumbing Code, these systems now require a plastic wrapping that's precisely the kind of non-conductive coating that makes a gas pipe useless as a grounding electrode.

In the case of the plate electrode, the code doesn't say anything about galvanizing or other types of protection for ferrous metals, and it doesn't specify a burial depth. Since a ferrous rod or pipe must be galvanized or otherwise protected, it seems likely that an inspector will insist on a plate electrode being similarly protected. It also would seem safe to bury it at least 2½ feet deep, the same as the pipe, rod, or #2 AWG copper wire ground ring.

The basic emphasis of the 1978 requirements was that the old traditional method of grounding the building electrical system via a single grounding electrode was no longer sufficient. At least two, and preferably more, grounding electrodes bonded together are now required to form a more effective grounding electrode system. These additional electrodes must be located no less than 6 feet apart. Keep in mind that if you're updating the electrical system on a pre-1978 building with an old grounding system still in place, you'll have to update the grounding system as well.

Overcurrent Protective Devices

Overcurrent protection is required as part of every branch circuit in every electrical installation, per *NEC* Section 210.20. This protection is provided either by fuses or by circuit breakers, most of which are mounted in the distribution boxes that form a major part of the service entrance equipment. The purpose of requiring overcurrent protection is to prevent damage to the conductors or other parts of the electrical system due to the overheating that results from an overcurrent. Overcurrent results from excessive current drawn through a circuit designed to carry only a certain amperage. It occurs for one of two reasons: overloading, or an unintentional and unexpected direct connection of power to ground through little or no resistance. The latter condition is called a *short*.

In Chapter 1, under Watt's Law, we had an illustration of overloading in which three appliances were connected to a 20-ampere appliance circuit. A 20-ampere circuit at 120 volts can deliver 2400 watts of power ($I \times E = P$). The three appliances in the example required 2800 watts. To deliver 2800 watts at 120 volts, the current will have to be 23.33 amperes. This is 3.33 amperes more than the wire insulation, the switches, or the receptacles were intended to carry. Any or all of them could be damaged by the excessive current. In this case, the required overcurrent protective device would shut off the circuit before any damage could occur.

Cases of overloading generally involve relatively small excesses of amperage over and above the rated capacity of the circuit. In contrast, shorts involve enormous excesses of amperage above the rated capacity

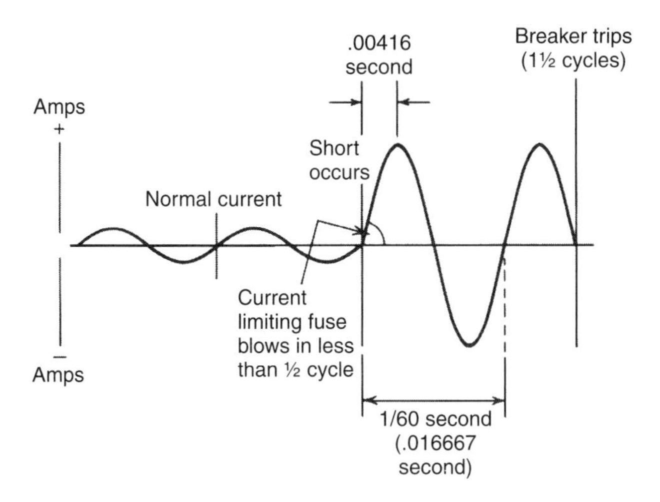

Figure 9-16
When a short occurs, the fuse responds faster than the breaker

of a circuit. Current flow can jump up to hundreds or even thousands of amperes at the first peak of the first half cycle after the short occurs, as shown in Figure 9-16. Consequently, overcurrent protective devices must react with incredible speed to avert serious damage to system components, since that first peak of the first half cycle occurs in 0.00416 second.

There are two basic types of overcurrent protective devices: fuses and circuit breakers. A fuse functions by melting a metal link in response to overcurrent. Since it has no moving parts, the fuse can open the circuit in less than one-half cycle. Breakers depend on mechanical movement, requiring about one and a half cycles to react. This gives the fuse a considerable advantage in terms of speed of response. But it has the disadvantage that, once blown, you usually have to replace it. In the case of some cartridge fuses, the fusible link can be replaced and the cartridge reused. The breaker reacts more slowly, but once it's tripped, you can easily reset it by simply moving a switch handle.

Fuses

In older electrical installations (more than 50 years old), the standard overcurrent protective device was the Edison base plug fuse, shown in Figure 9-17. This fuse has a threaded base like an incandescent light bulb. The top is transparent, leaving the fusible metal strip visible for easy inspection. When this type fuse is blown, the transparent top will be considerably discolored so you can instantly identify a bad one.

Figure 9-17
Edison base plug fuse

Plug fuses are available in 15, 20, and 30 ampere sizes, all of which fit into the same socket. Over the years, this has proven to be a very bad feature. It allows those who aren't familiar with fuses to overload a 15-ampere circuit and blow even more fuses. Then, without removing the overload, they try to fix the fuse trouble by simply using a larger fuse. This larger fuse will pass more current than the circuit components can handle safely. The end result can be a disastrous fire.

The *S type* fuse was developed to head off this problem. S type fuses consist of two parts: the fuse and a threaded adapter into which the fuse is screwed; see Figure 9-18.

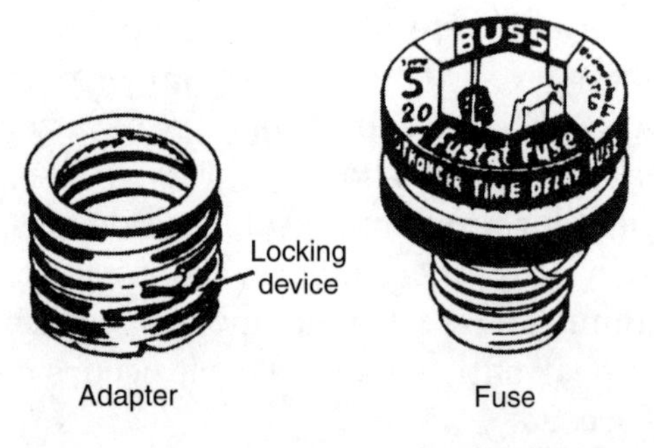

Figure 9-18
S type fuse includes adapter

This combination then screws into the standard fuse socket. Each fuse size has an adapter with a different inside thread, and all adapters are designed to be nonremovable once placed in a fuse socket. So, once a 15-, 20-, or 30-ampere adapter has been put into a socket, only that size fuse will fit that socket. If you're working on an older home, replace all standard Edison base plug fuses with S type fuses to prevent accidental overfusing (replacement of a blown fuse with one of too high a rating). In many jurisdictions, the inspection departments have already required that all old type plug fuses be replaced with S types to stop people from making sizing mistakes with their fuses.

Many electrical devices draw a considerably higher current when starting than when operating. As a momentary surge, this starting current presents no danger to the electrical system. But due to the very rapid response characteristic of fuses, it will often blow a fuse. To counteract this problem a *time-delay* fuse was developed that compensates for starting surges without blowing the fuse. The time-lag plug fuse, shown in Figure 9-19, is a dual-element device that can withstand the moderate overcurrents of starting surges, but will blow if a short causes a large overcurrent to occur.

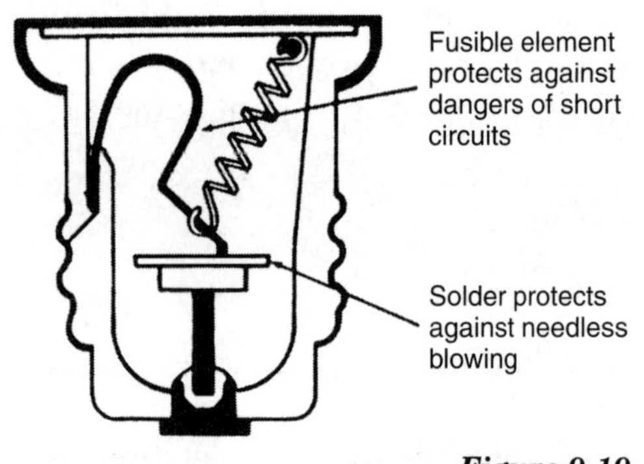

Figure 9-19
Time-lag plug fuse

To fuse circuits over 30 amperes you need to use one of two types of cartridge fuses: the ferrule contact fuse or the knife blade contact fuse. Both types are shown in Figure 9-20. You'll find these fuses in the main system disconnects and the fused disconnects for large equipment, such as a central air conditioning system. Either fits into the appropriate fuse clip. You need a fuse puller to insert and remove these fuses. It wouldn't be smart to put your fingers or metal pliers into a cartridge fuse holder.

Blown cartridge fuses can't be visually detected the way a bad plug fuse can. They must be tested. Pull and test a suspect fuse with an ohmmeter. If there's continuity through the fuse, it's good. It can also be tested in the circuit with the voltmeter. With the circuit on, put one probe at each end of the fuse. If the line voltage shows on the meter, the fuse is blown. If the meter reads 0 across the two ends of the fuse, but reads 120 volts (or whatever the line voltage is) from either end to ground, the fuse is good.

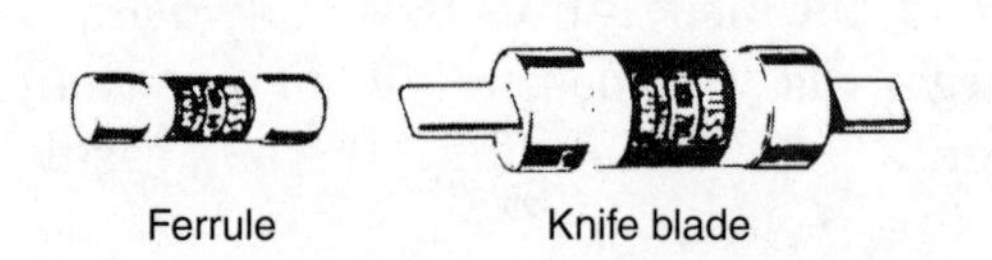

Figure 9-20
Cartridge fuses

Like regular fuses, you have to replace most cartridge fuses when blown. However, the type shown in Figure 9-21 has removable ends that allow access so you can replace the fusible link inside. While the cartridge length and diameter are increased with amperage, there can be more than one amperage at any particular cylinder size. This, as with plug fuses, opens the possibility of replacing a blown fuse with one the wrong size. Be particularly careful to replace a blown fuse with another of the proper size.

Figure 9-21
Replaceable link cartridge fuse

Circuit Breakers

A circuit breaker provides overcurrent protection and can also serve as a simple power switch. It has the advantage over fuses in that, when tripped by overcurrent, you can reset it almost immediately — providing you've found and corrected the cause of the overcurrent. If the overcurrent was caused by an overload that you haven't corrected, it'll usually go back to the *ON* position after a few minutes of cooling, but then trip again. If you don't correct the cause of a *short* that creates an overcurrent, the breaker absolutely won't hold in the *ON* position.

The heart of a circuit breaker is the bimetallic strip, shown in Figure 9-22, which is made of two different metals fused together. When heated, these two metals don't expand at the same rate, which causes the strip to bend. This strip acts as a latch, holding two spring-loaded contacts together. When a current greater than the breaker was intended to carry flows, the bimetal strip heats, bends, and releases the latch. This separates the contacts and interrupts the current flow.

The time it takes for the bimetallic strip to heat allows it to act in much the same way as a time-delay fuse; it carries momentary starting overloads without tripping. The handle on the breaker also allows you to use it as a switch — providing an easy way to shut off a single circuit for repairs without disturbing the others.

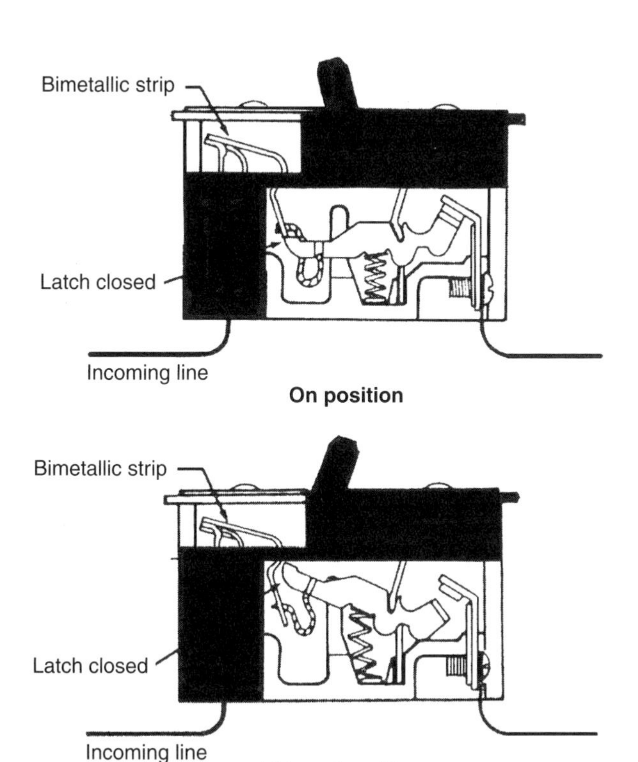

Figure 9-22
Typical circuit breaker mechanism

STUDY QUESTIONS

1. **What is the service drop?**
   A) The wires that drop from the weatherhead
   B) The wires that connect the meter to the main disconnect
   C) The wires that connect the meter to the weatherhead
   D) The wires that come from the street power pole to the weatherhead

2. **How much clearance is required for power wires from a street pole that pass over a roof with a slope of 4 inches in 12 inches, if their voltage is less than 300 volts?**
   A) 8 feet
   B) 6 feet
   C) 3 feet
   D) 8 inches

3. **When an overhead service entrance drop ends on the side of the building rather than on the roof, the attachment must be at least how high above grade?**
   A) 8 feet
   B) 10 feet
   C) 12 feet
   D) 16 feet

4. **The *NEC* specifically permits the residential service disconnecting means to consist of which of the following arrangements?**
   A) It's always just a single switch
   B) A single 240-volt breaker
   C) Up to four switches or breakers
   D) A single main breaker or up to six switches or breakers

5. **Where may the step-down transformer be mounted for underground service?**
   A) On a concrete pad at ground level
   B) On the power pole with a conduit run down the pole and underground
   C) In an underground vault beneath the building
   D) Lateral to the service entrance

6. What's the disadvantage of an underground service entrance?

A) It's not as safe as an overhead service entrance
B) It's more difficult to replace a buried service lateral than an overhead service drop
C) The service laterals must terminate inside the building
D) It requires more maintenance than an overhead service entrance drop

7. When the service entrance wires reach the meter box, where do the two hot leads go?

A) To the line terminals on the meter socket
B) To the load terminals on the meter socket, then to the required disconnect breaker
C) To the service disconnecting means
D) One goes to the top and one to the bottom terminal on the meter socket

8. What size breaker box do you need for a new single-family residence supplied with 100-ampere, 3-wire service?

A) A 6-breaker box
B) A 6-breaker box with 6 subpanels
C) A 12-breaker box with subpanels as needed
D) A 24-breaker box

9. According to the *NEC*, what must a grounding conductor consist of?

A) A copper or galvanized steel bar, with a minimum diameter of ½ inch, driven no less than 2½ feet into the ground
B) A continuous length of copper, aluminum or copper-clad aluminum wire
C) The metal frame of the building
D) A 2-foot-diameter ground ring buried no less than 2½ feet deep

10. What's the advantage of a fuse over a circuit breaker as an overcurrent protection device?

A) You can convert a circuit to handle a slightly-larger load by changing to a larger fuse
B) The top of a fuse becomes transparent when blown
C) Fuses are easily replaced when blown
D) A fuse opens the circuit faster

10

ROUGH WIRING

With plans and calculations completed and the building framework in place, you can begin the rough wiring based on the information given on the cable plans. The sequence of the work will vary depending on the materials being used, so we'll discuss cable, flexible metal conduit, EMT (thinwall conduit), and surface raceways (Wiremold) separately.

Concealed rough wiring in new work, as described below, is normally done with at least one side of each wall open to provide the electrician with access to framing members and the internal voids between them. This allows you to mount boxes conveniently, and place and secure box-to-box wire runs. When you've completed the work, the open wall allows the electrical inspector to easily see that cabling, Greenfield, or conduit work has been correctly installed and fastened in accordance with all *NEC*® and local code requirements.

Nonmetallic Cable — NMC

We've discussed the uses of NM cable as authorized by code, as well as various code-imposed limitations that must be observed when using this material in building wiring. And we've mentioned the use of the cable stripper to cut the outer (exterior plastic protective) sheathing to access conductors inside the NM cable. Now let's look at a wiring project using NM cable.

Mounting Boxes

All wiring projects begin with the installation of the boxes where receptacles, switches and fixtures will be placed, and where all splices and connections must be made, per *NEC* Section 300.15. You can use either plastic or metal boxes for NM wiring. Since plastic is less expensive, it's the type used most often.

Taking one circuit at a time, first determine from the cable plan how many boxes, along with their types and sizes, you'll need. For example, look at the bedroom 1 and hall circuit as cabled in Figure 8-59, back in Chapter 8. There are five duplex receptacles, each needing a 2 x 4 device box. The 4-way and two 3-way switches call for three more, making a total of seven. The ceiling light in the center of bedroom 1 mounts on a 4-inch octal box, and the two single-pole switches by the door go in a two-gang 4 x 4 box. The three recessed lights don't need boxes; they're self contained. So, you need the following boxes for this circuit:

2 x 4 device	7
4 x 4 (two gang)	1
4-inch octal	1

You need to check a few more details before settling on the list above. There are five 2-wire cables entering that two-gang switch box. The code limits the number of conductors in boxes based on their cubic volume. If you look at *NEC* Table 314.16(A) in Figure 10-1, you'll see that a 4-inch-square by 1¼-inch-deep box is approved for only nine #14 conductors; a 4-inch-square by 1½-inch-deep box is good for 10 #14 conductors; and a 4-inch-square by 2⅛-inch-deep box is good for 15. There are five two-conductor cables coming in, giving a count of 10, plus five grounds, which together count for one more, making a total of 11. The two switches count as two more, raising the total for the box to 13. So, a 4-inch-square box is fine, as long as you use one that's 2⅛ inches deep. The shallower ones are too small for that number of conductors.

Now check the 2 x 4 device box for the 4-way switch. Two 3-wire cables come to six conductors — one more for the two grounds makes it seven. When you add in the switch, the total becomes eight. There isn't room for #14 wire in any of the 2 x 4 device boxes listed in *NEC* Table 314.16(A). Now look at *NEC* Table 314.16(B) in Figure 10-2. It shows that the *NEC* requires 2 cubic inches per conductor when the wire size is #14. This means you need a box with at least 16 cubic inches of internal space. Single-gang plastic boxes, like the one shown in Figure 10-3, are available with internal volumes of 18 and 20 cubic inches. To avoid any possible confusion or doubt about the box size, check the volume on the box. It should be clearly marked on the inside.

Table 314.16(A) Metal Boxes

Box Trade Size			Minimum Volume		Maximum Number of Conductors* (arranged by AWG size)						
mm	**in.**		**cm^3**	**in.^{3}**	**18**	**16**	**14**	**12**	**10**	**8**	**6**
100 × 32	(4 × 1¼)	round/octagonal	205	12.5	8	7	6	5	5	5	2
100 × 38	(4 × 1½)	round/octagonal	254	15.5	10	8	7	6	6	5	3
100 × 54	(4 × 2⅛)	round/octagonal	353	21.5	14	12	10	9	8	7	4
100 × 32	(4 × 1¼)	square	295	18.0	12	10	9	8	7	6	3
100 × 38	(4 × 1½)	square	344	21.0	14	12	10	9	8	7	4
100 × 54	(4 × 2⅛)	square	497	30.3	20	17	15	13	12	10	6
120 × 32	(4 11/16 × 1¼)	square	418	25.5	17	14	12	11	10	8	5
120 × 38	(4 11/16 × 1½)	square	484	29.5	19	16	14	13	11	9	5
120 × 54	(4 11/16 × 2⅛)	square	689	42.0	28	24	21	18	16	14	8
75 × 50 × 38	(3 × 2 × 1½)	device	123	7.5	5	4	3	3	3	2	1
75 × 50 × 50	(3 × 2 × 2)	device	164	10.0	6	5	5	4	4	3	2
75 × 50 × 57	(3 × 2 × 2¼)	device	172	10.5	7	6	5	4	4	3	2
75 × 50 × 65	(3 × 2 × 2½)	device	205	12.5	8	7	6	5	5	4	2
75 × 50 × 70	(3 × 2 × 2¾)	device	230	14.0	9	8	7	6	5	4	2
75 × 50 × 90	(3 × 2 × 3½)	device	295	18.0	12	10	9	8	7	6	3
100 × 54 × 38	(4 × 2⅛ × 1½)	device	169	10.3	6	5	5	4	4	3	2
100 × 54 × 48	(4 × 2⅛ × 1⅞)	device	213	13.0	8	7	6	5	5	4	2
100 × 54 × 54	(4 × 2⅛ × 2⅛)	device	238	14.5	9	8	7	6	5	4	2
95 × 50 × 65	(3¾ × 2 × 2½)	masonry box/gang	230	14.0	9	8	7	6	5	4	2
95 × 50 × 90	(3¾ × 2 × 3½)	masonry box/gang	344	21.0	14	12	10	9	8	7	4
min. 44.5 depth	FS — single cover/gang (1¾)		221	13.5	9	7	6	6	5	4	2
min. 60.3 depth	FD — single cover/gang (2⅜)		295	18.0	12	10	9	8	7	6	3
min. 44.5 depth	FS — multiple cover/gang (1¾)		295	18.0	12	10	9	8	7	6	3
min. 60.3 depth	FD — multiple cover/gang (2⅜)		395	24.0	16	13	12	10	9	8	4

*Where no volume allowances are required by 314.16(B)(2) through (B)(5).

Figure 10-1
Volume allowances in metal boxes

Table 314.16(B) Volume Allowance Required per Conductor

Size of Conductor (AWG)	Free Space Within Box for Each Conductor	
	cm^3	**in.^{3}**
18	24.6	1.50
16	28.7	1.75
14	32.8	2.00
12	36.9	2.25
10	41.0	2.50
8	49.2	3.00
6	81.9	5.00

Figure 10-2
Volume allowances per conductor

Figure 10-3
Plastic box with internal cubic volume marked inside

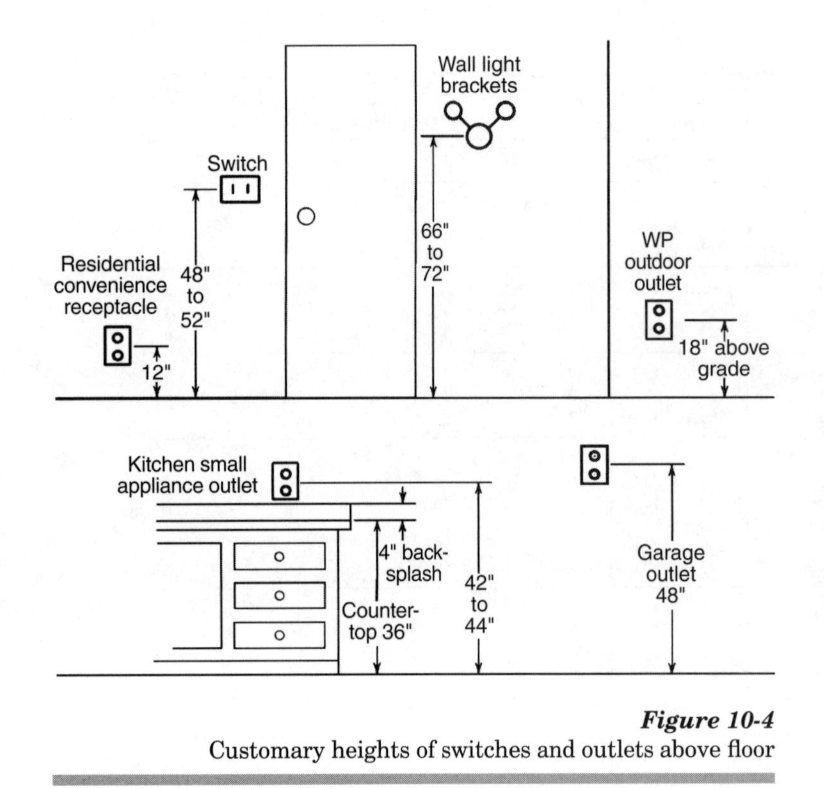

Figure 10-4
Customary heights of switches and outlets above floor

With the correct quantities and sizes of boxes now in hand, you can attach them to the building frame at the locations given on the floor plan. As mentioned earlier, the dimensions for box positions aren't given on the plans. However, the plans are drawn to scale; so you should be able to get very close to the desired placement using a scale ruler. Most residential and other small buildings commonly use a scale of ¼ inch = 1 foot. A triangular ruler, called an *architect's scale*, is available at any drafting supply house. This ruler has, in addition to the ¼-inch scale, all the other scales an electrician is likely to encounter. It's not an expensive item, and it's very durable (I've used the same one for at least 30 years), so it's well worth having.

What's the reason that exact dimensions for the locations of electrical boxes aren't given on the plans — except in special cases? Since pinpoint accuracy in box location isn't really necessary for proper operation of the system, it's actually very handy for the electrician to have the freedom to move a few inches one way or another to mount on an existing stud or joist. This freedom of movement saves a great deal of unnecessary labor. For example, *NEC* Section 210.52(A)(1) requires that you locate the first convenience outlet no more than 6 feet from a doorway. So, the last stud that's less than 6 feet from the door will meet code — the exact dimension is unimportant. Or, if the plan calls for a pair of switches beside a door on the handle side, it doesn't matter whether they're 3, 5 or 6 inches away from the door, as long as they're close and on the correct side.

The code doesn't dictate the proper heights for mounting boxes above the floor, but customary usage does. Figure 10-4 gives the heights on center normally used for most boxes in a residence. Special purpose boxes for TV antenna or telephone outlets are usually set, like convenience outlets, 12 inches on center off the floor. A special purpose outlet for a window air conditioner is usually 12 inches high as well. Range, refrigerator, disposal, dishwasher, or compactor is variable. The height of outlets for these units is a matter of convenience, since none are visible. Dryer and laundry outlet heights also vary. Most of the time they're placed just low enough so the machines hide them. Other times they're raised above the machines for accessibility. Remember, while the code says nothing about box height from the floor, *NEC* Section 314.20 is very definite about the box depth in relation to the finished wall surface — they should set back no more than ¼ inch.

Attach the boxes to studs and joists with nails. Most plastic boxes are supplied with nails. Mount metal boxes with brackets and secure them to the stud with 1-inch roofing nails. Large nails aren't necessary, as there's no strain on a box once it's in place. It's better to knock out the knockouts you'll be using *before* you mount each box, rather than after it's up.

When the plans require mounting a box in an exact location or dimension (the exact position for a ceiling chandelier in a dining room, for example), then you may need to use either a metal or a plastic box on a bar hanger. This will allow you to center the box precisely at the desired point.

Running and Fastening Cables

With all the boxes properly placed, you can install the wire runs as shown on the cable plan. When you use metal boxes, the NM cable must enter through a box connector that also serves as a clamp. The clamp firmly secures the cable to the box at the entry point, as required by *NEC* Section 314.17(B). Some boxes, both plastic and metal, have internal cable clamps. Whenever you use a clamp to secure the cable to the box, you must also secure that cable to the building frame within 12 inches of the box (top and bottom) using a staple or strap; see *NEC* Section 334.30. If you use a plastic box with knockouts, secure the cable within 8 inches above and below the box.

Inside the clamp in metal or plastic boxes, or within the knockout in a plastic box without clamps, strip the cable sheathing to leave at least 6 inches of the conductors free for splicing or connecting to devices, per *NEC* Section 300.14. The cable sheathing must extend into the box no less than ¼ inch inside each cable clamp or plastic box knockout, per *NEC* Section 314.17(C). And, for maximum ease when making splices and working in boxes, the sheathing shouldn't extend into the boxes *more* than that required ¼ inch. Since at least 6 inches of every conductor in every box must be clear of sheathing (per *NEC* Section 300.14), strip it before fastening the cable in place rather than after — it's much easier this way.

> ***IMPORTANT!*** When counting the number of conductors in a box, you must count an *internal* cable clamp as one conductor. The combination box connector and clamp that fits in the round knockout of a metal box does *not* count as one conductor.

Outside the box, the first staple or strap to secure the cable, whether at 12 inches or at 8 inches, will probably be attached to the same stud or joist on which the box is mounted. From there, until within the required stapling distance of the next box, secure

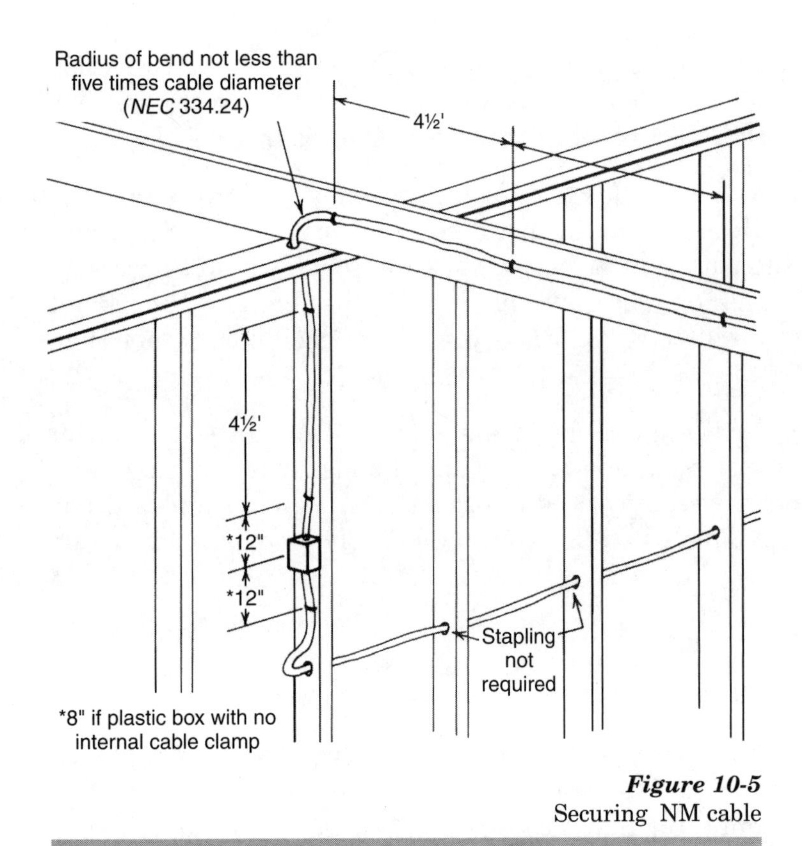

Figure 10-5
Securing NM cable

the cable every 4½ feet, as shown in Figure 10-5. Attach the cable along the joist, vertically through walls and horizontally across ceilings or under floors. You don't need to strap or staple cable when it runs horizontally through holes in studs or joists.

When you run a cable across a crawlspace at an angle to the floor joists rather than parallel to one of them, you don't need to drill the joists. Staple the cable to the bottom of the joists. If, however, you run a cable across an attic at right angles or on a diagonal to the joists, the fastening requirements depend both on the intended use of the attic and how it will be accessed.

While the code doesn't require drilling when making a diagonal cable run across an accessible attic, it's recommended. Otherwise, putting in the required guard strips and staples is as much or more labor than drilling — plus you have the additional material cost of the guard strips. And, if the owner wants to add an attic floor in the future, with drilling, the tops of the joists will be clear. Granted, diagonal cables between the joists will cause some inconvenience for the fellows who install attic insulation; but the advantages of drilling over using guard strips still outweigh the disadvantages.

Rough wiring is complete when you've installed the properly sized boxes, run all the cables connecting those boxes, and correctly fastened them in place.

BX Cable — Type AC (Armored Cable)

You can only use metal boxes with BX cable. In other respects, the box mounting procedures for BX are very much the same as those for NM cable. Find the horizontal distances by scale measurements from the floor plans. The heights are conventional unless otherwise noted. The same regulation regarding the mounting position of the box front and finished wall surface also applies.

Box Mounting

REMOVE THE KNOCKOUTS you'll be using *before* mounting the boxes. Remember, any knockouts you remove by mistake must be plugged, per *NEC* Section 110.12(A).

The basic work sequence is again to first properly mount the boxes, and then install the required box-to-box cable runs. Connect each cable to the boxes at both ends using box connectors that secure the cable with a screw clamp. Because of its metal armor, BX cable is much less flexible than NM cable. Fortunately, there are both straight and right-angled box connectors available to give you the necessary flexibility to bring cables in and out of boxes within the limited space.

Running Armored Cable

BX cable is easy to cut with an ordinary hacksaw. Cut between loops in the metal armor at approximately a 45-degree angle, then twist the armor to break it apart. Since BX already contains its insulated conductors, your primary consideration when cutting is to avoid damaging that insulation. It'll take a bit of practice to develop the knack of successfully cutting the armor without damaging the insulation. Be surprised if you *do* get it right the first time — not if you don't.

After cutting BX cable, install the required fiber bushing that slips inside the cut end before you clamp on the cable connector. This bushing is required because the sharp edges left from the cut might damage the insulation on the conductor and you won't see the damage or know it's there. The bushing protects the conductor insulation from direct contact with any of the armor that could damage it. The inside opening of the BX box connector is enlarged so the electrical inspector can easily see whether that bushing is in place.

Unlike NM cable, BX cable doesn't contain a grounding conductor of the same size and material as the circuit conductors. Instead, it's grounded through its steel armor, supplemented by a thin *bonding strip*, often of aluminum. After you've installed the fiber bushing, bend the bonding strip back to lie between the armor and the cable connector, making a good electrical contact with both.

The requirements for fastening BX cable are similar to those for NM cable. Fasten it within 12 inches of each box, and every 4½ feet thereafter, until it's within 12 inches of the next box. Like NM cable, you can pass it horizontally through holes in the wall studs without fastening every 4½ feet. The bending radius of BX cable can't be less than five times the diameter of the cable, measured from the inside of the bend, per *NEC* Section 320.24.

Since they require the same protection, the same drilling recommendations made for NM cable runs in an attic apply to BX cable as well. While code stipulations regarding NM and BX cables are generally very similar, there's one significant difference. You can embed BX in plaster or other types of masonry and you absolutely can't do that with NM cable. When BX is embedded in masonry, you don't need any fasteners to secure it.

In a BX cable containing the usual copper conductors, the potential difficulties due to dissimilar metals are obviously considerable. In some jurisdictions, using BX cable is limited or prohibited, for the following reasons:

- the proximity of dissimilar metals
- the ground is considered inadequate
- the danger of damage to conductor insulation when cutting the cable

Unless you're sure that BX is currently used and approved in your immediate area, before you install any of it, check with your electrical inspector to make certain. You don't want to find out later that you can't use it.

Flexible Metal Conduit — FMC (Greenfield)

Flexible metal conduit, or *Greenfield*, has spiral metal armor like BX cable. Also like BX, you can only use metal boxes with it.

Box Mounting

Use the same procedures we've already discussed to mount boxes for Greenfield. Scale the measurements from the plans to locate them horizontally, and place them at customary heights above the floor. As we advised earlier, remove any knockouts you'll need before nailing the boxes in place. Greenfield box connectors thread inside the spiral armor of the conduit.

Running Greenfield

When the boxes are installed, run the flex conduit from box to box, as called for in the plans. Cut the lengths between the loops with a hacksaw, just as you would with BX armor. Before cutting, be sure you

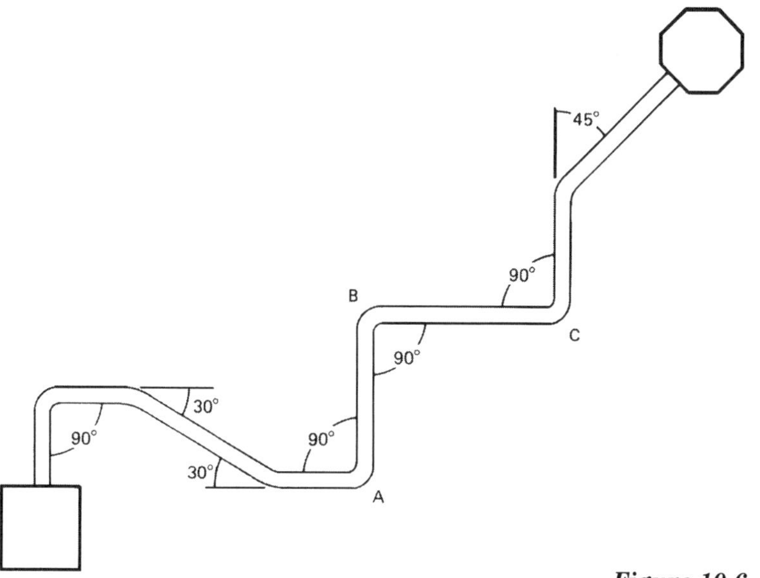

Figure 10-6
Too many bends between boxes in a FMC run

leave enough slack for fastening. *NEC* Section 348.30(B) requires you to secure flex conduit no more than 12 inches from each box, and then every 4½ feet between boxes, just the same as for both NM and BX cables.

The only limitation on bends in NM or BX cables is the radius of any single bend. That's not the case with Greenfield. *NEC* Section 348.26 limits bends to no more than 360 degrees in any one run from box to box or between conduit bodies, which can also be used with Greenfield. It requires careful planning at times to conform to this limitation. Greenfield is an empty armor that you have to pull wiring through — and the more bends, the more difficult this becomes. Figure 10-6 shows an example of a wiring run with excessive bends.

When you have several bends in a run, it's easier to fit and cut the flex conduit without stapling it into place. Then take it back out, lay it down and pull the wires through. Make sure you leave at least 6 inches of wire at each end (as required) for making connections when you cut the wires. Replace the Greenfield, tighten the box connectors and staple the conduit in place per code requirements. How to use a fish tape, and the proper techniques for pulling wires in conduit, will be discussed in the EMT section of this chapter.

Table 1 Percent of Cross Section of Conduit and Tubing for Conductors

Number of Conductors	All Conductor Types
1	53
2	31
Over 2	40

FPN No. 1: Table 1 is based on common conditions of proper cabling and alignment of conductors where the length of the pull and the number of bends are within reasonable limits. It should be recognized that, for certain conditions, a larger size conduit or a lesser conduit fill should be considered.

FPN No. 2: When pulling three conductors or cables into a raceway, if the ratio of the raceway (inside diameter) to the conductor or cable (outside diameter) is between 2.8 and 3.2, jamming can occur. While jamming can occur when pulling four or more conductors or cables into a raceway, the probability is very low.

From the *National Electrical Code* © 2007, NFPA

Figure 10-7
Percent of conduit that can be filled

NEC Chapter 9, Table 1, shown in Figure 10-7, gives the percent of the conduit that you can fill with conductors. This limits the number of conductors allowed in a conduit. When counting conductors to conform with box limitations and the conduit limitations in Table 1, don't forget that FMC isn't approved as an adequate ground path when the length of the path is over 6 feet. This means that, along with the required circuit wires, you have to pull a ground wire through each box-to-box run. This adds one to the conductor count in every Greenfield run and in every box.

Liquidtight Flexible Metal Conduit — LFMC

The code requirements relating to the use of liquidtight conduit are similar to those for Greenfield, except that it can be used where it'll be exposed to moisture. Its use in residential wiring is limited because it's an expensive choice. You'll most often use liquidtight conduit for very short final connecting runs to equipment, or in locations or applications where you need a conduit that can withstand vibration, water or both. It usually doesn't need fastening, and it's not likely to have bends that exceed 360 degrees in a run. And, since most runs will be less than 6 feet, it probably won't need a grounding conductor. It'll rarely carry anywhere close to its maximum conductor capacity.

> ***FOR ROUGH WIRING,*** cut liquidtight conduit with a sharp, fine-tooth hacksaw to avoid leaving a ragged edge. Then be careful to assemble the connectors correctly. Make sure you cinch them down tight so that the conduit will, in fact, be liquidtight and resistant to loosening due to vibration.

Electrical Metallic Tubing — EMT (Thinwall Conduit)

Producing clean, accurate, professional work using EMT requires much more skill and practice than any of the other materials used in rough wiring. While its use is limited in residential applications, it's used extensively in industrial and commercial wiring. Many industrial jobs are done entirely in EMT.

Box Mounting

Determine your box locations from the cable plan. Since one of the primary reasons for using EMT is that it provides good protection for exposed wiring, much of the time it's run exposed. As a result, the boxes where the runs terminate are usually surface mounted as well. The only boxes used with EMT are metal, and when surface mounted, you fasten them through the holes provided in the backs of the boxes.

If you're mounting a box on drywall at a point where there's a convenient wood stud, you can nail it. Most often, however, the boxes on drywall or masonry are placed at locations where nailing is impossible. In these cases, you'll need to use other types of fasteners, such as toggle bolts, or plastic or lead expansion anchors, like those shown in Figure 10-8. All of these fasteners require drilling before you insert the fastener. Use a slow-speed drill motor and carbide-tipped masonry drills.

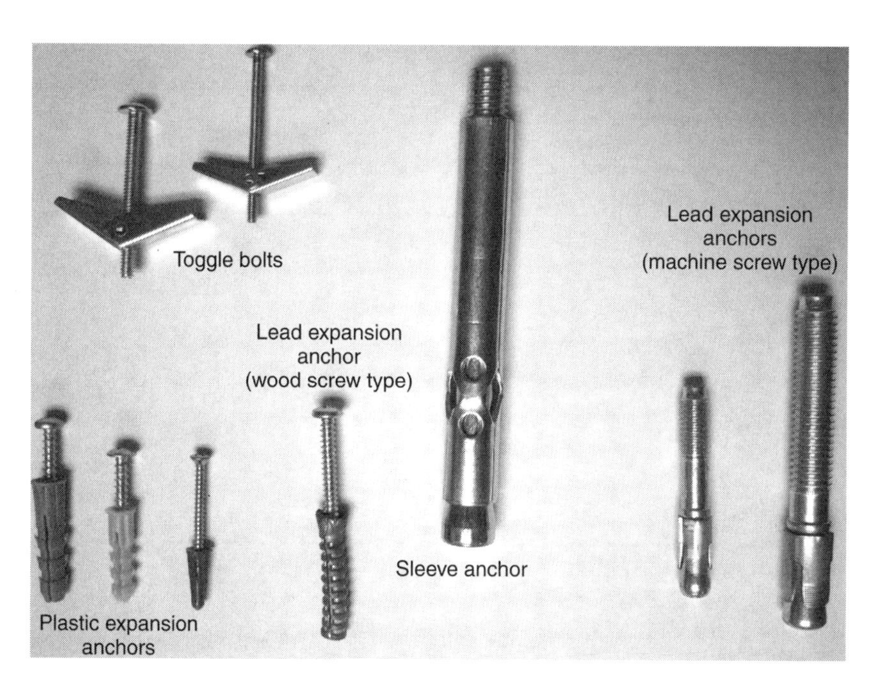

Figure 10-8
Masonry fasteners

Cutting EMT

EMT comes in 10-foot-long pieces that you can easily cut to length with an ordinary hacksaw. Use a 32-tooth, but nothing coarser than a 24-tooth blade. Otherwise, the saw will bind badly about halfway through. The most commonly used EMT size is ½ inch, which you can cut easily and cleanly with a tubing cutter. For larger sizes, you'll need a pipe cutter.

IMMEDIATELY after you cut a piece of EMT, *ream it* using a pipe reamer. Don't stop for coffee, lunch, or even the bathroom! Ream that pipe end before doing *anything* else. The inside edge of a cut pipe is very sharp. Anyone who fails to make it an absolute rule to ream every cut *at once* will eventually forget one.

If you pull wires through a conduit run with an unreamed cut, it's very likely that the insulation on one or more conductors will be damaged, if not completely torn, allowing wires to short out to each other or to the pipe. Only careful reaming of *absolutely every* cut will prevent this.

Bending

One big advantage that EMT has over other rigid conduit is the ease with which you can bend the smaller trade sizes to follow irregular contours. The majority of EMT work is done using ½-inch and ¾-inch trade sizes. You can easily bend these sizes using a conduit bender. Regardless of the manufacturer, any ½-inch conduit bender will be shaped to a curve matching the code-required conduit bend radius given in *NEC* Chapter 9, Table 2 (Figure 10-9) for ½-inch conduit, which is 4 inches. All ¾-inch benders will match the code-specified 5-inch radius. The radius of bend specified in the table depends on the trade size of the conduit. You can bend ½-inch conduit with a ¾-inch bender, but ¾-inch conduit can't be bent with a ½-inch tool. You can bend 1-inch and even 1¼-inch EMT using a hand tool, but anything larger will require a power bender.

Table 2 Radius of Conduit and Tubing Bends

Conduit or Tubing Size		One Shot and Full Shoe Benders		Other Bends	
Metric Designator	Trade Size	mm	in.	mm	in.
16	½	101.6	4	101.6	4
21	¾	114.3	4½	127	5
27	1	146.05	5¾	152.4	6
35	1¼	184.15	7¼	203.2	8
41	1½	209.55	8¼	254	10
53	2	241.3	9½	304.8	12
63	2½	266.7	10½	381	15
78	3	330.2	13	457.2	18
91	3½	381	15	533.4	21
103	4	406.4	16	609.6	24
129	5	609.6	24	762	30
155	6	762	30	914.4	36

Figure 10-9
Radius of conduit bends

The number of bends in a single continuous run between boxes or conduit bodies, such as ells (*L*s) or tees (*T*s), is limited to a total of 360 degrees, per *NEC* Section 344.26. There can be four 90-degree bends between boxes, or eight 45-degree bends, or any combination of angles (90-degree, 45-degree, 30-degree, 60-degree, etc.) as long as the total number of degrees in bends between boxes doesn't exceed 360.

Look back at the conduit run shown in Figure 10-6. This is an example of an excessive number of bends between boxes, and wouldn't be acceptable. A single 90-degree ell would have to replace the three bends at A, B, and C to make this run meet code.

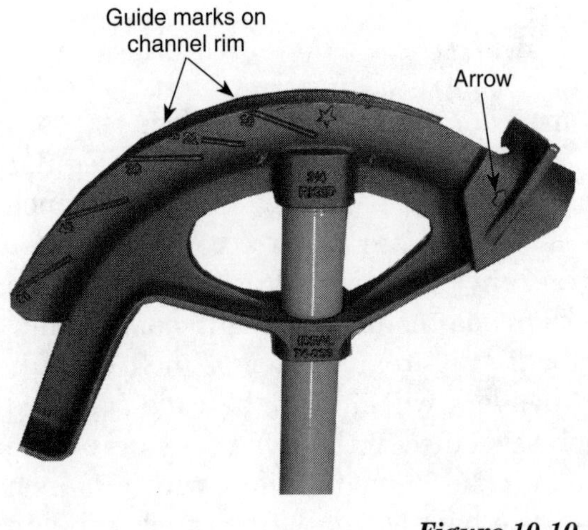

Figure 10-10
Conduit bender

The Conduit Bender

There are several different manufacturers of conduit benders, and you'll find as many models with differing details as there are manufacturers. As we pointed out earlier, basic benders are all very similar. The major differences are in the bending guides and aids that the manufacturer provides to assist the electrician in making accurate bends. One very helpful aid on the conduit bender in Figure 10-10 is an arrow close below the hook, plus two additional guide marks along the channel rim. The arrow is standard. The other two guide marks are specific to this manufacturer. To help you select a good tool, read the instructions accompanying it.

One manufacturer provides a pair of built-in levels to indicate 45-degree and 90-degree bends precisely. Another provides a system of guidelines cast into the tool to indicate 10-degree, 22-degree, 30-degree, 45-degree, and 60-degree bends, as well as right angles.

How to Make a Right-Angle Bend

1. First mark the conduit at the point where the bend is to start.
2. Lay the conduit on the floor.
3. Slide the bender over the conduit until the mark on the conduit is aligned with the start mark on the bender (see bender instructions to identify start mark).
4. Step firmly on the foot pedal and pull back on the handle simultaneously until you reach the 90-degree guide point.
5. Check the bend with a carpenter's square and adjust it with the bender, if necessary.

Other Angles

Follow the guidelines above to bend an angle for which the bender has a specific guide (such as a 45-degree level or an incised guideline), stopping at the appropriate guide point. If you need to make an angle for which the bender doesn't have a guide, use a common carpenter's tool called a "T bevel" to set the desired angle, and then match the bend to the T bevel.

Offset Bends

You're required to support EMT within 3 feet of each box or fitting and every 10 feet thereafter, per *NEC* Section 358.30(A). So, most EMT is run tightly against the wall. When it gets to the box, the knockout where it will enter must be ¼ inch (more or less) away from the wall. This requires an offset bend at the end of the conduit run, as shown in Figure 10-11.

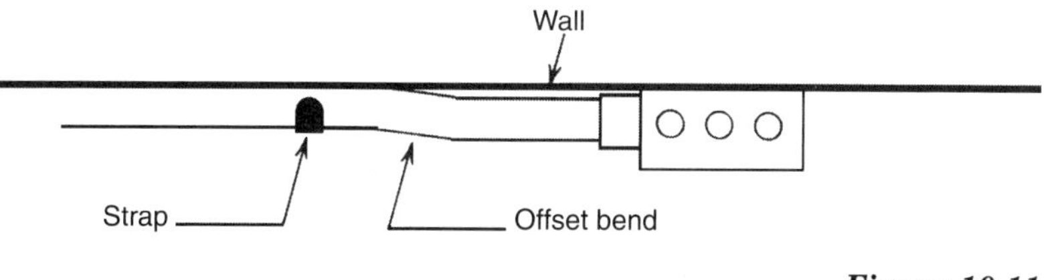

Figure 10-11
Offset bend into box

Figure 10-12 (a), (b) and (c) show how to make a shallow offset of this type. Measure from the conduit end about 4 inches and then make two marks 2½ inches apart (a). Place the bender on the second mark and bend up only 10 degrees (b). Now rotate

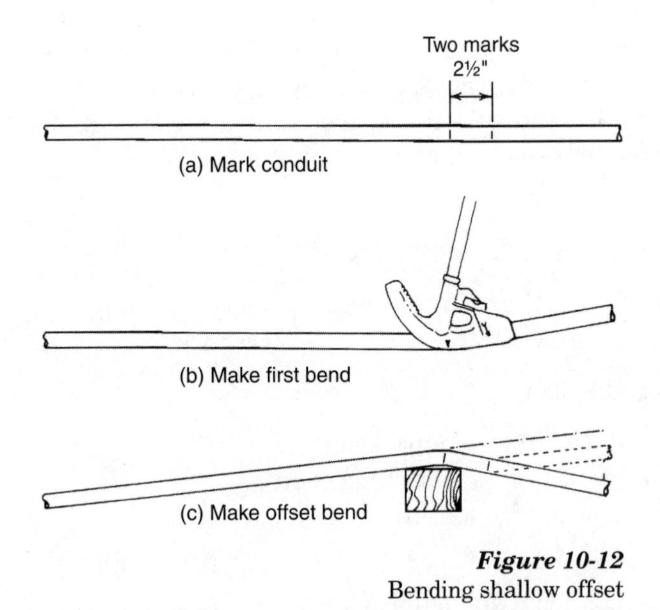

Figure 10-12
Bending shallow offset

the conduit 180 degrees and place it on a 2 x 4 on the floor so the first bend is pointing down. Move over to the first mark and bend up so the stub end is now parallel to the original run of the conduit but offset to the side (c).

Be careful to rotate the conduit exactly 180 degrees between bends. Otherwise, the offset won't be straight. If the offset is either too shallow or too deep to align exactly with the knockout, adjust it by slightly increasing or decreasing the bends.

To bend around obstructions, you may need to make much deeper offsets. To do this, increase both the bend angles as well as the distance between the bends. The table in Figure 10-13 lists the distances between bend marks when you're making offsets of various depths using 45-degree bends.

For various other bend angles and offset depths, the formula in Figure 10-14 will provide you with the distances between bends.

Using the formula in Figure 10-14, let's find the distance between bends for an offset depth of 7 inches using bend angles of 30 degrees.

Offset Depth	*Mark Spacing*
3"	4¼"
4"	5½"
5"	7"
6"	8½"
7"	9¾"
8"	11¼"
9"	12½"
10"	14"
11"	15½"
12"	16¾"
13"	18¼"
14"	19¾"
15"	21"

Figure 10-13
Distances between bend marks for offset bends

Constant Multiplier Formula

Offset Depth × Constant Multiplier for Bend Angle = Distance Between Bends

7 inches × 2 = 14 inches apart

Angle	*Multiplier*
10 degrees × 10 degrees	6
22½ degrees × 22½ degrees	2.6
30 degrees × 30 degrees	2
45 degrees × 45 degrees	1.4
60 degrees × 60 degrees	1.2

Use the constant multipliers above in the formula for the five common bend angles listed.

Figure 10-14
Constant multiplier

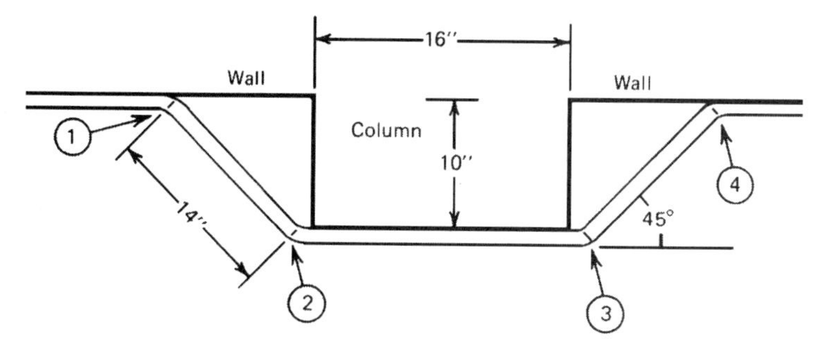

Figure 10-15
4-bend saddle

To pass an obstruction such as a beam or a column, you need to make a *four-bend saddle,* like the one shown in Figure 10-15. A four-bend saddle consists of two identical offsets with a connecting section between. For an offset depth of 10 inches using 45-degree angles, if we didn't have the distance already, the table in Figure 10-14 would supply us with a distance between bends of 14 inches ($10 \times 1.4 = 14$). That takes us to the front of the column. The same 14 inches with 45-degree bends will take us back to the wall on the far side of the column. The only information lacking to complete the saddle is the distance between bends (2) and (3) in Figure 10-15. After making the bend at (2), hold the conduit up to the column and measure between mark (2) and the column. That measurement doubled and added to the 16-inch width of the column is the distance between marks (2) and (3).

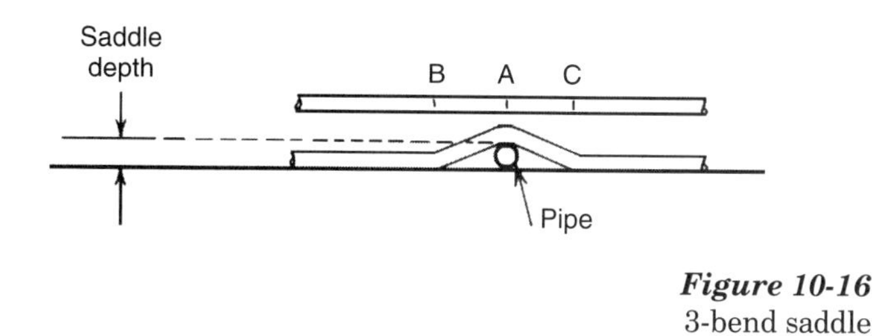

Figure 10-16
3-bend saddle

The most difficult part of making a four-bend saddle is keeping the whole thing straight. After the first bend, you must rotate the conduit *exactly* 180 degrees three times to make the other bends; otherwise, the piece will come out cockeyed. A pair of guidelines down opposite sides of the conduit will help immensely.

Saddle Depth	*Distance from Center*
1"	2½"
2"	5"
3"	7½"
4"	10"
5"	12½"
6"	15"

Figure 10-17
Distance marks from center for 3-bend saddles

To go around a smaller object, like a pipe, use a three-bend saddle, as shown in Figure 10-16. Make a 45-degree bend at (A), and two 22½-degree bends, one at (B) and one at (C). You need a special alignment mark on the tool for the center bend at (A) when making three-bend saddles. First make your 45-degree bend, then base the identical distances to the 22½-degree bends at (B) and (C) on the saddle depth. Figure 10-17 shows distances for some common depths. Use the usual start mark on the bender for the bends at (B) and (C). These bends are both made with the hook end of the bender pointing toward the center of the saddle. Again, as with the four-bend saddle, a pair of guidelines on opposite sides of the conduit will be a big help in keeping the piece straight through all three bends.

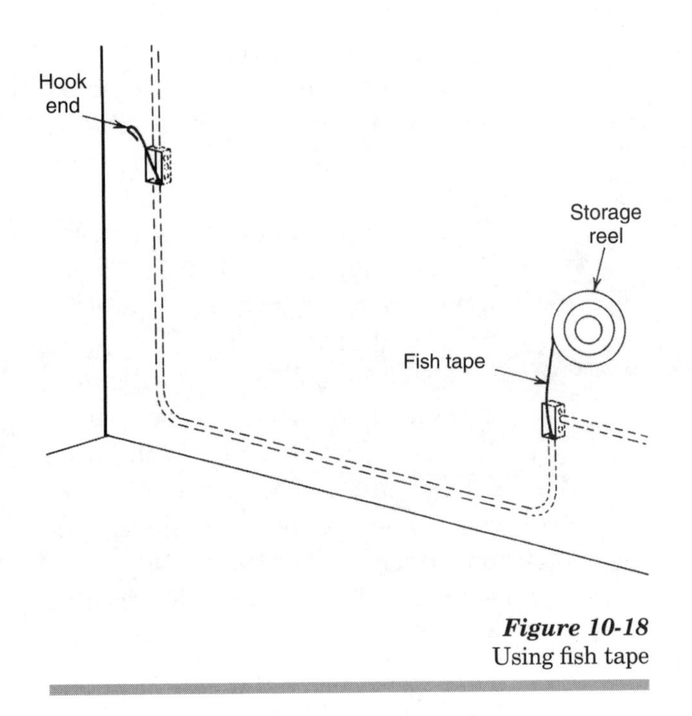

Figure 10-18
Using fish tape

Pulling Conductors in Conduit

Use a fish tape to pull conductors into conduits. Feed the tape from its reel through the conduit from one box to another; see Figure 10-18. There, attach the conductors to the hook end of the tape and pull them back through the conduit. It's easier to do this with two people. One person pulls on the fish tape while the second feeds the conductors into the pipe.

When you're pulling several conductors at once, which is usually the case, only one of them is passed through the hook on the fish tape. It's then doubled back on itself somewhere between 6 and 12 inches. Tape the other conductors to be pulled to the first one, and to each other, using plastic electrician's tape at intervals of about 3 to 6 inches for a distance of 1½ to 2 feet. The more conductors there are, the closer the tape intervals need to be and the longer the distance you need to tape them. This will insure that none of them fall out of the bundle and get left behind as you're pulling them through.

If it's a long pull or there are a large number of conductors being pulled (or both), before you start pulling, unroll enough conductor length from each wire spool to adequately cover the lengths you need. Stretch them out together on the floor and, with one man at each end, slap them hard against the floor several times. This will take the kinks out and straighten them, making the conductors easier to pull. Also, when there are many conductors or bends, have the man feeding the wires into the conduit lubricate them as he goes. There's pulling grease available for this purpose. It's a horrible yellow stuff, but it sure makes the job a lot easier for the guy at the pulling end.

When making a long pull that passes through several boxes (or a combination of boxes and *T*s or *L*s), at each pull point — whether a box, *T*, or *L* — you'll need to separate out of the taped bundle any conductors that are terminating at that point. Pull through lengths of each of the others sufficient to reach from that point to their final destinations. Then pull the entire lot through to the next pull point, and repeat the process.

At each box, separate the conductors that'll connect to devices or splices from those that simply pass on through. Don't cut any conductors that aren't required for connections. Leave a loop of each of them around the bottom of the box, and pass them on into the next conduit. Leave a foot of slack in each of the conductors that are to be cut, so that after cutting, each piece will be the 6-inch length required by *NEC* Section 300.14.

Surface Raceways — Wiremold

Where it isn't possible or desirable to conceal wiring, but where exposed EMT and its metal boxes are visually unacceptable, use a surface raceway such as Wiremold. It has an almost flat design, and though it may not be the most aesthetic material, it's still more appealing than EMT for exposed wiring.

Wiremold supplies a very complete selection of boxes, *T*s, inside elbows, outside elbows, flat elbows, twisted elbows, couplings, fastening straps, and adapter fittings for connection to standard concealed boxes, conduit, and other forms of conventional wiring.

Since you use slide-in flanges to connect this raceway and its various fittings to each other, start at a box and attach raceway lengths and fittings consecutively, in the same way you'd connect a run of threaded pipe. You can cut Wiremold with an ordinary hacksaw. After cutting, file the rough inside edges of every cut to prevent damaging the insulation when you pull the wires through. Each Wiremold fitting consists of two parts: the backing plate, which is flanged to slip into the raceway and contains the holes where you attach it to the wall; and the finished cover. Except for long runs or irregularities in the walls, only the backing plates are attached to the walls. The lengths of raceway are held in place by the backing plate flanges. Don't screw or snap the finished covers in place until all the wiring is pulled and the connections are completed.

Like EMT, but unlike Greenfield, Wiremold is its own ground. You don't need to pull grounding conductors through. As with all types of conduit, the number of conductors that you can pull through Wiremold is limited by the size of the specific raceway, and the size and insulation type of the conductors you're using. The limitations are given in Figure 10-19. Use fish tape to pull conductors through surface raceways in the same way as through EMT. You don't need any special equipment or procedures.

Identifying Rough Wiring

When rough wiring is completed, each box should contain conductors at least 6 inches in length, as required by *NEC* Section 300.14. When the time comes to complete the finish wiring, the conductors in most boxes should already be adequately identified simply by the color code used on the insulations.

However, in boxes containing several switches, particularly with a combination of single-pole and 3- or 4-way switches, some confusion can develop as to which conductors go where. This mix-up is further

Type of Raceway	Wire Size Gage No.	Number of Wires: Types RHH, RHW	Type THW	Type TW	Types THHN, THWN
No 200 (½ in., 11/32 in.)	12		2	3	3
	14		2	3	3
No 500 (¾ in., 17/32 in.)	8			2	2
	10	2	2	3	4
	12	2	3	4	7
	14	2	4	6	9
No 700 (¾ in., 21/32 in.)	6				2
	8		2	2	3
	10	2	3	4	5
	12	2	4	6	8
	14	3	5	7	11
No 5700 (¾ in., 21/32 in.)	8		2	2	3
	10	2	3	5	6
	12	2	4	6	9
	14	3	5	8	12
No 1000 (1 5/16 in., 15/16 in.)	6	2	3	3	5
	8	4	5	7	9
	10	6	9	13	16
	12	7	12	17	25
	14	9	14	22	34
No 1500 (1 9/16 in., 23/32 in.)	6				2
	8			2	3
	0	2	3	4	5
	12	2	3	5	7
	14	2	4	6	10
No 1900 (13/16 in., 9/16 in.)	12			3	3
	14			3	3
No 2000† (1 9/32 in., ¾ in.)	12			3* 3	3
	14			3* 3	3
No 2100† (1 1/4 in., 7/8 in.)	6	2	4	4	6
	8	4	6	8	10
	10	7	10	14	17
	12	8	13	19	28
	14	10	15	24	37
No 2200† (2 3/8 in., 3/4 in.)	6	5	7	3* 7	11
	8	8	11	7* 14	19
	10	13	19	10* 26	32
	12	15	23	10* 34	51
	14	18	29	10* 44	69
No 2800 (2 7/32 in., 23/32 in.)	6	2	3	3	5
	8	4	5	7	9
	10	6	9	12	15
	12	7	11	16	24
	14	9	14	21	33

†Figures for Nos. 2000, 2100, and 2200 are *without receptacles*, except where noted.
*With receptacles.

Figure 10-19
Conductors permitted in different sizes of Wiremold

compounded when, as often happens, the finish wiring isn't done by the same person who does the rough. The electrician who does the rough can save the electrician doing the finish a lot of time and irritation if he makes small sketches of all boxes that could be confusing to someone following behind him. It only takes very simple sketches, like the one in Figure 10-20, to make the job easier. You need a few lines to show where on the box each cable or conduit enters and a few words to identify where each one goes. This drawing, along with the cable plan, will be a great help to the finisher. Be kind to the finisher — remember, when the time comes, the person they send back just might be *you*.

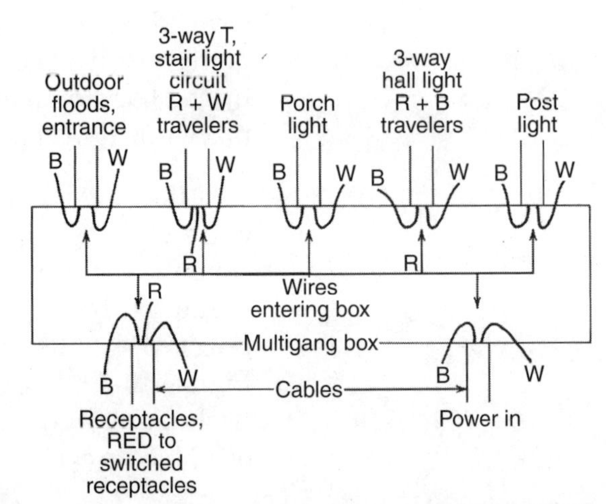

Figure 10-20
Sketch to record rough wiring of a large box

STUDY QUESTIONS

1. What do you need to determine the type and number of boxes required to wire a circuit?

A) The floor plan
B) The cable plan
C) The wiring diagram
D) The load calculation

2. How do you determine the number of conductors that can enter a box?

A) By the height of the box
B) By the width of the box
C) By the number of knockouts
D) By the box's internal cubic volume

3. The *NEC* requires that the first convenience outlet in a room be located how far from a doorway?

A) No less than 3 feet
B) No more than 4 feet
C) No more than 6 feet
D) No less than 10 feet

4. How far back should NM cable sheathing be stripped when entering a box?

A) 3 inches
B) 4 inches
C) 5 inches
D) 6 inches

5. What's an advantage of drilling when making a diagonal cable run across an accessible attic?

A) It leaves the joists clear so a floor can be installed later
B) It saves time on the installation of insulation
C) None, it's better to staple the cable run to the joists
D) Drilling is *only* an advantage when the cable runs perpendicular to the joists

6. What's the most difficult part of cutting BX cable?

A) Finding the correct saw blade
B) Not damaging the insulation
C) Cutting at the proper angle
D) Installing the fiber bushing inside the cut end

7. What's the major difference in the code regarding NM and BX cable?

A) They require different protection
B) The fastening requirements are stricter for NM cable
C) You can embed BX in plaster or masonry materials
D) The box mounting procedures

8. What's the major difference in the code regarding BX and Greenfield cable?

A) The limitations on bends
B) The fastening requirements
C) You can only use metal boxes with BX cables
D) The box mounting procedures

9. What's the *least* likely application for liquidtight conduit?

A) Connecting runs to equipment
B) Long residential runs
C) Runs subject to vibration
D) Runs in wet locations

10. Which type of material requires the most skill and practice to produce clean, accurate professional work?

A) BX cable
B) Wiremold surface raceway
C) Nonmetallic cable
D) EMT

11 FINISH WIRING OF NEW WORK

Finish wiring is one of the last tasks completed on a job. When the other trades are substantially finished, and all the walls are closed and finished in the manner specified in the plans, it's time for the electrician to return to install switches, fixtures and outlets, and wire in any direct-wired fixed appliances. Obviously, at this stage, all electrical work must be done with extreme care and neatness to avoid damaging the now-completed decorative finishes. Since the roughing in of the boxes and completion of the box-to-box wiring, all the installations have been inspected and approved. Now it's time to remove the drywall mud, paint globs, and other debris from those boxes, pull out and clean off the conductors, and prepare to complete the electrical system.

Receptacles

We'll start with the receptacles; the typical 120-volt common receptacles, plus the 240-volt receptacles needed for many of the fixed appliances in a home.

120 Volt

A typical 120-volt receptacle is shown in Figure 11-1. It consists of two three-hole sockets mounted on a metal strap called the *yoke*. Each socket will accommodate a standard 120-volt plug. These receptacles are called *duplex* receptacles, since they take two plugs. There are also single and triplex receptacles available, but they're not used as often.

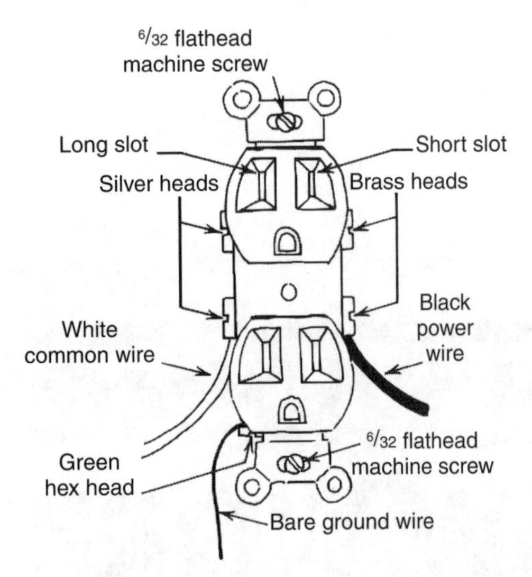

Figure 11-1
Typical 120-volt (end-of-the-line) receptacle

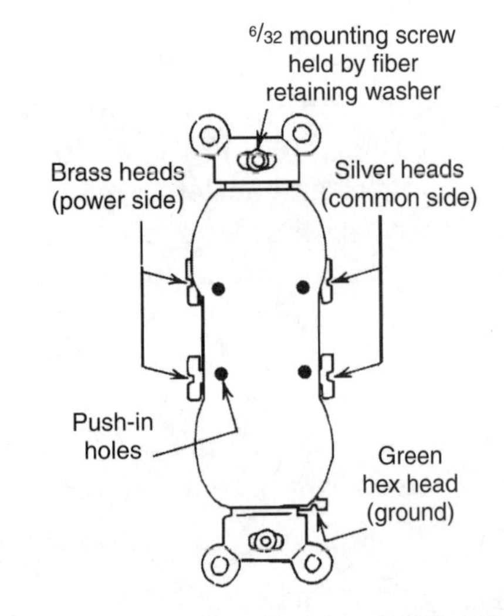

Figure 11-2
Back of 120-volt duplex receptacle

To connect the conductors to a receptacle, first strip ½ to ⅝ inch of insulation off each conductor. Most 120-volt duplex receptacles are furnished with two screw terminals (brass heads) on the power side, and two (silver heads) on the common side, as shown in Figure 11-2. The two brass screws connect to the power slots of the two sockets and to each other via a removable link.

The silver-headed screws similarly connect to the common sides of each socket and to each other via another removable link. So one connection on each side ties both power slots and both common slots into the circuit. Any end-of-the-line receptacle is connected like the receptacle shown in Figure 11-1. Although a receptacle is furnished with four terminal screws, you *only need to use two*. The ground wire (bare or green) connects to the hexagonal green screw at one end of the receptacle. If the wiring is EMT or Wiremold, you don't need a ground wire, as either one is code approved as its own ground.

A middle-of-the-line receptacle is connected between one line coming from the direction of the power source and another going on to another receptacle or other use point. Figure 11-3 shows the black power wires from both lines spliced to each other, as well as to a short black pigtail that's spliced to the brass power screw on the receptacle. The two white commons are similarly spliced to each other with a pigtail to a silver screw; and the two grounds spliced with a pigtail to the green hex-head screw.

Although there are two brass and two silver screws on the receptacle, only one of each is being used. The reason is that, by directly splicing the incoming and outgoing power and common lines with pigtails to the receptacle, subsequent damage to the receptacle, or even complete removal, won't interrupt the continuity of the rest of the circuit.

Most 120-volt receptacles are supplied with both terminal screws and holes for push-in connections. You can see both in Figure 11-2. At the risk of being considered old fashioned, it's my opinion that the screw terminal makes a significantly more positive connection than the push-in. While the push-in does reduce labor time when installing receptacles, it's electrically inferior to the traditional screw terminal.

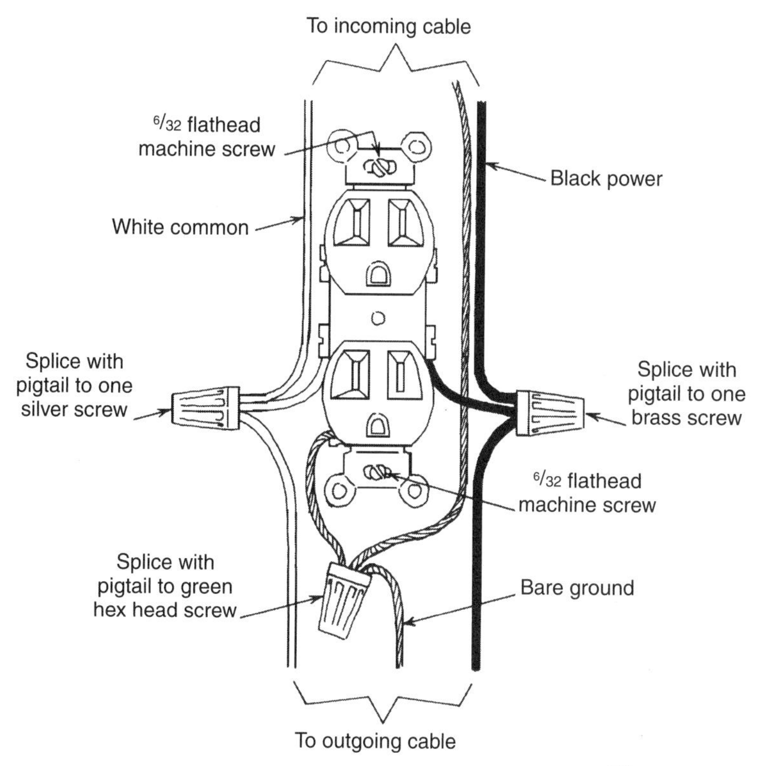

Figure 11-3
Middle-of-the-line receptacle

When you complete the splices and connect the pigtails, push the splices down to the bottom of the box to clear a space for mounting the receptacle. Get the grounds down first so they're below everything else. This minimizes the possibility of a stray loop of ground wire accidentally contacting a power terminal. Assuming that you had the proper 6-inch length of each conductor prior to making the splices (per *NEC*® Section 300.14), you'll have no difficulty separating the splices by pushing each wire nut down into a different corner, leaving only the pigtails standing up and available to connect to the receptacle.

Each new receptacle comes with two 6 x 32 flathead machine screws, which are held in the slots at the yoke ends by small square fiber washers. These screws fit the threaded holes provided for them in the device boxes. When you attach a receptacle to a *metal box, make sure you remove* these fiber retaining washers so that the receptacle yoke can make a good positive metal-to-metal contact with the box and insure proper grounding. In the case of a plastic box, the washers won't do any harm.

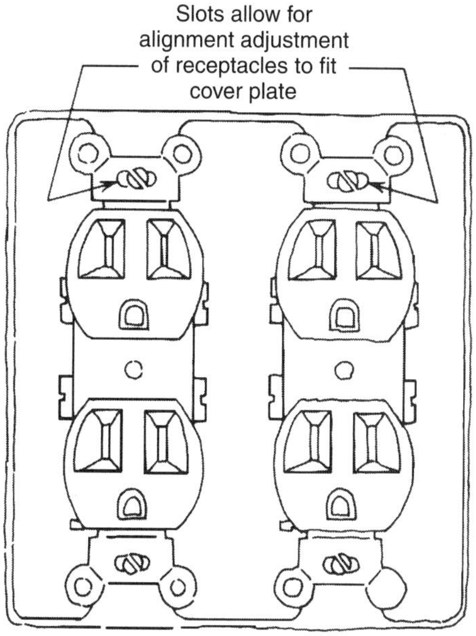

The slots for the mounting screws are there to allow you to align the receptacle, in case the box itself is somewhat cockeyed. After you firmly tightened the receptacle in place, complete the installation by adding the cover plate. Incidentally, the cover plate isn't a decorative option; it's required by *NEC* Section 314.25.

Slots allow for alignment adjustment of receptacles to fit cover plate

Figure 11-4
Two duplex receptacles in a 2-gang box

When you mount two or more duplex receptacles in a multigang box, as shown in Figure 11-4, you must correctly align them with each other or the required multigang cover plate won't fit.

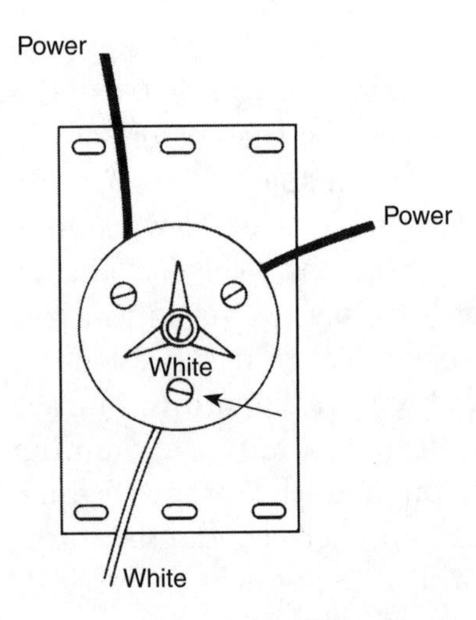

Figure 11-5
Back view of a 240-volt receptacle

240 Volt

The front faces of the 240-volt receptacles used for ranges and dryers differ; however, the connections that are made at the backs of both are the same. See Figure 11-5. In a normal installation, the screw terminal marked "white" will be down. Connections are made by inserting the circuit wires straight into the holes below the screw heads, and simply tightening down. As long as the white common conductor goes to the "white" terminal, either power wire may go to either of the other terminals. While 120-volt duplex receptacles are supplied with appropriate mounting screws for box attachment, the 240-volt range and dryer receptacles aren't. The electrician must furnish his own. The same cover plate requirement mentioned earlier also applies to box-mounted receptacles.

Weatherproof Receptacles

Where plans call for a weatherproof receptacle (you'll see them marked WP on the drawings), the receptacle part itself is the same as one used in any other location. They can be either 120- or 240-volt receptacles. The weatherproofing is provided by its attached cover plate. This plate has a neoprene gasket to seal the seam between it and the front edges of the box. The box also has spring-loaded covers with neoprene gaskets that fit over the receptacle openings. A weatherproof outlet is normally mounted in a weatherproof box. However, you can use any standard non-waterproof box for a weatherproof outlet if that box is embedded in an exterior wall so that the waterproof cover, when installed, will adequately seal it.

Switches

You can get single-pole, 3-way and 4-way switches with either push-in or screw terminals, or both. In my opinion, push-ins aren't any more desirable on switches than they are on receptacles. Regardless of how the conductors are physically connected to the switch terminals, the switching circuits will be based on one or other of those shown in Chapter 7.

The first step in finish wiring a switch box is to splice all the grounds together, and if it's a metal box, to clip a pigtail from that splice to the box to ground it out. If there are commons in the box, splice them to each other as well. Remember, commons never go to a switch. Just as in a receptacle box, push those splices down in the bottom of the box and out of the way before connecting the switch or switches. Then make your connections according to the wiring diagram. While you can't go too far wrong on a single-pole switch, the same isn't true of 3-ways or 4-ways, so always follow the diagram exactly.

> ***A WEATHERPROOF SWITCH,*** whether single-pole or 3-way, is wired just like any other switch. However, it comes already mounted on its cover plate with a weatherproofing neoprene gasket to seal the seams between that cover plate and the box on which it's mounted.

Switches have the same type of slot at the yoke ends as found on receptacles, and for the same reason — to allow you to adjust its position in the box. Like receptacles, most switches are furnished with 6 x 32 flathead mounting screws. Since there's no need to ground the switch yoke, you don't need to remove the fiber retaining washer when mounting in metal boxes, as you do with receptacles. When you've securely fastened the switch in place, place a cover plate over it to complete the installation according to code (*NEC* Section 314.25).

Photo-controls

Photo-controls are most commonly used to turn on various types of lighting at nightfall. Since they're normally activated by daylight, or the absence of it, your primary concern is to insure that they're oriented correctly for that purpose. They need to be shielded so that they won't be tripped out by other lighting during the night. Properly installed photo-controls are extremely reliable and extremely durable as well. They're generally mounted on outdoor boxes. Screw them into one of the threaded openings, and make connections to the wire pigtails furnished with the photo-control device.

Timers

A timer switch uses screw terminal connections. The terminals are color-coded with brass heads for power and silver heads for common. *Be sure to check the diagram* pasted inside the hinged cover to locate the correct input and output terminals. You would think that this step is too simple to require any mention, but it's astonishing how many times these devices are initially connected backwards. When this is done, the first time the timer switch cycles *OFF*, it not only shuts the load off but also shuts off its own clock, stopping everything!

Appliances

Appliances can be either direct-wired or plug in, depending on the type. Many appliances require 240-volt receptacles, and some require both a 120- and a 240-volt receptacle.

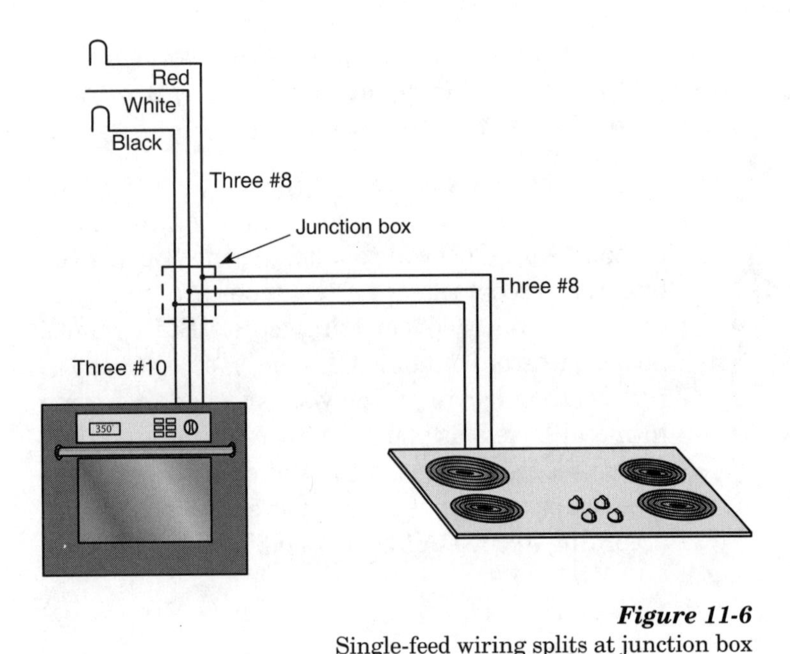

Figure 11-6
Single-feed wiring splits at junction box

Ovens/Ranges

Free-standing or drop-in type oven/ ranges are usually connected to the power supply by a plug inserted in a range receptacle. However, when the equipment consists of a counter-mounted cooktop with a separate wall-mounted oven, the two units are often directly wired to a common junction box or load center and fed from the main power panel. You can also run separate feeds for the cooktop and oven from the power panel — a 30 ampere with #10 wires to the oven, and a 40 ampere with #8 wires to the cooktop.

When you furnish separate power supplies for these appliances, each goes back to an appropriate 240-volt double breaker in the main panel. When the two are supplied by a single feed, it splits in two directions and a junction box (J box) is used, as shown in Figure 11-6. In the 4 x 4 box, splice the #6 or #8 feeders from the power panel to a set of #8s going on to the cooktop and a set of #10s connecting to the oven. The overcurrent protection at the power panel will be either a 40-ampere or a 50-ampere 240-volt double breaker.

WHEN THE COOKING appliances are gas-fired, you'll still need to furnish a 120-volt outlet to power accessories such as the spark ignition, oven light, timer, and clock.

If you're using a load center, it will be a small breaker box containing two 240-volt double breakers like the one in Figure 11-7. One will be a 30 ampere, for the oven, and the other will be a 40 ampere, for the cooktop. In this case, the subfeeders from the power panel feed the two breakers in the load center, which then individually protect the two appliances. The advantage of this arrangement is that, since each appliance has its own breaker, a problem occurring in one of them doesn't knock them both out of service.

Dryers

Many dryers plug into a typical dryer receptacle like the one shown in Figure 11-5; some, however, are directly wired. In either case, the wire size will be #10 and the circuit will be 30 amperes at 240 volts. You can use EMT, flexible conduit (Greenfield), or NM cable to carry the supply.

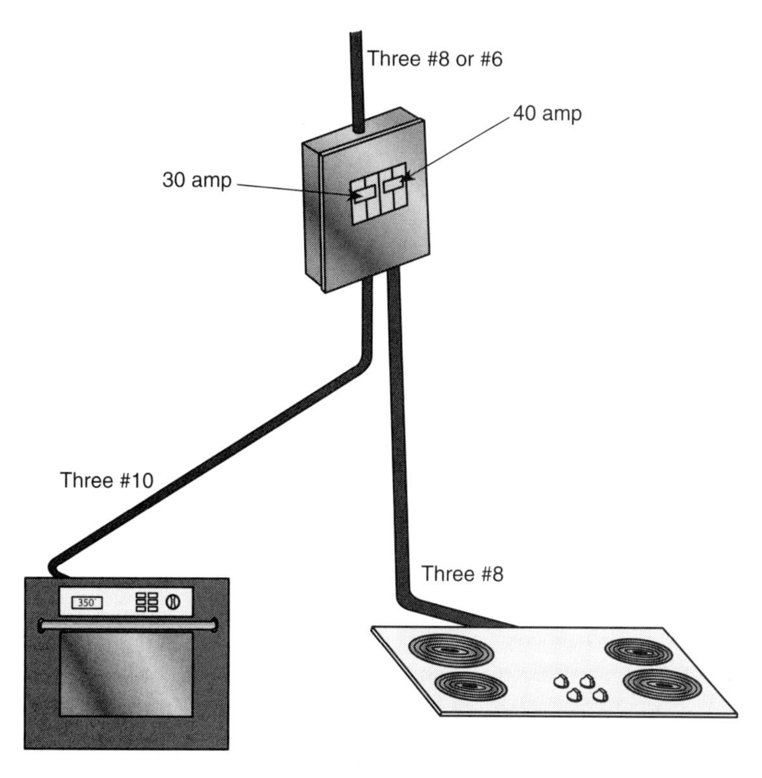

Figure 11-7
Power fed through two breakers in load center

If you use EMT, make a transition either at a junction box or through a transition coupling, so that the final connection to the dryer is through 3 or 4 feet of loose flex. This allows the dryer to be moved out for cleaning or servicing without disconnecting it. When you use cable or flexible conduit to make the entire wire run from the power panel for a direct connection, leave a loose loop of a few feet of cable or conduit at the dryer for that same purpose.

A gas dryer also needs a 120-volt supply to power its timer. You can use the same receptacle that operates the washer. Also, the gas pilot lights on newer gas dryers have been replaced with an *intermittent ignition device* (IID). This electrically-operated unit ignites the gas, saving a considerable amount of energy, as the pilot is only on momentarily.

Small apartment-size dryers are 120-volt units. The plug fits into any 120-volt convenience outlet. However, the heaters draw 10 or 12 amperes, and the motors a few more, so they'll approach, and in some cases even exceed, the capacity of a 15-ampere lighting circuit. To avoid difficulty, connect them to 20-ampere appliance circuits only. Any 20-ampere small appliance circuit will do, as well as the required 120-volt laundry circuit.

Water Heaters

An electric water heater requires another 240-volt circuit. To determine wire and breaker size, check the nameplate on the tank. The power requirements for the heating elements will be given in wattage. Water heaters have two heating elements, but the control is designed so that only one operates at a time. If the heating elements aren't the same size, use the maximum demand of the larger heating element.

Let's calculate the demand for a water heater with two elements rated at 2800 watts and 3600 watts respectively. The circuit will be designed for a demand of 3600 watts at 240 volts.

Using Watt's Law:

$$A = \frac{W}{V} = \frac{3600}{240} = 15 \text{ Amperes}$$

NEC Section 422.13 states that the water heater branch circuit rating must be no less than 125 percent of the nameplate rating of that heater.

125 percent of 15 = 18.75

You'll need a 20-ampere, 240-volt circuit using #12 wire for this water heater.

Water heaters are direct-wired, not plugged. On the top of the tank you'll find a small removable metal plate, held by one or two screws, containing the stamp for a standard ½-inch electrical knock-out. Under this plate are the wires to which you connect the incoming power. NM cable, armored cable (BX), or flexible conduit can be brought through the knockout, and secured to the plate with the usual box connector, as shown in Figure 11-8. However, if you want to use NM cable, first check with your local electrical inspection department. Some departments don't like this use of NM.

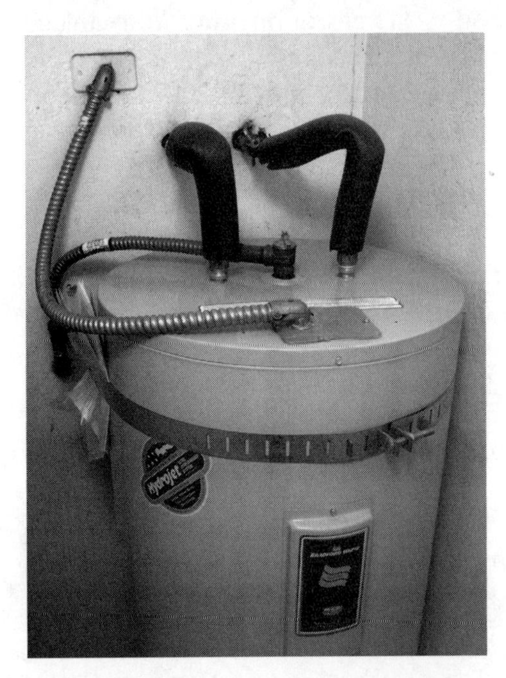

Figure 11-8
Top of water heater showing electrical connections

Disposals

Disposals are 120-volt appliances and can be either plugged or directly wired. While the code states in *NEC* Section 210.52(B) that a refrigerator can be connected to one of the two required small appliance receptacle circuits, the disposal isn't specifically mentioned here. However, in some jurisdictions, the disposal is permitted to be connected to a small appliance circuit. In other jurisdictions, a disposal is on a circuit by itself, or sometimes doubled up with a dishwasher. A disposal is considered an appliance, so despite the fact that it draws little power, and does so very infrequently, it still goes on a 20-ampere circuit.

For a plug-in connection, place a box containing a receptacle in the cabinet under the sink near where the disposal will be mounted. When you direct wire a disposal, use a cable or a short piece of liquidtight conduit. The cable or conduit enters the appliance through a standard ½-inch knockout and requires the usual box connector.

Dishwashers

The only finish wiring required for the dishwasher is the installation of the receptacle in the electrical box that was rough wired at the back of its undercabinet compartment. If someone blundered and put the box on the side of the compartment, move it to the back, or the appliance installers will never be able to get the machine in place.

Trash Compactors

As with the dishwasher, the only finish electrical work that a trash compactor requires is the installation of a receptacle in the box provided for it during the rough wiring.

Garage Door Openers, Attic Exhaust Fans, and Forced-Air Furnace Fans

Assuming the *roughing in* was completed correctly, the only finish work required for any of these appliances is, again, the simple installation of a receptacle.

Light Fixtures

Light fixtures can be recessed, surface mounted on a wall or ceiling, or hung from a ceiling. Each installation requires a different type of mounting box. We'll look at each type.

Incandescents — Surface-Mounted and Hanging Types

Where incandescent fixtures are going to be surface-mounted on a wall, or mounted or hung on a ceiling, octal metal or round plastic boxes are installed during the rough wiring. These boxes have two mounting holes on the front threaded for 8-32 machine screws. See Figure 11-9. To allow for the variety in fixture design, each fixture comes with an adapter plate. This plate is attached to the threaded holes in the box through the slots. The fixture is then secured to the adapter plate using its threaded holes into which fasteners furnished with the fixture are fitted.

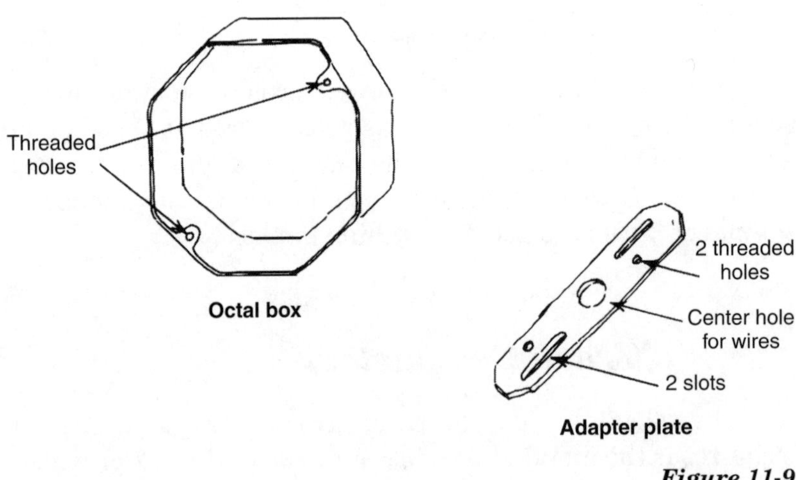

Figure 11-9
Octal box with typical fixture adapter plate

The reason for the slots in the adapter plate is that, although the roughing-in is generally done using standard 4-inch fixture boxes, from time to time 3-inch fixture boxes are used. The slots permit you to attach the adapter plate to either one. Adapter plates can vary in other ways, depending on the requirements of the fixture design. For instance, the spacing between the threaded holes on the adapter can differ for different fixtures. In some cases, the center hole is threaded for a 3/8-inch nipple and the two small threaded holes are omitted. In others, the two small threaded holes are replaced by one small hole in the center.

Each fixture is accompanied by a drawing or assembly instructions, or both. Believe me, it's *not* a sign of weakness or stupidity to read and follow these instructions. There's a distinct possibility that the people who made that fixture also know how to put it together and install it. Admittedly, instruction sheets are sometimes poorly written and confusing, but at least give them a try before striking out on your own!

Fixtures are often supplied with a pair of wire nuts considerably smaller than the smallest size used in the circuit wiring. This is because the fixture wires will be #16 or #18 stranded, and must be securely attached to a #14 solid. The smaller wire nut is necessary in order to make a tight connection. When splicing solid wires to each other with wire nuts, it isn't necessary to twist the wires together. When splicing stranded to solid, as when installing fixtures, first twist the stranded wires around the solid, then cap with the small wire nut. Make sure they both turn in the same direction. Twist the stranded wire around the solid wire clockwise, then turn the wire nut down on the two of them clockwise as well.

Ceiling fixtures are normally not too heavy to be supported by a correctly-installed fixture box. However, in the case of a large chandelier, you'll need to reinforce the box support. Instead of using the usual offset bar hanger to hold the box, nail a 2 x 6 crossways between the ceiling joists, then attach the box to that.

Recessed Incandescent Fixtures

No matter how elaborate or expensive it may be, a recessed incandescent light fixture, when reduced to its essentials, is simply a tin can facing down containing a very hot, not-very-well-ventilated, light source. It's important to install these fixtures to code primarily to reduce the danger of fire resulting from overheating.

Recessed light fixtures contain their own junction boxes, so no fixture boxes are provided for them in the rough wiring. When the plans call for power to feed from a switch to a single recessed fixture, there are no special requirements relating to the fixture junction box. However, when the plans call for the power to pass through several recessed lights in sequence, then the boxes on those lights must be approved for *through-wiring*, see *NEC* Section 410.21. Use only boxes identified as approved for through-wiring for this purpose. If it's not approved, *do not* use it for this purpose.

In order to prevent the temperature of recessed fixtures from reaching dangerous levels, *NEC* Section 410.116 requires that recessed parts should be spaced at least ½ inch from any combustible material, such as wood ceiling joists. In addition, there should be 3 inches of clearance between the fixture and any thermal insulation that's placed above the ceiling. This will prevent heat from being trapped in the fixture.

While trimming thermal insulation 3 inches around each recessed fixture will encourage heat dissipation from the light when it's operating, the resulting gap in the insulation will also allow cold air to enter when the light is off. The alternative is to use Type IC fixtures. These are identified and approved for contact with insulation, per *NEC* Section 410.116(A)(2). However, any fixture not specifically identified as Type IC must be installed with the proper clearances given in *NEC* Section 410.116(A)(1), Non-Type IC.

Fluorescent Fixtures

In response to growing energy conservation efforts, a good deal of incandescent lighting is being replaced with fluorescents. Furthermore, many more initial installations are being done using fluorescents than in the past. While fluorescents generate less heat than incandescents, a fact often overlooked is that they're also vastly more temperature sensitive.

A fluorescent fixture contains a *ballast*, which is a type of transformer. The ballast normally has a service life of about 60,000 hours. However, if its operating temperature exceeds 90 degrees C (194 degrees F), that service life will be reduced. The higher the temperature, the greater the reduction.

IF YOU INSTALL a recessed fixture in a clothes closet, check the clearance requirements between the fixture and storage space given in *NEC* Section 410.16.

At the other end of the spectrum, the tubes are designed for optimum performance in an air temperature of about 25 degrees C (77 degrees F). While dropping off somewhat, light output remains good down to about 10 degrees C (50 degrees F). Below that point, the light output decreases rapidly with decrease in temperature and, depending on variations in humidity, line voltage, fixture design, and other factors, the lamp may not light at all unless you're using special cold weather ballasts, lamps, or both.

So, when you do the finish wiring for fluorescents, pay particular attention to the effect the location of the fixture mounting may have on the operating temperature. The more distance you have between the ceiling and the fixture, the cooler its operating temperature. For example, by hanging a fixture 1 inch below the ceiling, the ballast will run 6 degrees C (10.8 degrees F) cooler than if it were mounted directly to the ceiling. At 2 inches below the ceiling, the temperature will be down by 14 degrees C (25 degrees F); and at 6 inches, the ballast temperature will be down 22.5 degrees C, or 40.5 degrees F, while the surrounding air temperature has remained unchanged. As you can see, in cases where the ballast operating temperature is close to or exceeding the recommended maximum of 90 degrees C, a small drop down from the ceiling can make a huge difference.

When the ambient temperature is too low, the solution isn't as simple. You can get special cold-weather ballasts that will start lamps at temperatures as low as –20 degrees F. These, coupled with 800 M.A., 1000 M.A., VHO, or Power Groove tubes, will often provide satisfactory low-temperature service. Another factor that'll help considerably is the shielding around the tubes. Totally enclosing the lamp in a wrap-around plastic shield can raise the ballast operating temperature by as much as 14.5 degrees C (26 degrees F).

Fluorescents, like recessed incandescents, need no junction boxes since the case is adequate. There's a significant difference between them, however, in that you can use any fluorescent for feed-through wiring.

Breaker Box

Conductors entering a breaker box should be left much longer than the 6-inch minimum required for junction or device boxes. You should leave them about 18 inches long, or more. The reason for this additional length, is that the distance between the point where a conductor enters a breaker box and its point of connection in that box can be considerable. In addition, it may have to go around a large block of breakers to get to that connection. Look back at the breaker box illustrated in Chapter 9, Figure 9-13. The commons in the two conduits entering on the right side of the panel box must go down to the bottom of the box, then across and all the way up the left side to reach the neutral bar, where they're finally attached.

All commons and all grounds go to the neutral bar, where you can attach them to any available terminal screw. Power leads (black, red, blue, etc.) only go to the breakers. Never attach more than one conductor per breaker, and be certain that the conductor is large enough to handle the amperage that breaker will pass. You can put #12 wire, good for 20 amperes, on a 15-ampere breaker; but never put #14, good for only 15 amperes, on a 20-ampere breaker. Keep all wiring neat and orderly so that if someone has to trace a problem or make changes or additions, he'll have no difficulty understanding the original installation.

NOTE The exception: if the ground-fault interrupter is included in one of the breakers, then, and only then, the branch circuit common wire (white or grey) goes to the terminal screw on the breaker marked with a white dot. The white wire furnished with the breaker then goes on to the neutral bar instead of the circuit common.

STUDY QUESTIONS

1. How many screw terminals do you need to connect to for an end-of-the-line receptacle connection?

A) One power terminal and one common terminal
B) One power terminal, one common terminal and one ground
C) Two power terminals and two common terminals
D) Two power terminals, two common terminals and one ground

2. How many screw terminals do you need to connect to for a middle-of-the-line receptacle connection?

A) One power terminal and one common terminal
B) One power terminal, one common terminal and one ground
C) Two power terminals and two common terminals
D) Two power terminals, two common terminals and one ground

3. What's one difference between installing 120-volt and 240-volt receptacles?

A) 240-volt receptacles are not provided with mounting screws
B) The terminal for the white common wire is white on a 120-volt receptacle
C) The terminal for the white common wire is not labeled on a 240-volt receptacle
D) Power wires can go to either terminal on a 240-volt receptacle

4. How does wiring for a switch box differ from wiring for a receptacle box?

A) Push-ins are more desirable in switch installations
B) You need to retain the fiber washer on the receptacle mounting screws
C) Commons never go to a switch
D) You can't weatherproof a switch

5. How would you wire for a separate cooktop and oven installation?

A) On a single 30-ampere circuit, with #10 wire to a 240-volt double breaker in the main panel
B) They are always wired separately, each to its own 240-volt double breaker
C) Subfeeders from the panel box feed a single breaker in the load panel, protecting both appliances
D) Each unit is directly wired to a common junction box fed from the main power panel

6. What is an IID?

A) An intermittent ignition device
B) An internal ignition device
C) A Class II dryer receptacle
D) A two-stage drying cycle

7. Which is true regarding clothes dryer installations?

A) A residential dryer may not be direct-wired
B) The washer and dryer may be on the same 40-ampere circuit
C) A gas dryer needs no electrical connection
D) A small apartment-size dryer can plug into a 120-volt outlet

8. How do you determine the wire and breaker size you need for an electric water heater?

A) Check the nameplate on the tank
B) Total the wattage demand of both heating elements
C) Use the wattage demand of the larger element
D) Calculate it on 125 percent of the combined wattage demand of both elements

9. The *NEC* specifically allows which appliance to be connected to one of the two required small appliance outlets?

A) A disposal
B) A dishwasher
C) A cooktop
D) A refrigerator

10. When installing fluorescent fixtures, what is one simple way to provide for longer ballast life?

A) Remove the ceiling insulation directly above the fixture to allow the heat to dissipate
B) Use a wrap-around plastic shield to insulate the fixture from variations in room temperature
C) Install the fixture a few inches below the ceiling instead of flush
D) Use Type IC fixtures

12

ADDITIONS AND ALTERATIONS TO OLD WORK

When doing new work, an electrician needs to know a good deal about the building structure in which he's working so he can determine how the power will be distributed from the entry location to the various required use points. His interest isn't, however, with the details of framing a wood stud wall. He simply needs to know how much wiring he'll need and where to run it. Once the framing is up, he'll run his wiring through whatever is there, all of which is open and accessible.

With old work, it's a different story. *Old work* refers to work in a building that's been finished. The term has nothing to do with the age of the building. The important difference is that the walls are now closed and finished. The framing, including any wiring or piping that's inside, is concealed and inaccessible. In order to add or change anything, you need to know exactly what's normally contained in that concealed structure.

Framing Types

Wood frame is the most common construction type and also has the most complicated internal structure. However, due to the requirements of the *International Building Code®* (*IBC®*), wood frame construction is also highly standardized. An electrician merely has to familiarize himself with the basic characteristics of two framing systems: platform or western framing, and balloon framing.

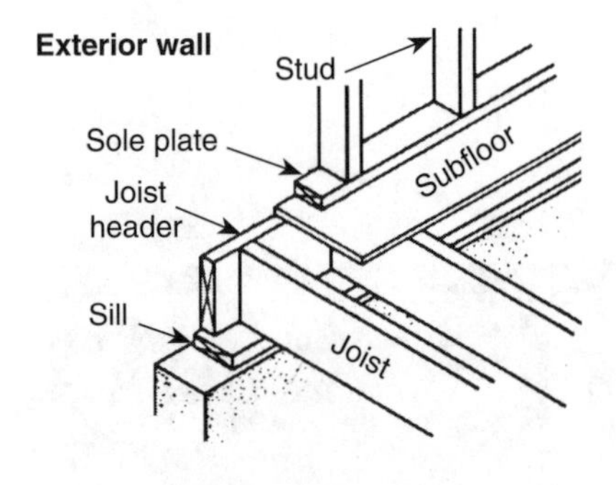

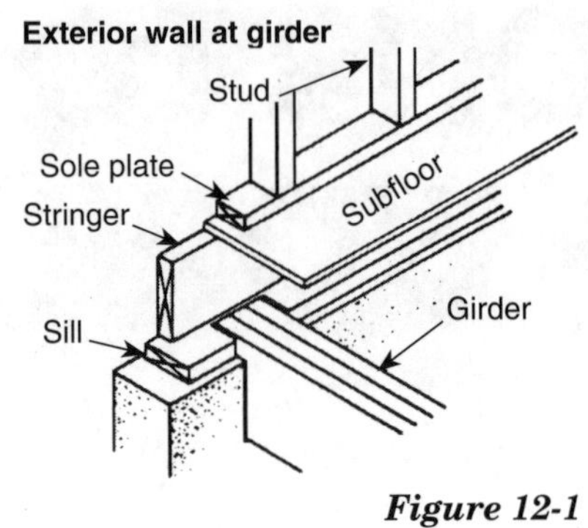

Figure 12-1
First floor platform framing

The basic parts — joists, studs, plates, headers, trimmers, and rafters — are common to both systems. Most of the material used to make these parts is of 2-inch nominal thickness (actual thickness is 1½ or 1⅝ inches), and of varying widths, the majority falling between 4 and 12 nominal inches.

Platform Framing

Both framing systems start with bolting a sill to the top of the foundation wall. Resting directly on the sill are the floor joists, which form the support for the subfloor. The header, placed across the joist ends, completes the framing of the platform, from which this system gets its name. The next step is to nail the subfloor to the top of that frame. Figure 12-1 shows the first floor assembly.

In Figure 12-2 you can see the perspective and a typical section of platform framing. The sole plate, which is the bottom framing member of the wall, rests on the subfloor. On

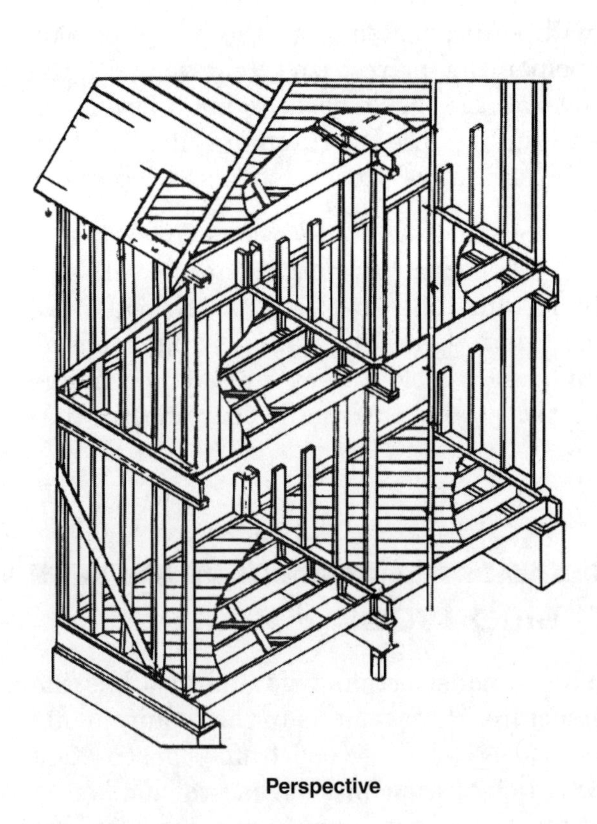

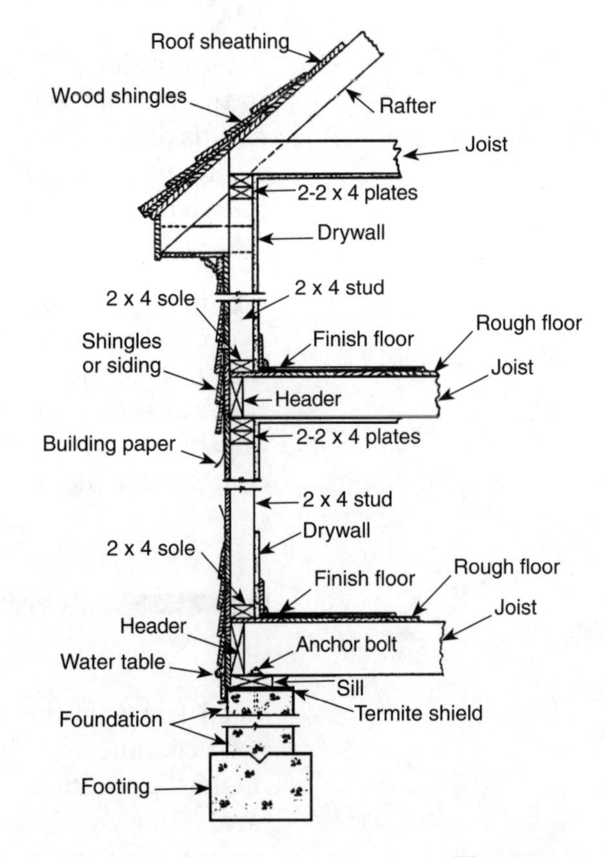

Figure 12-2
Typical section platform framing

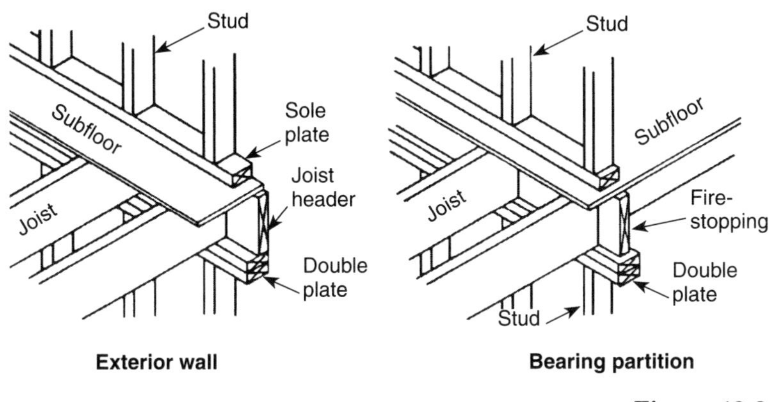

Figure 12-3
Second floor platform framing

the sole plate stand the wall studs, where drywall, plaster, or paneling will be attached on the inside, and any variety of surfaces on the outside. Placed above the wall studs is a doubled top plate. This doubled plate forms the support for the joists of the ceiling or next platform level, whichever comes after. In the case of a second floor, the subfloor placed on those joists forms the base for the sole plate, and then the wall studs of that next floor. The second floor details are shown in Figure 12-3. Each floor platform and wall frame is a repetition of the last one.

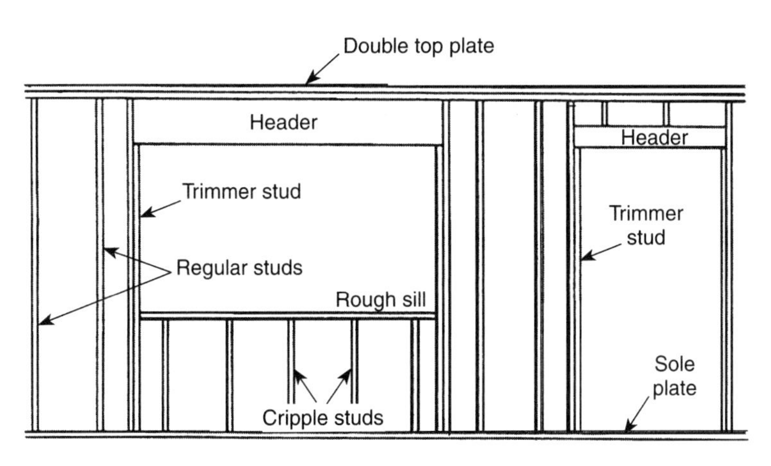

Figure 12-4
Framed openings in platform construction

The basic frame wall in platform construction, shown in Figure 12-4, consists of regular 2 x 4 wall studs spaced 16 inches on center. Door or window openings are framed with a regular stud on each side of the opening, which is doubled with a trimmer stud the height of the opening. The header sits on the trimmer, forming the top of the opening. In some cases, the header is made thick enough to completely fill the space between the opening and the lower side of the top plates. This saves the labor of cutting and installing the short stud pieces called *cripples* that would otherwise provide support between the header and the doubled top plate above.

To conserve energy, the spaces between *exterior* wall studs are filled with one of several types of insulation. The type and density of the insulation varies depending on the climate conditions where the building is located. So, when planning additions or alterations, it's important to remember that the spaces inside exterior walls are *not* going to be empty. This complicates the job of wiring in these walls.

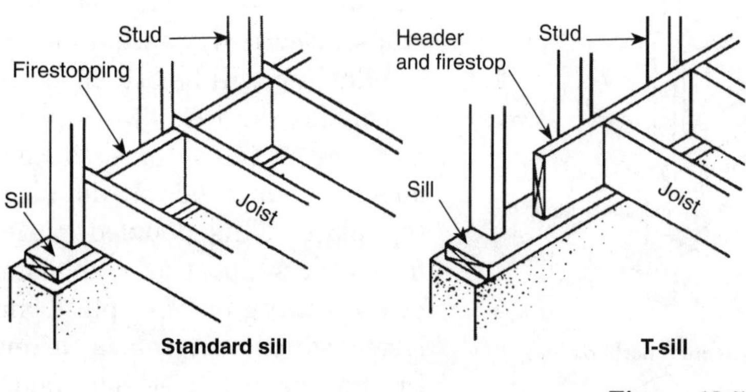

Figure 12-5
First floor balloon framing

Balloon Framing

Balloon framing also starts with a sill bolted to the top of the foundation. The floor joists rest on the sill, but in balloon framing, the wall studs also rest on that sill. See the illustration of the first floor level in Figure 12-5. In a *standard sill* arrangement, the joists and studs overlap and are nailed together. If a *T sill* is used, the joists and studs abut. In either case, the subfloor ends at the studs, instead of passing under them as it does in platform framing.

At the second floor, a piece called the *ribbon* is set into notches cut in the studs. The ribbon must be cut into the studs to support the second floor joists, because the studs continue uninterrupted up to the roof plate. The second floor joists rest on the ribbon and are nailed to the studs. See the typical section shown in Figure 12-6. Notice in the second floor details,

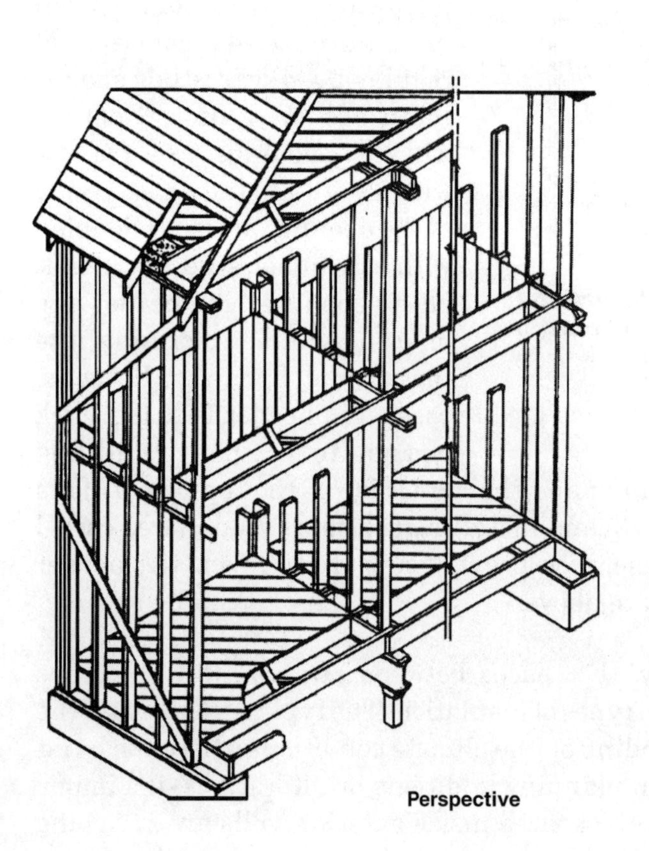

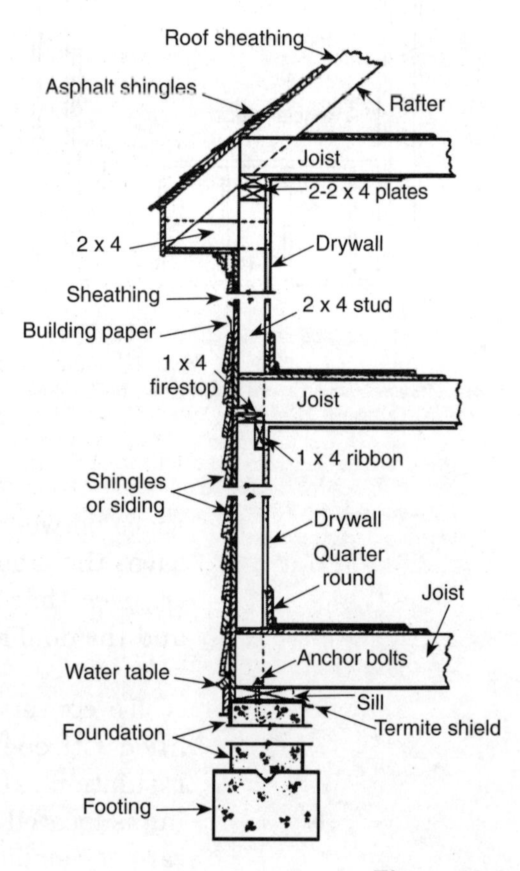

Figure 12-6
Typical section balloon framing

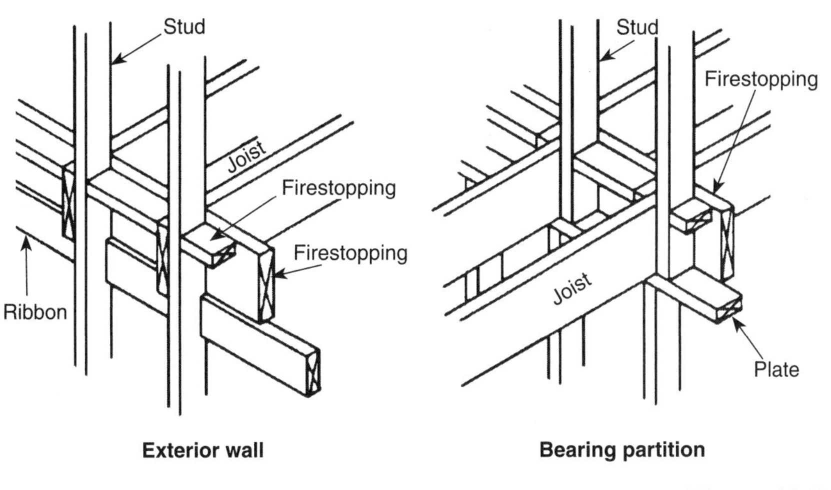

Figure 12-7
Second floor balloon framing

shown in Figure 12-7, there are firestops inserted between both studs and floor joists. These will be in the way of fishing wiring between or under floors. Firestops aren't needed in platform framing, except at bearing partitions, because the wall top plates and joist headers between floors perform that function.

The normal spacing of studs in balloon framing is 16 inches on center, the same as in platform framing. However, the openings for doors and windows are handled differently in balloon construction; see Figure 12-8. Instead of adding studs, trimmers, and a header cut to fit the required opening, the header extends beyond the required opening to the next standard-spaced stud. The opening is then rough framed from the header down, as shown.

The need for insulation in the exterior walls also applies to balloon framing. The same insulation materials are used for either construction method. The types of interior and exterior finishes are also the same for both platform and balloon framing.

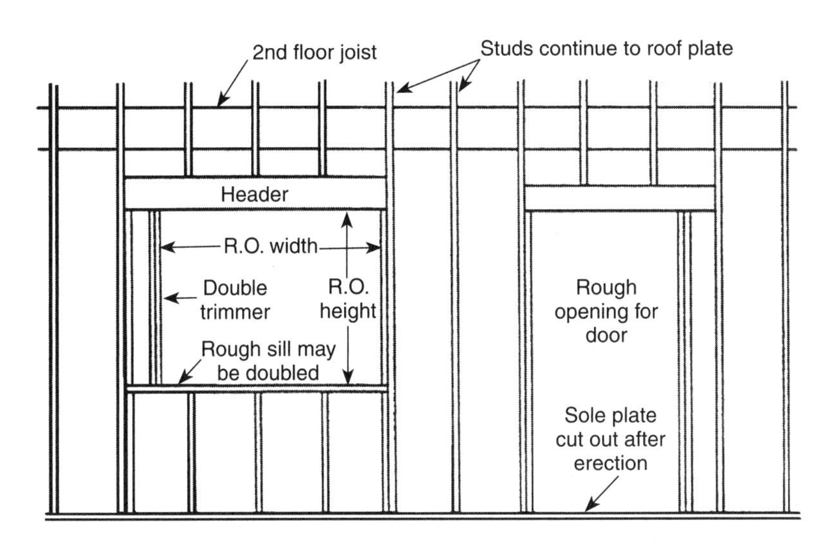

Figure 12-8
Framed openings in balloon construction

Concealing Electrical Additions or Alterations

The most commonly requested minor changes to an electrical system, other than replacing light fixtures, are adding receptacles, switches, and lights in new locations; or new switching for pull chain or self-switching fixtures, such as ceiling fans. In utility areas, you can use exposed EMT for these purposes, but most people don't want exposed cables in their living areas. They prefer to have you conceal new wiring in walls and ceilings or under floors, just like the original wiring. Since the walls are now closed and finished, this requires fishing wiring through inside voids and installing cut-in boxes.

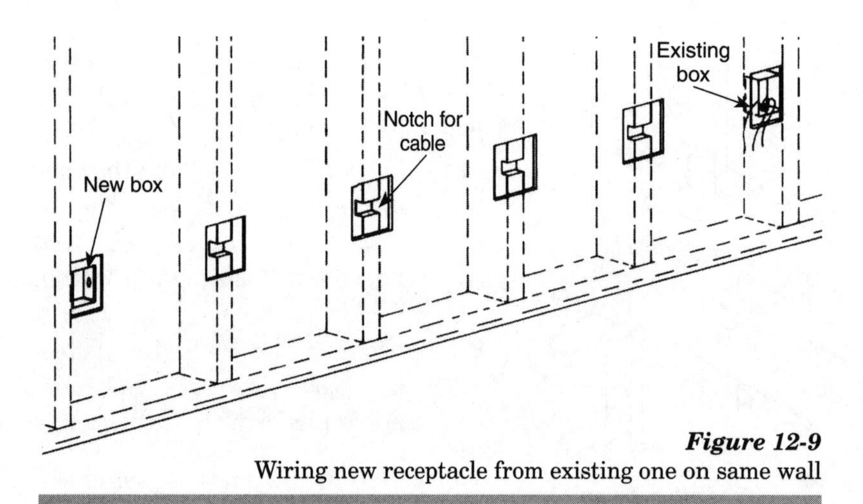

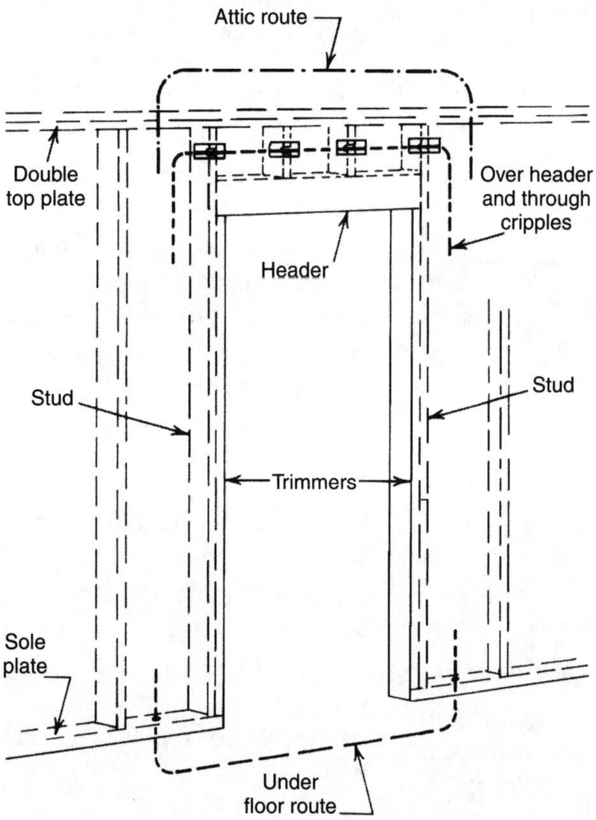

Figure 12-9
Wiring new receptacle from existing one on same wall

Adding a Receptacle

To add a receptacle powered from an existing one, first make a hole for a cut-in box at the desired location. With a stiff wire, feel back behind the wall for the first stud. Mark that distance on the wall to guide you in making a neat rectangular hole, about 4 inches wide and about 3 inches high, exposing the stud. At intervals of 16 inches on center, there should be studs all the way back to the existing box. Make a neat rectangular hole to expose each of them, as shown in Figure 12-9. Cut a notch in each stud to provide an open path for wiring from the existing box to the new one. Before you insert and secure the new box in its mounting hole, attach the box connectors and the new cable (Greenfield or BX) to it. Once it's in place, only its interior is accessible. When the new wiring is in place, either you, or whoever is designated in the contract, can patch the holes over the studs and repaint the wall.

Figure 12-10
Wiring around a doorway

If there's a doorway between the existing box location and the new one, it'll be necessary to go over or under it (Figure 12-10). When there's a basement or crawl space, making it possible to drill up through the sole plates, that's the shortest and easiest way. A slab floor makes this route impossible, so you'll have to go over the door, using the attic route. If you can't access an

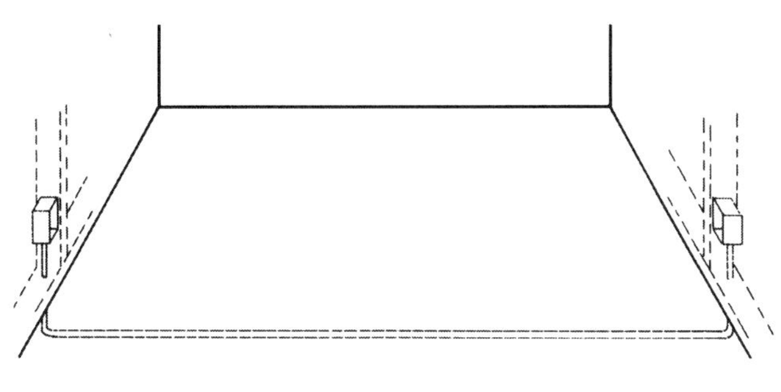

Figure 12-11
Running wiring under the floor

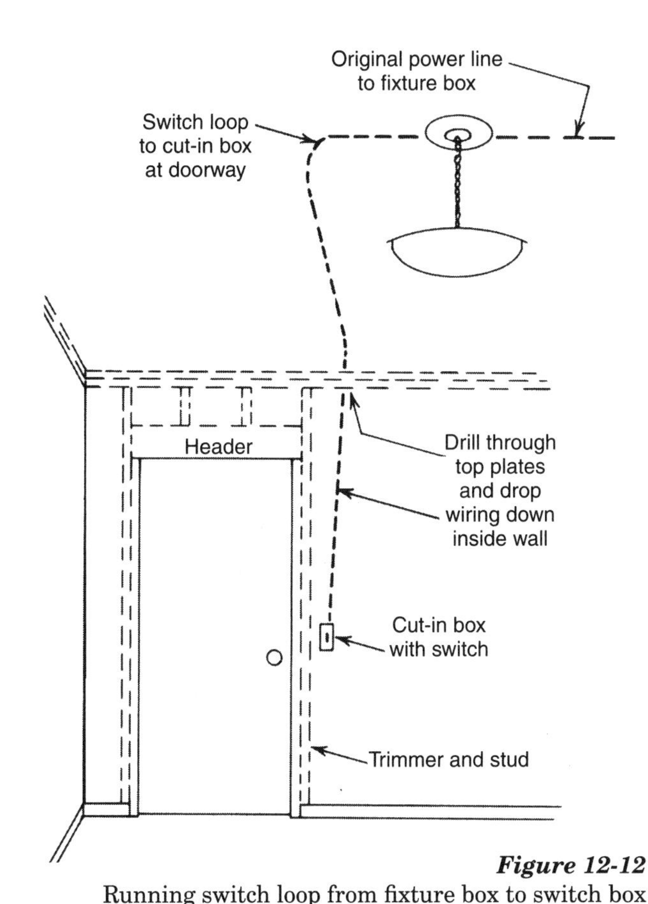

Figure 12-12
Running switch loop from fixture box to switch box

attic, the only other alternative is over the header and through the cripple(s) above it. In a case where a full header goes all the way to the top plates (not an uncommon situation), you'll have to go back to the drawing board and start again; you can't get through. A tip I learned the hard way: before you start any cutting, tap above the doorway *first* to determine whether a path exists.

If the new box location is across the room from the existing receptacle, and there's a crawlspace or basement, it may be easier to run the wiring under the floor to the new box. See Figure 12-11. This will again require drilling holes through the sole plates at the base of the walls on both sides. This should be done from below. If the floor joists run in the same direction as the wire run, then the wiring can be secured to the side of a joist. Otherwise, it'll have to be secured to the bottoms of joists, or passed through a series of holes drilled through them. Where the wiring is passing over a usable basement area, taking the time to drill holes in the joists is probably worthwhile; if it's only a crawlspace, then certainly not.

Adding a Switch for an Existing Fixture Location

In older buildings, you'll sometimes find light fixtures with self-contained switches. When modernizing, you'll generally want to replace these with fixtures that require wall switches. You can easily do this by running a switch loop from the fixture box to a new switch box, usually located on the latch side of a doorway, as shown in Figure 12-12. After you drop the wiring down into the wall from the hole in the top plates, pull it out and connect it through the box hole beside the door. If it's hard to find the wiring in the wall, make a hook from a wire coat hanger to catch it and pull it through.

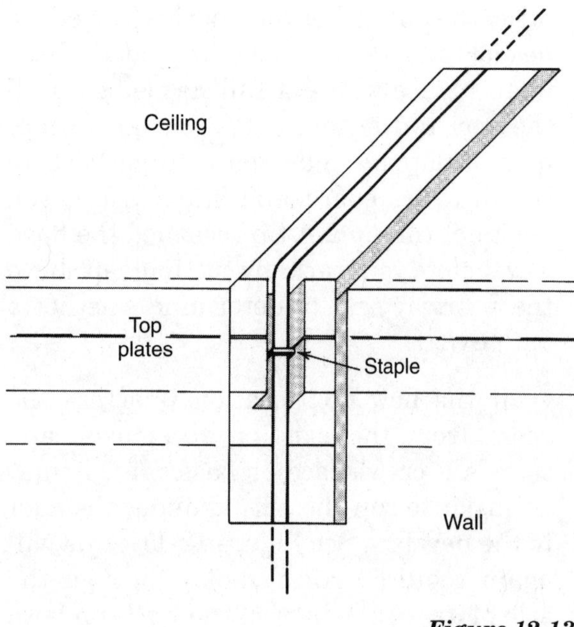

Figure 12-13
Cut access hole at top of wall to ceiling

If it's impossible to drill through the top plates from above, you can cut an access hole into the wall and out on the ceiling. See Figure 12-13. After notching the top plates, pull the wiring to and from this access hole with the fish tape. Then secure the wiring in the notch and patch the holes in the wall and ceiling.

Adding a Switched Fixture

To add a switched fixture in a new location, you need to install a new box for the switch. You also need a box on which to mount the fixture, *unless* it's a recessed fixture with its own junction box, or a fluorescent fixture, where the case is an acceptable junction box. Next, connect the switch to the fixture with a 2-wire run. The last step is connecting to a power input at one end or the other.

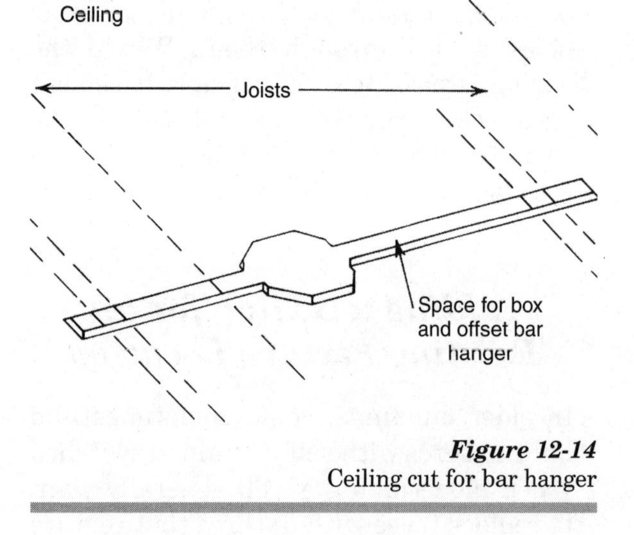

Figure 12-14
Ceiling cut for bar hanger

A cut-in box will readily fit the switch. For the fixture box, if the wiring is in NM cable, you can use a round plastic cut-in type box as well. If your wiring requires a metal box, octal cut-in boxes aren't available. You'll have to mount the metal octal from the attic above. If attic mounting isn't possible, make a hole for the box from below, then cut ¾-inch-wide slots in the ceiling from that hole to the joists on either side so you can insert an offset bar hanger to support the box. See Figure 12-14. After you've installed the bar hanger and box, patch the slots in the ceiling.

Run the wire from the fixture to the switch box across the ceiling, and then down inside the wall. If you can't go through the wall top plates to get inside the wall, use the access hole technique illustrated in Figure 12-13.

With both switch and fixture boxes installed and the 2-wire run connecting them in place, it's time to connect to the power input. If you installed the fixture box and connecting wiring from above, you might be able to locate a junction box containing unswitched power and run a power input from that to the fixture box. From there, complete the job by using the 2-wire switch box connecting run as a switch loop.

If you installed the fixture box from below, or if you can't find a suitable power connection in the attic, you'll have to pick up power from a receptacle box and use whatever route is necessary to take it to the switch box. If both boxes are on the same wall, then notching the studs along the wall between them, as shown in Figure 12-9, will get power from one to the other. Should you have to cross a doorway, then run the wiring around the door as we discussed; see Figure 12-10. Or, if the boxes are on different walls, the best wiring method might be the underfloor route illustrated in Figure 12-11.

Adding 240-Volt Circuits — Dryer, Heater, Air Conditioner, or Other

You need 240-volt circuits only when there are devices that require considerable power. Your first step before installing such a circuit is to determine whether the incoming service can handle the proposed additional load. This involves calculating the present loads on the system, adding the proposed load (or loads), and then comparing the total obtained to the rated capacity of the existing service using the methods we explained in Chapter 8.

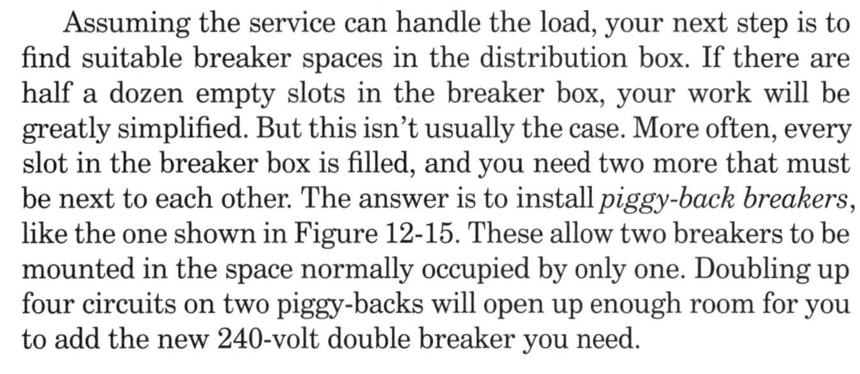

Assuming the service can handle the load, your next step is to find suitable breaker spaces in the distribution box. If there are half a dozen empty slots in the breaker box, your work will be greatly simplified. But this isn't usually the case. More often, every slot in the breaker box is filled, and you need two more that must be next to each other. The answer is to install *piggy-back breakers*, like the one shown in Figure 12-15. These allow two breakers to be mounted in the space normally occupied by only one. Doubling up four circuits on two piggy-backs will open up enough room for you to add the new 240-volt double breaker you need.

Figure 12-15
Piggy-back breaker

Both the rough and finish wiring of the circuit will be the same as for any other 240-volt circuit previously described, except that physically wiring and placing the boxes is going to be more difficult because you're doing the work in a completed building.

Adding a 120-Volt Lighting Circuit

Remodeling work that includes room additions, space conversion, or changes in the size, shape or use of an existing room usually requires additional lighting and convenience outlet circuits. Sometimes you can supply added rooms by using available capacity from one or more circuits supplying adjacent areas; but not always. More often, you have to add a new circuit.

The design of a new circuit will be the same as any of the ones we discussed in Chapter 8. It'll most likely be a 15-ampere circuit, #14 wire, and it'll be limited to a computed connected load of no more than 80 percent of its capacity, or 1440 watts. You do both the rough and finish wiring in the conventional manner. The only problem you're likely to have is that there may not be any additional space in the distribution box. However, that's easily solved by using a piggy-back.

Exposed Additions and Alterations

Sometimes you have to make electrical system alterations or additions in living areas with exposed wiring. While using EMT will make a really ugly job of it, Wiremold is generally quite satisfactory.

Wiremold surface raceway comes in 10-foot lengths, the same as EMT, and in several different cross-sections. The most commonly used sizes are 200, 500, and 700 raceways. The larger raceways (No. 1000 and higher) are used in office and commercial installations, and the saddles (No. 1500 and No. 2600) are for floor runs in offices, showrooms, stores, etc. There's a considerable variety of surface-mounting fixtures, devices, and switch boxes available with this raceway, adequate to meet most normal requirements. Unlike EMT, Wiremold raceway sections don't bend. You make changes in direction by cutting and joining raceway sections with elbows and *T*s, like the ones in Figure 12-16.

The raceway and the various boxes and connecting units that you use with it all consist of two parts: a base of galvanized steel, and a painted steel cover. The covers on all the boxes screw onto the base plates. On *T*s and elbows, the covers snap onto the base plates. The base plates of the fittings are punched for fasteners, and each length of raceway comes with a connector, which is also punched for fasteners. The bases of all fittings (boxes, *T*s, and elbows) have curved tongues that fit behind the base strips of the raceways.

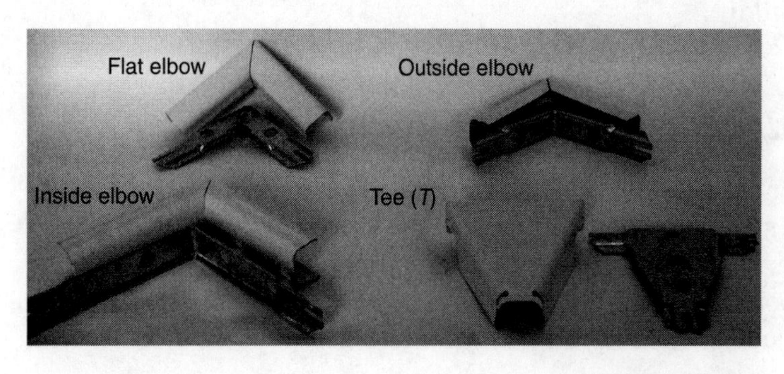

Figure 12-16
Wiremold connectors

All the parts of this system assemble progressively. For example, first fasten the base plate of a fitting, device, or fixture box to the wall. Then slip a precut length of raceway onto one of its tongues. Slip the base plate of another fitting into the far end of that raceway, and then fasten that to the wall. The raceway section is now securely held between the base plates of the two fittings. Next, connect another length of raceway to the second base plate. That raceway will be fixed in position by the base plate of another fitting that will be placed at its far end. Repeat this progression as many times as necessary for your installation.

Cut the raceways to size with a fine-tooth hacksaw. Since they can't be reamed, clean the cut ends with a small file to remove burrs from the sawing. There's a fiber bushing, similar to the one required for the ends of BX cable, available for use in Wiremold raceway ends. However, it's been my experience that light filing is the normal procedure, and the bushings are seldom used.

Those who've been reading carefully may have noticed that, so far, I've said nothing about putting covers on any of the boxes or fittings. What's on the walls and/or ceiling at this stage is a number of base plates for various fittings, connected to each other by lengths of empty raceway. That's because, before you install any covers, you need to pull the wiring through. Use the same fish tape that you use for EMT to do this job. Through the shorter raceway lengths, the fish tape won't be necessary, since #14 wire, the smallest that you can use, is still stiff enough to slide through without it. As with other wiring methods, the runs should be uninterrupted from box to box.

After the wiring is pulled, leave the usual 6-inch tails for making connections on all wires in all boxes. As when wiring in EMT or Greenfield, any conductors that merely pass through a box without making connections don't get cut. The finish wiring of switches and fixtures is the same as concealed wiring. The finish wiring may only differ when wiring the grounds on receptacles.

A correctly installed surface raceway wiring run, as just described, provides continuous metal-to-metal contact — from the tongues on the galvanized fitting base plates to the galvanized backing strips in the lengths of raceway joining them. This continuous metal-to-metal contact is considered by code an adequate safety grounding path — so you don't need to pull any grounding conductors in surface raceways.

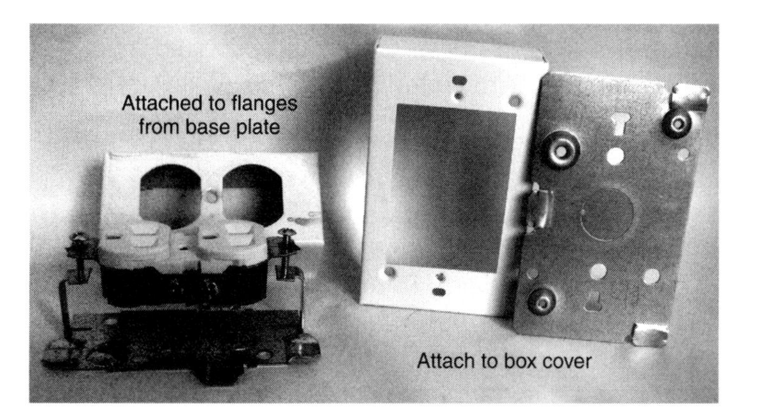

Figure 12-17
Wiremold receptacle boxes

Install receptacles on either of the two types of boxes shown in Figure 12-17. When you mount the yoke of the receptacle on the flanges that are part of the base plate, removing the fiber washers on the receptacle mounting screws will permit a direct metal-to-metal contact. This forms a positive ground path from the yoke to the base plate. However, when you mount a receptacle on a painted metal

Figure 12-18
Base plate and surface-mount adapter

cover plate, connect a wire pigtail from the ground screw on the receptacle to one of the fasteners holding the base plate to the wall to insure a good ground.

When installing switches or receptacles into these types of boxes, you must attach the box cover before mounting the device. Machine screws come with the box for this purpose. Be careful that they're not lost in the process of doing other preliminary work, as they're special screws and difficult to replace.

Adding Surface-Wired Receptacles

To add surface-wired receptacles powered from an existing box, start by removing an existing receptacle. Attach the base plate of a surface-mount adapter box in its place using the same mounting holes used on the old box. See Figure 12-18. To one or more of the connecting tongues on that base plate, attach the raceways and whatever fittings you need to reach the location, or locations, of additional outlets. Fish new wiring through the raceways and splice it to the existing wiring in the old box, along with pigtails to reconnect the old receptacle. With the raceways attached to the box base plates, remove one or more knockouts from each box cover before you mount those covers. Remember, per *NEC*® Section 110.12(A), you can't leave any unused openings in a box. Be careful to break out *only* the correct openings, because there are no plugs available for these boxes.

Adding a Switch for an Existing Fixture Location

Surface-mount fixture boxes, like surface-mount device boxes, are available in both standard type, with a full base plate, and adapter type, with a large round hole in its base plate. To add a switch to an unswitched fixture box, start with the adapter-type box.

Working from a connecting tongue on the adapter base plate, attach raceways and fittings to reach the location for the new switch. Pull a pair of wires for the switch loop, and complete finish wiring in the usual manner. If you want to install additional wall receptacles as well, pull a third conductor for the common down to the switch box. Connect the raceways from there to the receptacle locations and complete as described above.

THE DIFFERENCE between the adapter box and the standard-mount device box is the large rectangular opening in the base plate of the adapter box. Because of this opening, you can only use the adapter when attaching to an existing box.

Adding a Switched Fixture at a New Location

To add a switched fixture at a new location, use the fixture box with the solid base plate. If you'll be bringing power into the circuit at the fixture box, that base plate has a standard ½-inch knockout through which the wiring can enter. Connect the raceways from fixture box to switch box as previously described.

If power will enter at the switch box, it'll probably have to come from a nearby receptacle. In this case, move the receptacle forward onto an adapter-type box. You can now run raceways from there to the switch location; and from there to the fixture location.

STUDY QUESTIONS

1. **Why is it important for the electrician to know the type of framing used when doing old work?**
   A) He has to work around door and window openings
   B) Only one side of the interior walls will be open and visible
   C) He needs to know what's normally concealed inside the walls
   D) In platform framing there are no open spaces between studs

2. **Which type of construction has the most complicated internal structure?**
   A) Concrete block
   B) Wood frame
   C) Structural clay tile
   D) Brick

3. **Which framing systems does a residential electrician normally need to be familiar with?**
   A) Platform, western and steel
   B) Platform and western
   C) Concrete block, balloon and platform
   D) Platform and balloon

4. **What is required in order to power a new receptacle from an existing receptacle?**
   A) The *NEC* prohibits power from an existing receptacle being run to a new receptacle
   B) That the circuit still be sufficient to carry 125 percent of the maximum anticipated load
   C) That there are no more than six existing receptacles on that circuit
   D) That there be an acceptable path for the wire run from the existing receptacle

5. When a slab floor and an inaccessible attic make it necessary to run wire over a door, what should you check for before beginning to cut into the wall?

A) That there are cripple studs installed so you can support the wire
B) That there isn't a full header all the way to the top plates
C) *NEC* Section 250-119(B) generally prohibits wiring installed over doorways
D) That it won't create a run longer than permitted

6. When replacing a light fixture with a self-contained switch with a wall-switched fixture, what wiring method is recommended?

A) Run a switch loop from the fixture box to a new switch box
B) Run a 3-wire line back to the main junction box, then wire as for a new fixture
C) Add a grounded junction box on the top plate over the latch side of a doorway, then connect the new switch
D) An entire new circuit is required for this type of alteration

7. Which factor must be considered when adding a 240-volt range or dryer circuit?

A) The 240-volt wiring must be enclosed inside the walls
B) There must be no fewer than two empty spaces in the breaker box
C) The existing service must be able to handle the additional load
D) There must be an existing receptacle circuit to run power from

8. How can you add a new 120-volt light fixture if all existing circuits are at capacity and there are no spaces in the distribution box?

A) Double up the 240-volt breaker using a single piggy-back breaker in its place
B) You can't do it unless there's an existing circuit to handle the new load
C) Increase the load of one 120-volt circuit to 1620 watts
D) Add a new 15-ampere circuit with a single piggy-back in the distribution box

9. How can you make exposed wiring acceptable in residential use?

A) Exposed wiring is not permitted for use in the living areas of residences
B) Enclose it in Wiremold raceway
C) It can only be used in basements and laundry areas where it's permitted to staple the wires directly to the walls
D) Use EMT colored to match the walls

10. In what way is Wiremold different from EMT?

A) It's a system of steel parts that you assemble
B) It bends more easily
C) It's available in many different lengths
D) It requires a grounding wire

13

TROUBLE-SHOOTING AND REPAIRS

Common electrical difficulties fall into one of two general categories: either some part of the permanent building electrical system fails, or some piece of equipment using power from that system breaks down. Our concern here is with the first category.

Because of this, we haven't discussed the servicing of small plug-in appliances, except to the extent that they cause the electrical supply system to malfunction. And, our concern for the larger appliances, such as ranges, dryers, water heaters, ovens, microwaves, disposals, air conditioners, and baseboard or radiant heating systems, is also primarily when failures in these appliances cause difficulties with the building electrical system. The electrician who installs or maintains the power wiring in buildings isn't generally an appliance repairman. Appliance repair is an independent, and very extensive, field of work — far too extensive for us to address as part of a chapter in a book on building wiring. So our discussion of appliance repairs here is deliberately very limited.

Shorts

All branch circuits must be safeguarded by overcurrent protective devices, per *NEC*® Section 210.20. These overcurrent protective devices can be either breakers or fuses, with a current rating no greater than the ampacity of the wiring in the circuit. So, if a #12 AWG copper

Figure 13-1
Checking power supply through receptacle with a test light

wire with THW insulation is rated for 20 amperes, the breaker or fuse protecting that wire can't be rated more than the same 20 amperes. If the breaker were smaller, that would be permissible, but it can't ever be larger.

In the event any piece of electrical equipment fails to function, the first step in troubleshooting is to verify that it's getting power. While that may seem too obvious, it's astonishing how often I've seen time wasted puzzling over equipment connected to dead power supplies. What makes this doubly frustrating is the fact that testing the power supply is so simple. You can test the power supply using the AC scales on a voltmeter, or, even more easily, with a small pocket test light.

Test a plugged appliance by pulling the plug out and inserting test probes into the receptacle in its place, as illustrated in Figure 13-1. If the test light or meter probes fail to make a good connection with the interior contacts in the receptacle, you can get a false indication that the receptacle is dead. In that case, make absolutely certain the receptacle is dead by removing the cover plate and testing the terminal screws directly, as in Figure 13-2. If they also produce no reading, move on to the next step.

Figure 13-2
Testing receptacle at terminal screws with test light

Shorts in Appliances

When a receptacle is dead, check the breaker box for a tripped breaker. If you find one, switch it to the *OFF* position, and then try resetting it to *ON*. If it won't stay, but immediately drops back to the *tripped* position, *don't attempt to hold it at ON*, and *don't attempt to reset it again*. Either attempt will be useless; and holding it *ON* could be dangerous. Somewhere there's a short circuit. This means that at some point, the ungrounded *hot* side of the line is in direct contact with either the ground or common, or both.

If you're working in an old building, the overcurrent protective devices might be fuses. In residential structures, these are usually plug fuses. When a fuse is blown, the discoloration of the transparent top is obvious. The standard old Edison base plug fuse is made in three sizes: 15, 20, and 30 amperes. As we discussed earlier, this is a potentially dangerous situation because of the ease with which a 15-ampere circuit can be over-fused with a 20- or 30-ampere fuse. To prevent over-fusing (or under-fusing, for that matter), replace all standard plug fuses with S type plug fuses; see Figure 13-3. Only the fusible inner part of an S type plug fuse is removable, and the adapter, once installed, will accept only one fuse size. So, once a 15-ampere adapter is in

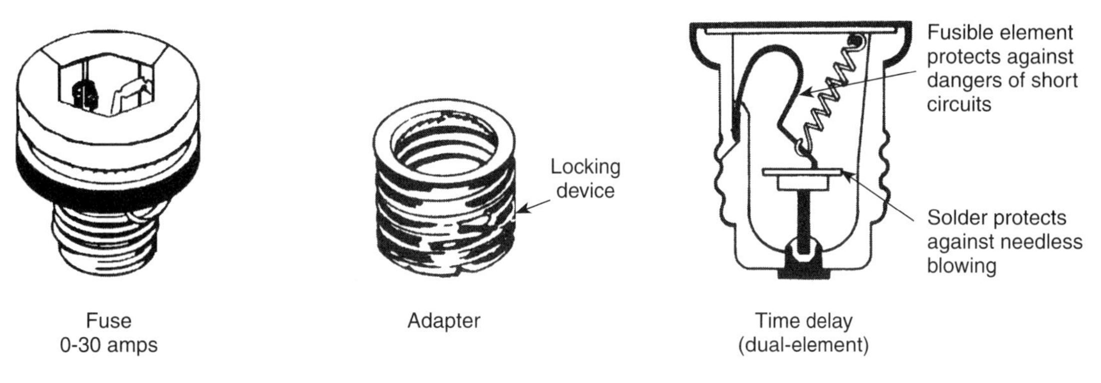

Figure 13-3
S type plug fuses

place, that circuit can no longer (either accidentally or on purpose) be over-fused with a 20- or 30-ampere fuse. Many jurisdictions now require all standard plug fuses to be replaced with S type.

In fused disconnects or fused switches, the fuses are generally cartridge types, with either ferrule or knife blade contacts, like those illustrated in Figure 13-4. With these fuses, there's no visible indication whether the fuse is good or bad. You have to test them if you think they're bad. You can use a voltmeter to test them in place with the power on; or you can test them separately with an ohmmeter. Testing in place with power on, the voltmeter will read zero volts across the two ends if the fuse is good. A blown fuse will show 120 volts difference between the two ends. A good fuse tested separately with an ohmmeter shows continuity (0 ohms resistance) between the two ends; a bad fuse will show no continuity (infinite resistance).

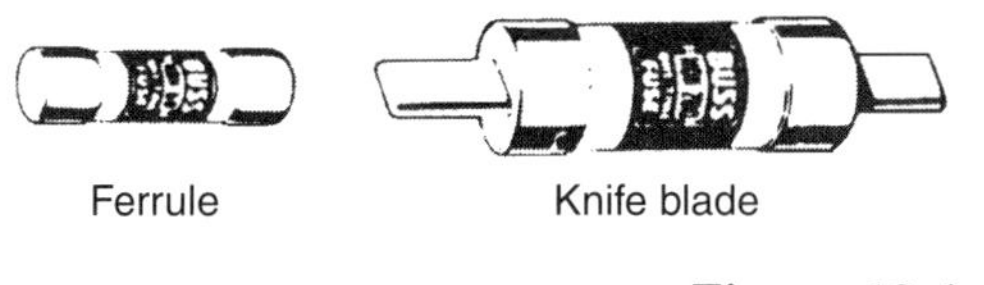

Figure 13-4
Cartridge fuses

After checking the breaker box, if it appears that there's a short, the next step is to try to isolate it as either in some piece of equipment or in the building wiring. Unplug or disconnect all readily-removable equipment from the circuit, turn all the building switches on, and try the breaker again. If it then holds in the *ON* position, the short is in one of the disconnected units. If it still won't stay on, the short has to be in the circuit wiring or in one of the permanently wired-in devices, such as a receptacle, a switch or a fixture. There's one last check you can make before believing absolutely that there's a short in that circuit. Disconnect the power wire from the output terminal of the breaker and, using an ohmmeter, test from the power wire to the ground buss in the breaker box. A zero reading eliminates breaker failure, and indicates the certain existence of a short somewhere out in the circuit.

If, instead of a breaker, the overcurrent protective device is a plug fuse, test the circuit by replacing the blown fuse with a new one only after disconnecting all power-using equipment, as just described. Of course, if the short is in the building wiring, the new fuse will promptly blow and follow the first into the trash barrel. You can avoid unnecessarily blowing out a second fuse by shutting off the main disconnect to the fuse box, and testing the circuit with an ohmmeter for continuity between the power wire coming from the fuse and the ground buss, as described earlier. If the meter reads a very low resistance, there's still a light bulb or two in place somewhere. If it reads absolutely no resistance, there's a short at some point in the circuit.

When, after the removal of the equipment from the circuit, the breaker holds at *ON*, or if it's a fused circuit, the fuse holds, the problem is a matter of appliance or equipment repair. That's beyond the scope of this book.

Building Circuit Shorts

Wiring problems in the run between boxes inside a wall, under a floor, or across an attic are possible, but unlikely. After examining the box carefully for any indications of defects, test every entering cable for shorts with the ohmmeter. To do this, you'll need to open all the splices so you can check each cable completely independent of the others. Usually, when you're testing cables this way, if there are two or three or more cables entering a box, one of them will show zero resistance between its conductors (indicating the short), and the others will indicate there's no problem. If you know or can determine the sequence of boxes in the circuit, tracing the short is a simple matter of following the shorted cable to its next box, and its next after that if necessary, until you find the short.

THE VAST MAJORITY of shorts occur inside an electrical box — either in the wiring that's been stuffed in, or in a wiring device or fixture mounted there. Tracing down the short requires going through the circuit from box to box, opening each one.

When you don't know or can't accurately reconstruct the circuit sequence, tracing a short can become a time-consuming and frustrating exercise. You have to continue in hit-or-miss fashion until you find the trouble — and that's just contrary to the way you'd normally like to do electrical work.

Overloads

The same overcurrent protective devices (breakers or fuses) that are activated by short circuits are also activated by overloads, but the indicators of overload are rather different from a short. In the case of a short, the tripped breaker absolutely won't reset until the short has been removed from the circuit. When an overload trips the breaker, it *will* reset after a few minutes cooling time. How long it will hold before it trips again varies, depending on the magnitude of the overload and the make of the breaker. It might last only a minute or two, or it might hold for 5 to 10 minutes. However, it's certain that as long as the overload remains connected to the circuit, the breaker will trip again, but there'll be a perceived time lag. The immediate and obvious difference between a short and an overload is that, although the breaker won't hold the *ON* position in either case, it can temporarily be reset when the circuit is overloaded — but not when there's a short in the line.

You'll notice the same time difference in a fuse-protected circuit. A short will instantly blow a fuse even as it's being inserted. An overload will permit a time lag, even if it's only quite short.

WHAT HAPPENS during an overload is that the circuit is required to deliver more power than the wires and the breaker are rated to supply.

Breakers, and some fuses, will permit temporary overloads — such as the starting surge of tungsten lights — but when the overload persists, the breaker trips or the fuse blows. Increasing the capacity of the breaker or fuse is *not* the proper solution.

Replacing a 15-ampere breaker or fuse with, perhaps, a 30-ampere is a deceptively simple *wrong* answer. While it's likely that the breaker or fuse will hold the circuit on, the wiring and its insulation will be operating under overloaded conditions, which will cause the wires to overheat. The wires won't get hot enough to melt the metal, but they can get hot enough to melt the insulation.

If that happens, you can have arcing between a now inadequately insulated power lead and a common wire or ground. Another possibility is that melted insulation will allow the hot wire to directly touch the common or ground, and this *will* short out the circuit. Neither of these are desirable scenarios, but of the two, the arcing is the more dangerous. A dead short will activate the circuit overcurrent protective device, which will shut off the power. In a situation involving arcing, the overcurrent protective device might not be activated, permitting the arcing to continue for some time. This situation can and has produced some very destructive fires.

The only real correction for an overload is to reduce the power demand on the circuit until the current being drawn falls within the range that the circuit is rated to supply. This means one or more items must be removed from the overloaded circuit and transferred to another.

In order to decide what load (or loads) to transfer, first you need to know the total amount of the overload. You can determine this directly with a meter, such as an Amprobe, or by calculation. If you have an Amprobe or other clamp-on field sensing ammeter, clamp it around the power lead coming from the breaker, switch the breaker on, and quickly read the amperage being drawn. The breaker will usually hold long enough for you to take a reading. Now, with the meter reading available, inspect the area served by that circuit, and list each load individually. Total the listed loads, then compare that total with the meter reading. If the sum of loads is significantly lower than the meter reading (0.5 ampere or more), you've missed a load. If the sum is significantly higher, you've obviously mistakenly included some load belonging on another circuit.

Loads & Wattage		*Amperage*
1 @ 200 watts =	200 watts	1.67 amperes
1 @ 350 watts =	350 watts	2.92 amperes
4 @ 100 watts =	400 watts	3.33 amperes
1 @ 250 watts =	250 watts	2.08 amperes
1 @ 900 watts =	900 watts	7.50 amperes
Total	*2100 watts*	*17.50 amperes*

Figure 13-5
Individual loads on the circuit

Once you've established the total actual load on the overloaded circuit — and you also have a breakdown of its component parts — you can easily correct the problem by disconnecting enough individual items (going by either wattage or amperage) to bring that total down within required limits. For example, suppose you have a 15-ampere breaker that repeatedly trips, resets, and trips again. Since this appears to be an overloaded circuit, let's first look at your meter reading. It shows a current of 17.7 amperes being drawn. Compare that to your direct examination and listing of the loads on that circuit shown in Figure 13-5.

The difference between the meter reading of 17.7 amperes and the total 17.5 ampere load you calculated by adding up the items on the circuit is only 0.2 amperes. That's an acceptable error. A 25-watt light bulb or some other small overlooked item can account for an error of that size.

To get your 15-ampere circuit down to the maximum load that it can carry, you'll have to remove at least 2.7 amperes. Look at the list in Figure 13-5. At first you might think that cutting out the one load of 350 watts would do it, but that's not enough. Remember, *NEC* Section 210.20(A) requires that a circuit be rated no less than 125 percent (5/4ths) of its load; or working as we are now with an established circuit capacity, the load can't be more than 80 percent (4/5ths) of our 15-ampere circuit.

This means that the total load can't just be reduced by 2.7 amperes to 15 amperes, but must actually be reduced by 5.7 amperes to 12 amperes (80 percent of 15 = 12). To accomplish this, you'll not only have to remove the 350-watt load, but the four 100-watt loads as well. Depending on where and how these items are located, it might be simpler to remove, or run a separate circuit to, the one 900-watt item rather than dealing with five different items.

If you don't have a suitable meter available, you can handle overload situations by calculation alone. Since you can't do the initial step of reading the actual load, start with the second step, and list each of the individual loads connected to the circuit that appears overloaded. It isn't important whether you list the size of each load in watts or in amperes. You can easily convert one to the other using Watt's formula:

$$\text{Volts} \times \text{Amperes} = \text{Watts}$$

or

$$\text{Watts} \div \text{Volts} = \text{Amperes}$$

To arrive at a proper total, convert all the individual loads to a common unit of measurement, either watts or amperes. If your total exceeds the rated maximum for the circuit, then you've clearly exposed the existence of the overload. Let's look at our 15-ampere circuit in the prior example using total watts instead of total amperes. First, find the maximum load in watts that the circuit can handle: 120 volts × 15 amperes = 1800 watts. So, if you were figuring your total loads in watts, anything over 1800 is an overload on the 15-ampere circuit. The 2100 watt total in Figure 13-5 is clearly 300 over what our circuit can handle.

The obvious disadvantage of determining the amount of an overload by calculation alone is the possibility that you miscount and leave out one or more of the loads. But as long as your count at least shows an overload, even if you leave something out, chances are the reduction of the known load to 80 percent of circuit capacity will deal with it adequately.

If your total of the loads connected to the circuit doesn't show an overload, you've probably overlooked something. This happens most often in an older building that's been remodeled, and the remodeling has included changes in circuitry. However, if you thoroughly investigate the system and still don't find any unidentified load or loads that could be tied to the problem circuit, the answer may be a defective breaker.

Defective Breakers

Breaker failure is extremely uncommon; so much so that it's often misdiagnosed when it does occur. I've never encountered a breaker that failed in the *ON* position and couldn't be shut off, but since it's a mechanical device, that type of failure is possible. I *have* encountered breakers that were impossible to get *ON* for several tries, and also breakers that would snap into the *ON* position but not hold under load.

A breaker makes its internal connection through a pair of contact points similar to, but larger than, the contact points in a car. Not only when a short occurs, but also any time a breaker is switched *OFF* or *ON*, there's arcing across those points when any appreciable load exists on the circuit. This arcing, sometimes gradual, sometimes sudden, ruins the points — creating a high-resistance spot through which the rated current won't pass or where enough heat is generated to trip the breaker quickly. If a breaker repeatedly trips, although measurements and/or calculations show it isn't overloaded, it's probably bad. So, if a breaker keeps tripping when you can't find an overload, try replacing it. If the new breaker holds *ON*, the old one was bad. If the new one still won't stay on, start looking for that overload again.

A breaker is tripped *OFF* by the heating and subsequent bending of a bi-metal strip. When a breaker operates with a current close to its maximum capacity for some time, that bi-metal strip may have been running hot and become weak. If this happens, the breaker may refuse to stay on, and must be replaced.

Receptacles

The vast majority of receptacles you'll be installing are the common duplex grounded 15-ampere, 120-volt type. Today, most of these are furnished with both terminal screws and holes for push-in connections. The outside receptacle case consists of two parts: a back shell and a front plate slotted to receive the standard three-prong grounded plug. Both parts are molded of a hard thermosetting plastic. Inside the case are metal contact strips for power and common, plus ground, which connects to the mounting yoke.

THE RECEPTACLE'S outside plastic case, though hard, is quite brittle and cracks easily on impact. The usual mounting height of 1 foot off the floor is intended to minimize damage caused by cleaning implements, such as brooms, mops, or vacuum cleaning equipment. While this mounting height reduces damage, it by no means eliminates it. A cracked receptacle is dangerous and should be replaced as soon as possible.

Besides cracks, other problems with receptacles are caused by poor contact between the plug and receptacle. These can be due to dirt in the slots, or the result of an oxide coating forming on the metal of the internal contact strips. In either case, it's easier to replace the receptacle than to try and clean the contacts. Occasionally, a short in an appliance or a loose wire in a plug can cause arcing that severely burns both the plug and receptacle. Again, you need to replace the receptacle.

Aluminum wire is another source of difficulties. Very few 120-volt duplex outlets are made for use with both copper and aluminum. Those that are marked for copper only, or for copper or copper-clad aluminum, can't handle bare aluminum. And, if you use those that *do* take bare aluminum, you must make attachments only at screw terminals — *never* use push-in contacts with aluminum wire.

Receptacles with larger amperage, 30-ampere and 50-ampere, are generally marked *Al-Cu*, which means you can use bare aluminum with them; I don't recommend it, however. A great many cases of arcing and burned receptacles in these sizes occur daily because of the oxidation problems we discussed in Chapter 4. If you use aluminum wire to feed these receptacles, for the power supply to dryers or ranges for example, you need to take steps to avoid problems. About once a year, completely remove the receptacle from its box, disconnect the aluminum wires, treat them with antioxidant, and then put them back in place. If this isn't done regularly, the chance of trouble with these receptacles is quite high. Unfortunately, in recent years, the high price of copper has made the use of aluminum wiring in range and dryer outlets more appealing, even when the rest of the residence is wired in copper. The owner often doesn't even know about these aluminum lines until the unhappy day they burn out.

Switches

The single-pole switches used in homes are generally the toggle-type that make or break connections internally with a pair of contact points. It's those contact points, illustrated in Figure 13-6, that are the vulnerable parts of the switch. Every time the switch is opened or closed, a slight electrical arc occurs across those points. The larger the load being switched, the stronger the arc will be. Repeated arcing gradually pits and deforms the points. Finally, they become so badly misshapen, and the area of contact between the points so small, that a current can no longer pass and the switch fails. There's usually a period of intermittent malfunction before the switch fails entirely. During this period, you may have to try operating the switch several times before it actually works.

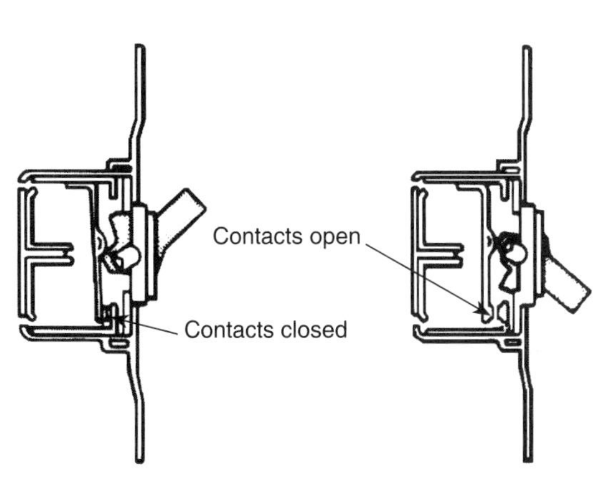

Figure 13-6
Contact points inside single-pole switch

Switches normally fail in the *OFF* position because the points are so badly damaged that adequate contact across them is no longer possible, but occasionally they appear to freeze in the *ON* position. Moving the toggle back and forth has no effect. The reason they appear to be frozen at *ON*, is simply because they *are*. At some time when the switch was being closed, a strong arc occurred that actually welded the points together. Even if flicking the toggle back and forth repeatedly succeeds in breaking the points apart, that switch will never operate properly again, and must be replaced. The same thing is true of switches that fail in the *OFF* position — you can't repair them — they have to be replaced.

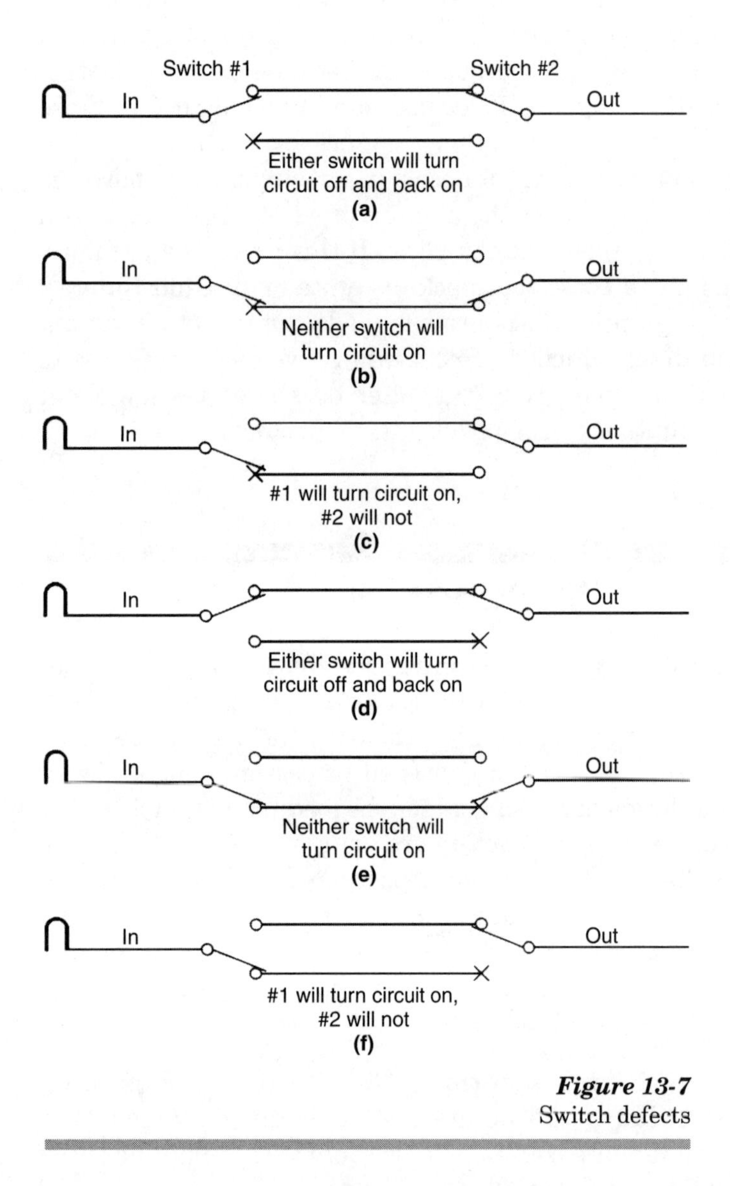

Figure 13-7
Switch defects

Like single-pole switches, 3-way and 4-way switches also make and break connections through contact points. These points are also subject to arcing, pitting, and eventual failure. When a single set of points in any 3-way or 4-way switch fails, it disrupts the entire multiple-switch circuit. This makes it very difficult to determine which switch has gone bad. Figure 13-7 shows why. In example (a), the defect is in switch #1; and in (d), switch #2 is bad. But in both examples, with the switches in the positions shown, the circuit will be on and either switch will turn it off and back on as though nothing were wrong.

With the defects in the same places, but switch positions changed, as in examples (b) and (e), neither switch will operate the circuit; but in each case, only one switch is bad. Finally, in examples (c) and (f), switch #1 will operate the circuit, while #2 won't. Replacing #2 will help in the (f) defect, but certainly not in the case of (c).

With two 3-way switches and one 4-way switch, all sorts of deceptive combinations exist. A few are shown in Figure 13-8. In example (a), any one switch will shut the circuit off and turn it back on again, although the 4-way is bad. In example (b), none of them will turn it on, but only one is bad. In case (c), switches #1 and #2 will turn the circuit on and off, but #3 won't, even though there's nothing wrong with it.

Since defective multiple-switch circuits show seemingly inconsistent malfunctions, the only way to isolate the bad switch is to test each one individually to see whether it will pass power in both positions of the toggle. Another, sometimes simpler, way of dealing with trouble in a multiple-switch circuit is the substitution method. With a 3-way switch circuit, replace either one of the switches. There's a 50 percent chance you'll get the right one. If luck was looking the other way and the circuit still doesn't operate properly, use the switch you removed to replace the second one. The logic behind the substitution approach is that, while you *don't* know which switch is bad, you *do* know that you most certainly have to replace one. By simply starting off with a replacement, in 50 percent of the cases, you'll be right the first time; and at worst, you only wasted five minutes or so changing the wrong switch. With two 3-ways and a 4-way, if changing both 3-ways fails to solve the difficulty, only the 4-way is left. The odds are 2 to 1 it will be one of the 3-ways, right?

Light Fixtures

As far as repairs go, you'll be dealing with both incandescent and fluorescent fixtures. We'll look at both types separately.

Incandescent Light Fixtures

There are basically three types of incandescent fixtures:

1. surface mounted
2. hanging
3. recessed

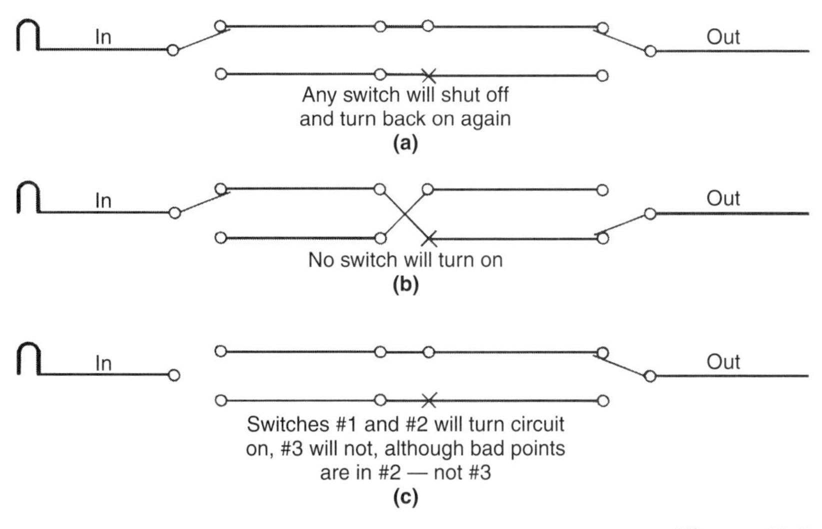

Figure 13-8
Defects using two 3-way switches and one 4-way switch

All three types are comparatively trouble-free. When an incandescent fixture fails to light, it's usually just a burnt-out bulb. Of course, in the case of fixtures with internal switches, the switch part of the fixture is subject to the same kind of failure, and for the same reasons, as any of the wall switches we discussed earlier.

In light of the current and ongoing interest in energy conservation, you'll find insulation installed in a great many formerly uninsulated attics. In addition, other homeowners are increasing the thickness of their existing attic insulation. As long as they take into consideration the 3-inch clearance around recessed fixtures required per *NEC* Section 410.116(B), difficulties with recessed lights in attic insulation installations are very unlikely. If not, at the very least, the fixture will burn out bulbs at an alarming rate. Operating at an excessively high temperature can also scorch the inside finish of the can, substantially reducing the light output. And, in those fixtures with plastic lenses, the lenses can be damaged. While the insulation on the wiring in these fixtures is rated to survive temperatures well above anything likely to be encountered in normal operation, a large bulb, burning base up in a fixture surrounded by insulation and that's not dissipating its heat adequately, can damage the fixture wiring.

Another common cause of overheating, resulting in damage to recessed incandescent light fixtures, is using bulbs larger than the maximum wattage for which the fixture is rated. Never use a bulb size larger than the wattage stamped on the inside of the fixture case. *NEC* Section 410.120 requires this number to be stamped, *in letters at least ¼-inch high*, where it can be seen when changing bulbs. Be sure to look for it.

Fluorescent Fixtures

FLUORESCENT BALLASTS and lights are far more sensitive to temperature extremes and voltage variations than incandescents.

Fluorescent lights are far more energy efficient than incandescent lights, so you'll be seeing more and more of these fixtures in residential use. Most homeowners now use screw-in type fluorescent bulbs, which for our purposes work exactly like incandescents. Any problems with these involve either just changing the bulb, or dealing with the fixture, which is just a standard one. But tube fluorescents are still common in kitchens and work areas — and the rule in commercial establishments. You'll be dealing with them, so we'll deal with them here.

Tube fluorescent fixture operation is considerably more complex than incandescent — they're subject to a good many more kinds of malfunctions. Some of these problems are caused by improper installation, while others result from unfavorable operating conditions. Ballasts are particularly subject to premature failure if they're operated at excessively high temperatures (above 105 degrees C or 221 degrees F), either due to inadequate ventilation or excessive line voltage (above 120 volts). And, the light output of fluorescent lamps decreases at temperatures below 10 degrees C or 50 degrees F. Depending on humidity, line voltage and lamp design, at varying temperatures below 50 degrees F, fluorescents will fail to light altogether.

There are three basic types of fluorescent fixtures: preheat (older fixtures only), rapid start, and instant start. Some parts of the troubleshooting drill are the same for all three types. When a fixture fails to light, the first step in all cases is to change out the tubes. In most instances, that'll fix the problem. In two-tube fixtures, one tube generally won't

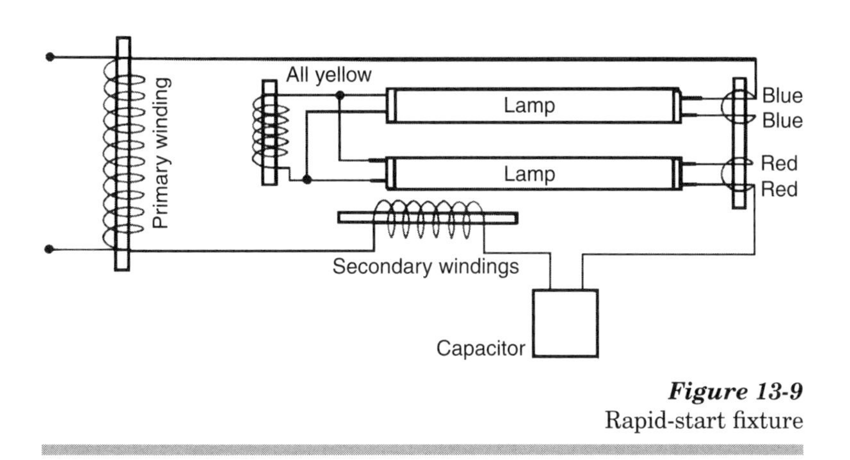

Figure 13-9
Rapid-start fixture

operate alone — if one goes bad, they both stop working. In four-tube fixtures, the tubes are usually wired in pairs, so if one tube is bad, its mate will also go out, but the other two will continue to operate. When changing tubes, before you install a replacement tube, check both sockets to make sure that all the lamp pins make good contact.

The majority of both new and existing fluorescent installations today are rapid-start fixtures. The older preheat-type fixtures have generally been replaced with either rapid-start or instant-start fixtures. If you come across one of these old fixtures, and the problem appears to be more than replacing a tube, replace the fixture.

For proper operation of rapid-start fixtures, it's very important that you securely ground these fixtures. If you don't, the result will commonly be not starting or else random starting. If the tubes are good and the fixture is properly grounded, but it still won't start, there may be inadequate cathode voltage to the lamps. Blue-blue, yellow-yellow, and red-red leads run to the filament windings in the ballast, as shown in Figure 13-9, and supply between 3.4 volts and 4 volts to the lamp cathodes. If you can't measure that voltage at the lamp sockets, the lamps aren't properly heated, and may not start, or if they do start, they may not operate correctly. Inadequate voltage can be caused by an improperly connected lead wire from the ballast, low power supply voltage to the ballast, or bad ballast winding.

Assuming the cathode voltage is satisfactory, check to see if the starting voltage is adequate. To read starting voltage, connect the meter from the highest reading red lead to a blue. The correct starting voltage varies with lamp type and size, temperature, and number of lamps. The table in Figure 13-10 shows the starting voltage for various lamp types in one- or two-lamp rapid-start fixtures.

What would cause rapid-start lamps to start slowly or fail to start under high humidity conditions when the cathode heater voltage and starting voltage are correct? It could be dirt on the lamps affecting the silicon coating, or it could be a poor job of silicon coating. If the installation has been in service for some time, try washing the lamps to get the dirt off. Small droplets of water should then form on the tubes, similar to the droplets that form on a newly-waxed car.

Rapid start — 430 MA lamp type	*Starting voltage (minimum @ 50° F)* *Single lamp*	*Two lamp*	*Filament voltage*
F14T12	108	162	7.5-9.0
F15T8	108	162	7.5-9.0
F15T12	108	162	7.5-9.0
F20T12	108	162	7.5-9.0
F22T9	185	—	3.4-3.9
F30T12/RS	150	215	3.4-3.9
F32T10	205	—	3.4-3.9
F40T12/RS	205	256	3.4-3.9
22-32 (circline comb.)	—	235	3.4-3.9
32-40 (circline comb.)	—	295	3.4-3.9

Rapid start — 800 & 1000	*Starting voltage (minimum)*					
	Single lamp			*Two lamp*		
MA lamp type	*50° F*	*0° F*	*−20° F*	*50° F*	*0° F*	*−20° F*
F24T12/HO	85	110	140	145	195	225
F36T12/HO	115	155	190	195	235	260
F48T12/HO	165	215	255	256	290	310
F60T12/HO	210	240	290	325	350	365
F64T12/HO	240	270	330	345	370	388
F72T12/HO	275	300	360	395	410	420
F84T12/HO	285	325	370	430	445	455
F96T12/HO	280	330	360	465	480	490

Note: Filament voltage 3.4-4.3

Rapid start — 1500 MA	*Starting voltage (minimum)*					
	Single lamp			*Two lamp*		
lamp type	*50° F*	*0° F*	*−20° F*	*50° F*	*0° F*	*−20° F*
F48PG, VHO, SHO	160	205	240	250	265	300
F72PG, VHO, SHO	225	270	310	350	360	400
F96PG, VHO, SHO	300	355	400	470	470	500

Note: Filament voltage 3.4-4.3

Figure 13-10
Rapid-start lamp types and starting voltages

Lamp type	*Starting voltage (minimum)**
F24T12	270
F36T12	315
F40T12	385
F40T17	385
F42T6	405
F48T12	385
F64T6	540
F72T8	540
F72T12	475
F96T8	675
F96T12	565

*For single lamp, measure voltage between red and white leads. For two lamp (series sequence), measure voltage between red and white, insert lamp in red and white position, then read voltage between blue and black. For two lamp (lead lag), measure voltage between red and white and blue and white leads.

Figure 13-11
Starting voltage table for instant starts

When it's a new installation, but random starting is occurring when the humidity is high, generally the line voltage is low or there's a poor silicon coating on the lamps. If droplets don't form when water is run on the tubes, the problem is the silicon coating.

Occasionally with a two-lamp rapid-start fixture, one tube will light properly but the other won't. Look at Figure 13-9 again. If the tube that lights is between the red and the yellow leads, look for a damaged yellow lead. However, if the lamp between the red and yellow leads won't light but the other does, the ballast is probably shorted and it'll have to be discarded.

Depending on the frequency with which it's switched on and off, you can expect a fluorescent tube to last between 15,000 and 20,000 service hours. As a tube reaches the end of its service life, the ends will gradually turn black before it fails entirely. Premature blackening and tube failure can be the result of inadequate cathode heating, which can result from inadequate voltage to the cathodes from the ballast, as mentioned earlier. However, inadequate heating can also occur when the ballast is fine. Other causes could be poor seating of the lamp in the socket, a broken socket, a broken lamp pin, a loose socket in the fixture causing socket spacing to be too great, or an internally damaged lamp cathode.

You can expect a ballast to provide around 60,000 to 100,000 hours of service — again, depending on the frequency of switching. If they're installed properly, so they're not running overheated, they're extremely reliable. However, they'll eventually fail and require replacement. When a ballast finally fails in a simple utility-type fixture, consisting only of ballast and lamp sockets mounted on a lightweight sheet metal case, it'll often cost more for labor and material to replace the ballast than to just replace the whole fixture.

The tubes used in both preheat and rapid-start fixtures are the familiar bi-pin type. Instant-starts are single-pin, found most commonly in 6-foot and 8-foot-long fixtures. Starting voltages for various sizes of instant-starts are considerably higher than for rapid-starts of comparable size. These voltages are given in Figure 13-11. You can determine the starting voltage using a voltmeter. Remove the lamp and connect the voltmeter between the respective primary and secondary leads of each lamp, as designated on the ballast label. For series-sequence ballasts, the red lead must be in position while measuring the starting voltage of the remaining lamp.

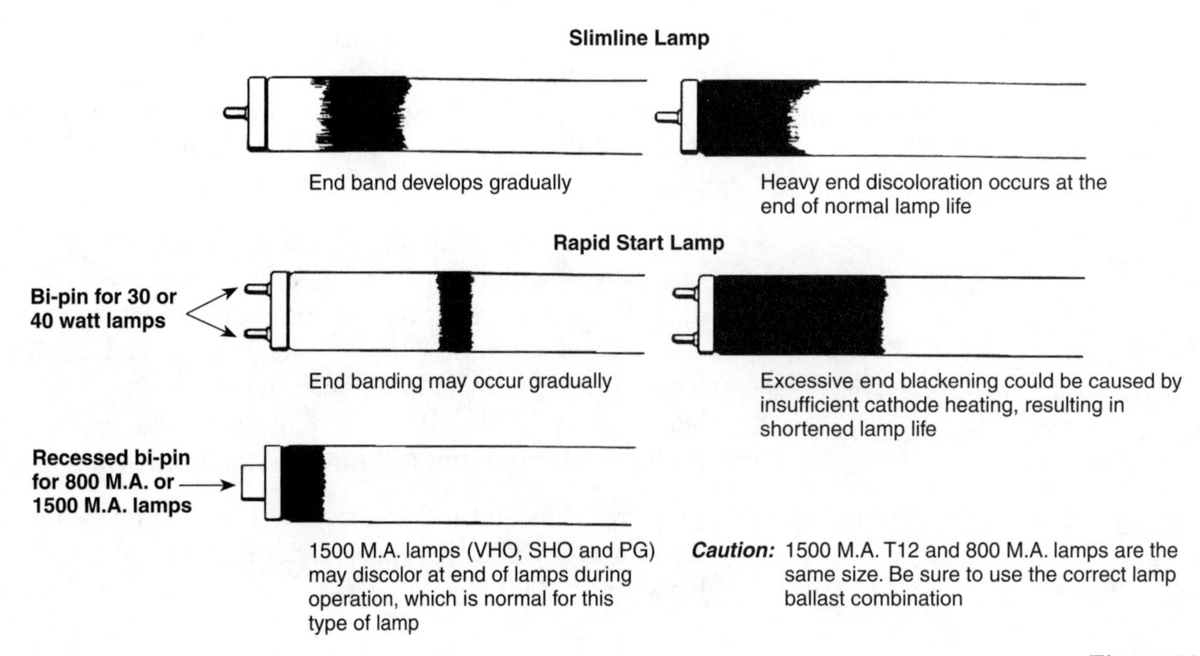

Figure 13-12
Normal darkening through age of slimline and rapid starts

The effects of aging and pending failure of instant-start (or slimline) tubes are similar to that on rapid-starts. These are shown in Figure 13-12. Dark bands near the tube ends may develop during normal service. As the tube approaches failure, there's a gradual increase in the development of the heavy blackening on the ends. However, since normal tube life is between 10,000 and 20,000 hours, end blackening after only a few thousand hours of operation is clearly a sign that something is wrong. Line voltage to the fixture may be low, lamp-to-socket contact may be poor, or the fixture could be miswired. Check all three. You can normally find the correct wiring diagram for the fixture on the ballast.

With slimlines or instant-starts, it's common for one tube of a two-tube fixture to light when the other tube has a defective filament. The defective one will flicker and blacken at one end very rapidly. *Do not continue* to operate the fixture with a bad tube. The ballast will overheat and fail.

The tube sockets at both ends of either preheat or rapid-start bi-pin tubes are exactly the same. Install the tubes by slipping the pins into the end slot in the socket, then rotating the tube 90 degrees to seat the pins in the contacts, as illustrated in Figure 13-13. Check for proper rotation and fit of the lamps in the socket. The base mark on the lamp must be aligned with the center of the socket. The lamps must seat in so that the spacing between the socket and the lamp is small enough to assure that proper contact is made, but free enough to prevent binding. Check for loose or broken sockets and dirty socket contacts when you change lamps.

Single-pin slimline tubes are suspended between two sockets that aren't the same. One end is fixed in position while the other has a movable spring-loaded contact. Install the tube by depressing the spring-loaded end to allow space to seat the single pins at both ends. In my opinion, the spring-loaded single-pin mounting system isn't adequately engineered. I've seen the spring-loaded socket repeatedly overheat, resulting in a sudden burnout, accompanied by the tube dropping out and shattering. This burnout occurs uniformly at the spring-loaded end, not the fixed end.

Checking New Wiring

After you bring new circuit cable back to the breaker panel and connect the power lead to its assigned breaker, test the circuit for shorts. This should be a routine procedure in your installation. First, make sure that there isn't any equipment connected or plugged into any outlet, and that there are no light bulbs in any fixtures. Now, with the breaker off, using the ohmmeter, test from the output terminal on the breaker to the ground buss in the box. The meter should read *infinity* (∞). If it reads a low resistance — 2 or 3 ohms — a light bulb or something else that shouldn't be there is in the circuit. If it reads 0, it's probably shorted.

Follow the procedure outlined early in the chapter, under *Building Circuit Shorts*, to trace down the short. For new wiring, this should be quite easy because you'll know the wiring sequence of the various outlets, switches and fixtures. That makes it possible to follow the power distribution from box to box through the circuit in an orderly manner.

Assuming there wasn't a short, or if there was one, you've located and corrected it, your next step is to turn on the power and check all switches, fixtures, and outlets. At this point you want to put bulbs in all light fixtures to test both fixtures and switches. If any fixture fails to

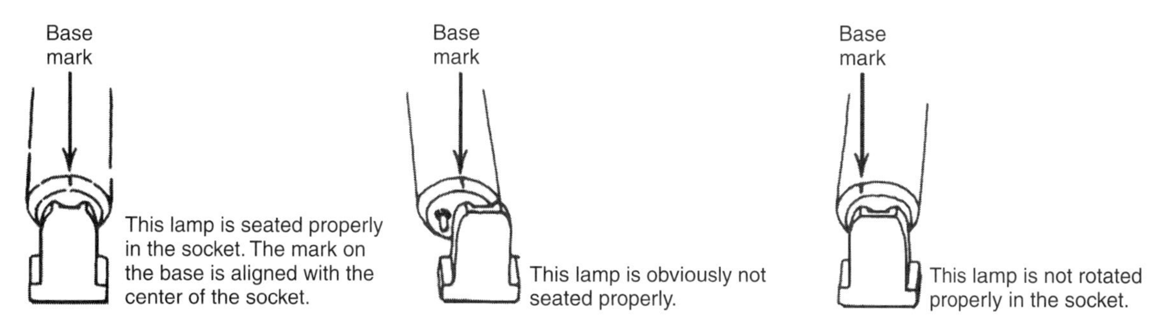

Figure 13-13
Seat bi-pin tube properly in socket

Figure 13-14
Using 120-volt outlet tester

light with the switch on, remove the switch from its box and, using a pocket test light, check from both input and output terminals to ground to make certain the power is getting into and out of the switch. If power isn't getting into the switch, work back toward the source. Somewhere there's probably a poor splice. If power is getting into, but not out of, the switch, replace the switch — it's defective.

If power is going through the switch, test at the fixture. If power is getting to the fixture and you know that the bulb is good, the trouble has to be in the common return wiring — either a bad terminal connection or a bad splice.

When the lights and switches are all operating properly, the next step is to test all the receptacles. You can use a special outlet tester for this purpose (shown in Figure 13-14), a voltmeter, or even a little pocket test light. Using an outlet tester, simply plug it into each receptacle. The way it lights up, or fails to light, will indicate the presence of any existing difficulty — such as an open ground, reversed polarity, open common, or open power lead.

Figure 13-15 (a) is a normal 120-volt receptacle. To test this with a voltmeter, first test across the slots (1). The reading should be approximately 120 volts. Then, test from the shorter slot to the *U*-shaped ground

Defects could be as follows:

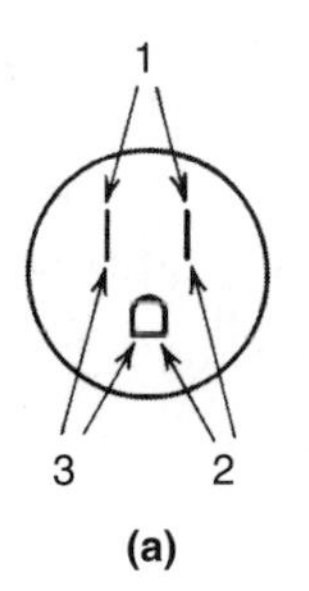

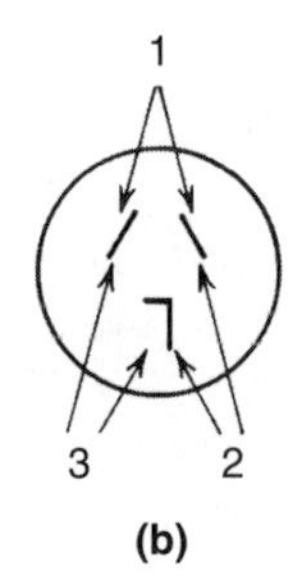

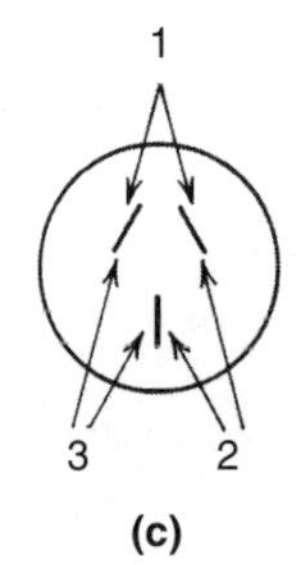

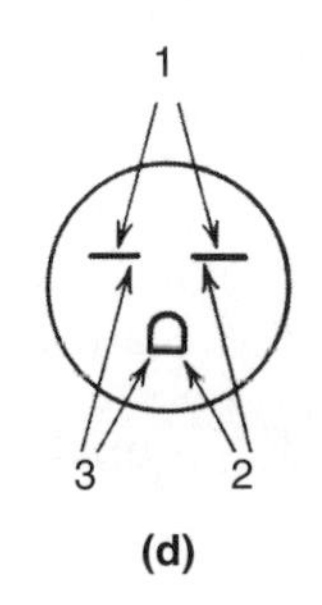

Figure 13-15
Test sequence for 120-volt and 240-volt outlets

1	*2*	*3*	*Defect*
120	0	120	Power and common reversed — polarity backwards
120	0	0	Open ground — ground continuity broken somewhere
0	120	0	Open common — common continuity broken
0	120	120	Power and ground reversed — ground is hot
0	0	120	Common is hot and power is open
0	0	0	Power is open — no power

Figure 13-16
Possible defects in a 120-volt outlet

1	*2*	*3*	*Defect*
120	120	0	Left side power line open
120	0	120	Right side power line open
240	0	0	Common open
120	240	120	Common and left side power line reversed
120	120	240	Common and right side power line reversed

Figure 13-17
Possible defects in a 240-volt outlet

hole (2), which should also read 120 volts. And last, test from the ground hole to the long slot (3), which should read 0 volts. Figure 13-16 shows the possible defective connections.

You can do the same tests with a pocket test light. But, in every instance where the meter would show 120 volts, the test light will just light up. What the test light won't show is whether the voltage itself is correct. If the voltage is running either too high or too low, a test light doesn't show it.

To test a dryer or range receptacle, see Figure 13-15 (b) and (c), use the voltmeter. When you test across the two top angled slots (1), the reading should be 240 volts. With one test lead in the bottom vertical slot, the readings to each of the angled slots (2) and (3) will both be 120 volts. Figure 13-17 shows the possible defects that these tests indicate.

Figure 13-15 (d) shows a type of 240-volt receptacle often used for small room air conditioners or space heaters. Both the tests and the defects applicable to the receptacles in (b) and (c) also apply to this type of receptacle.

STUDY QUESTIONS

1. What overcurrent protective device could you use to safeguard a circuit wired with #12 AWG copper with THW insulation rated for 20 amperes?

A) A circuit breaker rated at more than 20 amperes
B) A 20-ampere fuse in a 15-ampere circuit
C) A circuit breaker or fuse rated at 20 amperes
D) A 30-ampere fuse in a 30-ampere circuit

2. What should you check first if an electrical device fails to function?

A) Check the circuit fuse or breaker
B) Test the receptacle or power supply to the device
C) Test the inoperative device for a short
D) Plug the device into a different receptacle

3. How should you deal with a blown Edison base plug fuse?

A) Replace it with the next larger size
B) Replace it with an S type fuse and adapter of the same size
C) Replace fuse with a time-delay type
D) Consider replacing the fuse box with breaker box

4. If you have a fused disconnect that uses cartridge fuses, how do you test the cartridge fuse to see if it's bad?

A) Test the fuse in place with an ohmmeter
B) Remove all equipment from the circuit and then test each one for a short
C) Remove the fuse and use an ohmmeter to test the circuit for a short
D) Test the fuse in place with a voltmeter

5. What's the first step in isolating a short?

A) Disconnect all power-using devices and try resetting the breaker
B) Test each power-using device for a short
C) Test all circuit receptacles for shorts
D) Remove power wire and test the breaker

6. What's indicated if, when you reset the breaker, it stays on briefly, but then trips again?

A) There's a short in the wiring
B) There's an overload, not a short
C) The breaker is bad
D) There's a short in a connected device

7. What should you do if aluminum wire is used to feed a receptacle marked *Al-Cu*?

A) Once a year, disconnect the wires in the receptacle box and treat them with an antioxidant
B) Nothing, this is a common usage
C) Replace the receptacle
D) Bare aluminum wire is never used to feed receptacles

8. When a single-pole toggle-type switch fails in the *ON* position, what is the first step to take in dealing with the problem?

A) Move the toggle repeatedly back and forth to free the fused contacts
B) Clean the contacts and treat them with an antioxidant
C) Replace the switch with one of slightly-higher amperage to prevent reoccurrence
D) Replace the switch with a new one of the same type

9. When a *fluorescent* light fixture fails to light, what should be the first step in fixing it?

A) Check to see if it's receiving excessive line voltage
B) Check the cathode voltage to the lamps
C) Change out the tubes
D) Replace the ballast

10. Using a voltmeter to test a normal grounded 120-volt outlet, what should the voltages be if you test first across the slots, then short slot to ground, and then the ground to long slot?

A) 120 volts on all three
B) 120, 0, 0
C) 0, 120, 120
D) 120, 120, 0

14 SUPPLEMENTARY SYSTEMS

In this chapter we'll discuss the three most common supplementary electrical systems that you'll be dealing with in residential wiring:

- Signaling and warning systems
- Communications systems
- Entertainment systems

Signaling and warning systems include doorbells, chimes and buzzers as well as smoke detectors and burglar alarms. Communications systems include telephone for outside communications, and internal communications via voice intercom systems. The entertainment area includes radio, TV and stereo, which may include wiring runs for antennae or remote speakers. Electricians who do power wiring aren't required to get involved in electronic installations, though it's probably a good idea to keep abreast of technology advances in cable TV, satellite dishes, Internet and DSL wiring, and video monitoring and alarm systems.

Signaling and Warning Systems

Most supplementary electrical systems are low voltage, meaning that they operate on anywhere from 6 to 24 volts, instead of the 120- and 240-volt circuits we've been dealing with. Doorbells, chimes and buzzers are uniformly powered by step-down transformers attached to the

120-volt supply. Smoke detectors and burglar alarms, with internal step-down transformers, are usually directly connected to the 120-volt line. In many cases, they're separately powered by batteries, or have battery back-ups that take over in case of failure of the 120-volt line.

When the same emergency that causes a fire also cuts off the electrical power, a battery-powered smoke detector or a battery back-up can literally be a lifesaver. Battery-powered or battery-backed-up burglar alarms continue to operate even if a thief shuts off the main electrical disconnect. So battery-powered warning systems independent of the building electrical system have distinct advantages.

The batteries used in these systems have a standard service life of about a year. Their very durability can be a disadvantage, since it's all too easy to forget when they should be replaced. Occupants should be strongly advised to follow the manufacturer's instructions regarding routine testing of these units, and encouraged to follow a rigid schedule of battery replacement.

Doorbells, Chimes, and Buzzers

The doorbell circuit starts with the bell transformer, shown in Figure 14-1. This is a step-down transformer that reduces the 120-line voltage to somewhere between 10 and 24 volts. The primary is furnished with two wire pigtails, a black and a white, that attach to the 120-volt line. Since all connections to devices or splices of conductors to each other must be made in an approved electrical box (*NEC*® Section 300.15), many transformers are supplied with a special bracket that fits a standard ½-inch metal box knockout. Using this bracket, you can mount the transformer on any standard metal box with ½-inch knockouts, so that the primary pigtails are inside ready for connection, and the secondary screw terminals are outside and available for the low-voltage bell wires that go on to the bell button and the bell to be attached.

Figure 14-1
Bell transformer with box mounting bracket

Although there are bell transformers with secondary voltages as high as 24 volts, a 10-volt secondary is adequate for most purposes. Some transformers have multiple taps that provide two or three voltages, such as 8, 16, and 24. When all else fails, check the instructions that came with the bell or chime. They should indicate the minimum voltage required by the device. There's no danger in erring on the high side. If the bell wants 8 volt, then a 10-, 12-, or even 16-volt supply won't cause a problem. The reverse might not be true: If the device wants 16 volt, it may not operate on 8 volt.

While connection of the transformer primary to the 120-volt supply must be made inside a box, the secondary terminals are outside, because low voltage connections don't require the protection of a box. Also, the wire size used is vastly smaller than the #14 minimum required for the 120-volt power circuits. Bell wires range from #18 down to #22. You can make splices at any point, no box required.

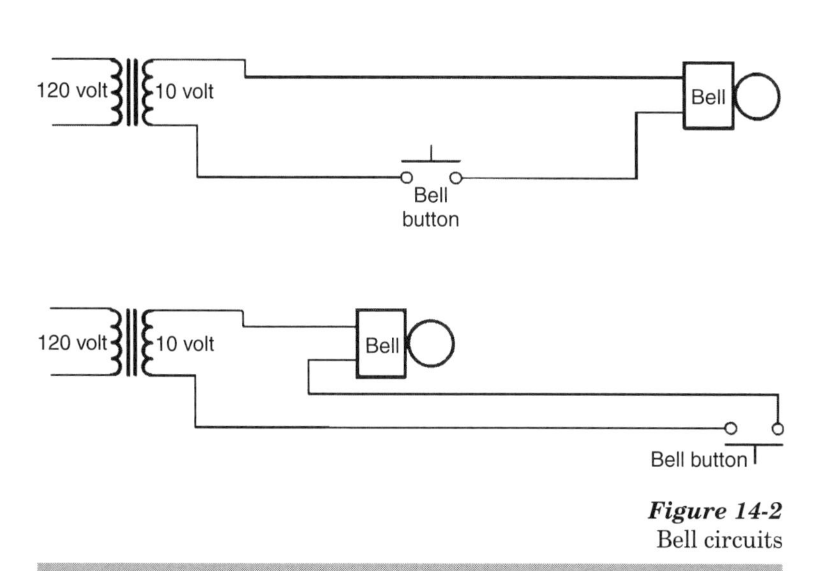

Figure 14-2
Bell circuits

Typical bell wire consists of two individually insulated conductors furnished in two colors, as a twisted pair. The colors commonly used are red and white. Different colors are used solely to help the electrician identify which is which. They have no other significance.

From the transformer forward, the circuit consists simply of a switch (the bell button) and a load (the bell). As always, one side of the line goes through the switch, and both sides go to the load. Depending on the physical locations of the parts, the wiring sequence can be transformer to button to bell, or it can be transformer to bell with switch loop to bell button. See the two examples in Figure 14-2. These two circuits are identical to the wiring in conduit shown in Chapter 7, Figures 7-8 and 7-9; they merely connect different components using much lighter wire.

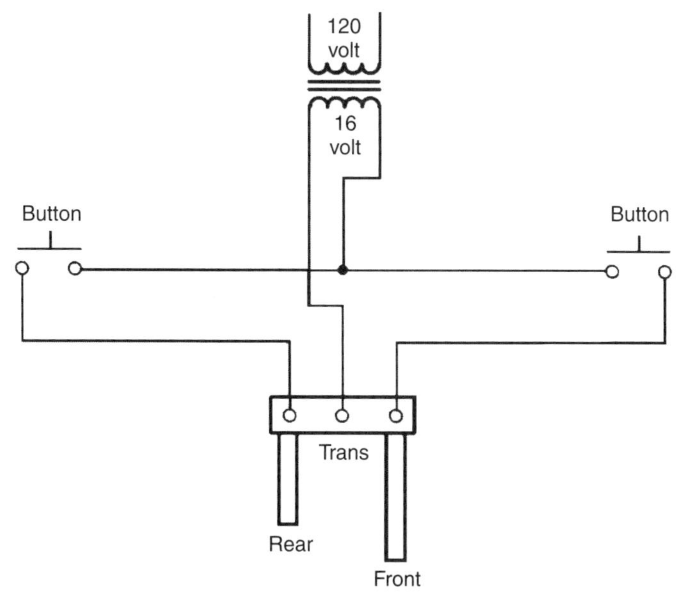

Figure 14-3
Two-tone chime circuit

It's very common to use two-tone chimes to indicate, with different sounds, which of two entrances is being approached. Figure 14-3 shows how you would normally wire these. A low-voltage pair runs from the transformer to the chime. One wire connects to the terminal marked either *trans* or *common*; the other splices to one side of each of two switch loops going out to bell buttons. The return side of one goes to the *front*, and the other to the *rear*, allowing each button to activate a different chime tone.

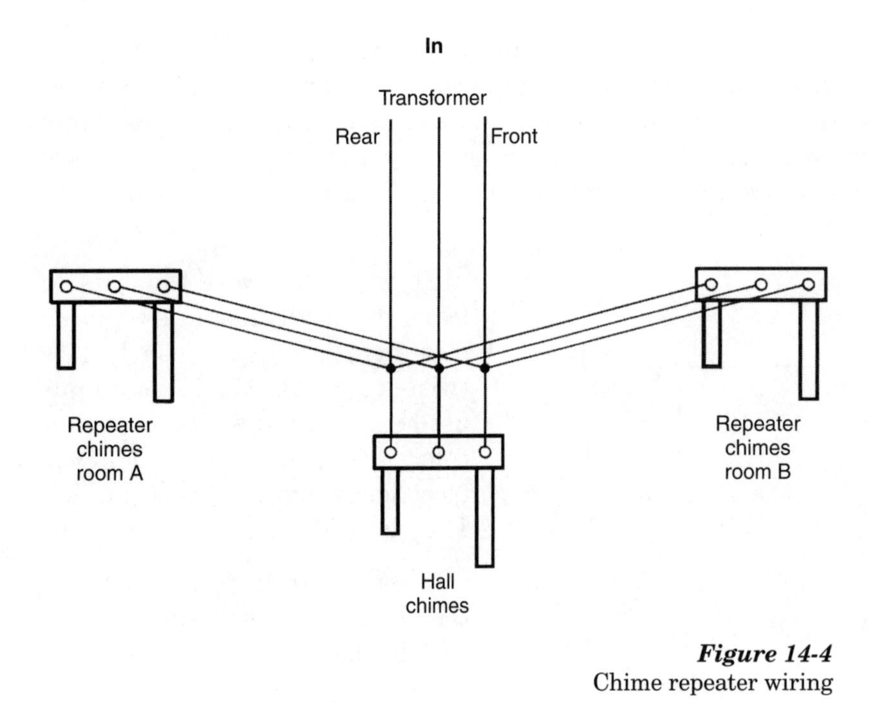

Figure 14-4
Chime repeater wiring

Because of the distances and shapes of halls and rooms, a single bell or chime is often inadequate. If so, you can easily connect repeaters to the initial unit, as illustrated in Figure 14-4. In order to adequately power the repeaters, you might need to increase the secondary voltage of the bell transformer if it's only supplying 10 or 12 volts.

With all due respect to our colleagues in manufacturing, doorbell buttons are generally not very well made. Also, because they're outdoors and continually exposed to the weather, difficulties with doorbell circuits are often caused by failure of the bell button. If a bell or chime fails to operate, start by pulling out the bell button. Then take off one wire and short it to the other wire. Generally, the bell will ring merrily, indicating that the button is bad. In that case, all you have to replace is the button.

If the bell doesn't ring when you test it in this way, use a voltmeter to measure across the wires at the bell. When the transformer is all right, you'll see a reading somewhere between 10 and 24 volts; but if the transformer is bad, you won't find any voltage. As a general rule, the bell itself is the last component to go bad; more often it's the transformer.

Smoke Detectors

There are two types of detectors commonly used: photoelectric and ionization. The photoelectric unit contains a light sensor that responds to a constant light in a small chamber. Smoke entering the chamber obstructs that light. The sensor detects the change and activates an alarm. This type of detector isn't highly sensitive to fires produced by comparatively clean-burning materials, such as gasoline or other highly-refined liquids.

Ionization detectors contain a small amount of radioactive material that causes ionization of the air in a test chamber fitted with two oppositely-charged plates. Due to the ionization, an extremely tiny current flows between those plates. Smoke entering this chamber reduces the flow. When that reduction is detected, the alarm sounds.

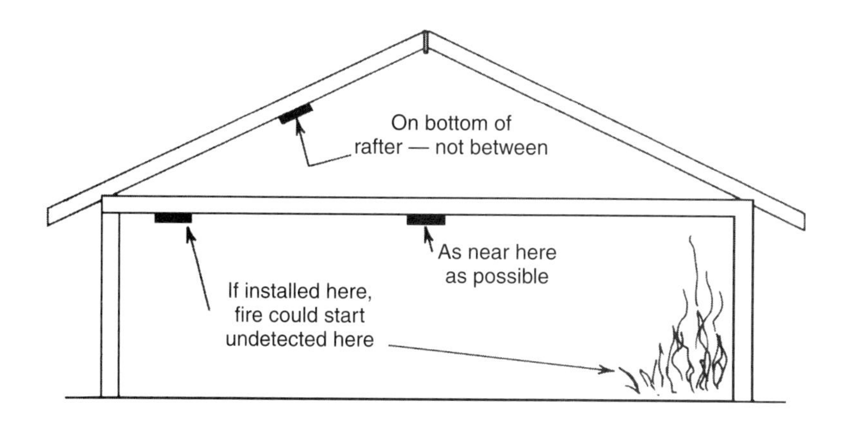

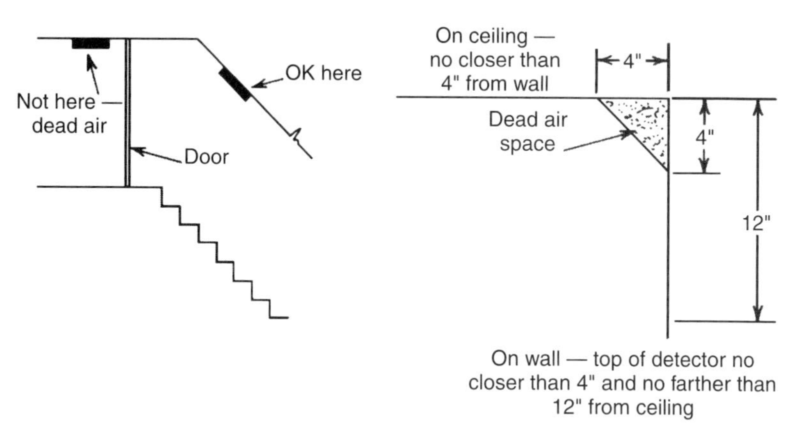

Figure 14-5
Smoke detector location

This type of detector does much better with clean-burning fires than the photoelectric type.

Some smoke detectors have black and white wire pigtails so you can connect them directly to the 120-volt line. You must mount this type directly on a box. Others are equipped with a cord and plug for connecting to a normal wall receptacle. Be certain it's an unswitched receptacle, and also be sure to place the clip on the cord under the screw that holds the receptacle faceplate in place. This is a safety device that prevents the plug from being accidentally pulled out of the socket and disabling the detector.

Still other smoke detectors are battery powered. These units don't require supply wiring, giving the installer the freedom to locate them in the most strategic places possible. Good quality batteries last approximately one year, but they should be tested regularly — in accordance with the manufacturer's instructions — to insure that the device is operating properly.

The exact location of a smoke detector is *critical* to its proper operation. Figure 14-5 illustrates many of these locations.

Following are some *dos* and *don'ts* for smoke detector installation:

- Where possible, install the detector on the ceiling to allow the smoke rising from a fire to enter the chamber.
- Install the detector as close to the center of a room or hallway as possible. If it's installed off to one side of a room, a fire on the other side could become quite intense before the detector picks it up.
- In an *unfinished* attic, install the detector on the bottom edge of a rafter or joist. The space in between rafters and joists is considered dead air space.
- Install the detector at the top of an open stairway. Don't put it in the dead air space that might be closed off by a door.

- If you must install a detector on a wall, place it so that its top is no more than 12 inches from the ceiling, but no closer than 4 inches (above that is dead air).
- When you must install a detector off to one side on a ceiling, place it no closer than 4 inches from the wall (again, dead air).
- Don't install detectors in bathrooms, laundry areas or other places where visible water vapor will be created.
- Don't install a detector near any forced air registers (either heating or air conditioning). Moving air can keep smoke from entering the detection chamber.
- Don't install detectors in kitchens where cooking smoke can trip a false alarm.

Burglar Alarms

A burglar alarm system consists of numerous small, inconspicuous low-voltage sensors placed so that the movement of a door or window will activate them. When activated, the sensor trips a relay, which sounds the alarm at the premises and, where the system is monitored, alerts the alarm service who can then contact the police. Some systems guarding commercial and high-end premises not only sound an alarm where the intrusion is occurring, but also send a signal to the local police station or to a commercial guard service.

A common sensor type consists of two metal contacts, one mounted on the door or window and the other on its frame. In the closed position, the contacts touch, completing the circuit. If the door or window is opened, the circuit is opened as well, tripping the alarm. On a window, the sensor can be combined with metallic tape; if the glass is cut or broken, that'll also open the circuit.

A burglar alarm system that operates on the main building power supply is usually equipped with a battery back-up, in case the supply either fails or is deliberately interrupted. The battery supply is normally maintained in the off position by a relay that's held open by the main power supply. When that fails, the relay is released, switching the battery circuit on.

Once a sensor is opened and the alarm tripped on, merely closing that sensor won't shut the alarm off. The relay that activates the alarm latches in the alarm position, and must be turned off manually.

As with smoke detectors, burglar alarms should be tested regularly. These use rechargeable batteries, but even rechargeables don't last forever. Most manufacturers recommend replacing them about every three to six years.

Communications

Residential communications systems consist of the telephone system, which very often won't be the electrician's responsibility, and intercom systems, which will.

Telephones

Most of the time, the electrician wiring a building has nothing to do with the installation of the telephone system. Sometimes, however, you might be required to provide positions for modular-type telephone sockets and routes for concealed wiring to those positions. The telephone company supplies the actual telephone wire, including all associated parts and fittings. They also pull their own wire and install all their own parts. What electricians are called upon to do is provide boxes, and to fish wires through the insides of finished walls, either up into the attic or down to a basement or crawl space, so the telephone people can get their wiring where they need it.

In commercial installations, the telephone company frequently requires that an electrician furnish a conduit system, through which it'll run its cabling. Occasionally, the phone company may also want conduit for a residential system, especially if the home has masonry walls.

Since telephones are low-voltage systems, you can use standard electrical boxes, but they aren't required. In some cases, you can use a plain plaster cover plate without the box, and in other cases, just a hole in the wall with a fish wire inside will do. Consult with both the building owner and the telephone company before you begin wiring to determine whether you'll be needed to make any provisions for telephone lines and, if so, what's required.

Intercommunication Systems

There are several different types of intercom systems on the market, each providing different capabilities. One simple system runs on the existing telephone system. The telephone company simply supplies extension phones equipped with switches and buzzers to permit communication from room to room through that existing system. You don't need to make any special provisions for this system in the building wiring.

Another method is an independent system of telephone handsets powered and wired separately from the outside telephone system. This system requires a power supply in the form of a step-down transformer connected to the 120-volt line and delivering 10 volts. The transformer is exactly the same type used for the doorbell system, and is mounted on any junction box in the same way. Although both circuits use the same type of transformer, don't put them both on one. Provide each with a separate transformer.

A third intercom system is considerably more versatile than the telephone handset type. It consists of a master control station connected to a number of remote stations, any of which can be called from the master, or can call into the master as well. The master generally includes an AM-FM radio receiver that can be piped through the system when it's not being used for communication. This is possible because each remote station contains a small loudspeaker through which messages are received from the master station, and which serves as the microphone for transmissions back to the master. Another important feature of this system is that a remote station can be switched to the transmit position and left that way, allowing someone at the master location to monitor that remote station. This feature is particularly useful for parents of small children, or in cases where it's important to hear signs of distress from someone suffering from a serious illness or who's incapacitated.

These systems are also low voltage, and usually require a step-down transformer connected to the building power, although some contain their own internal step-downs and plug directly into the 120-volt line. If drawings indicate an intercom in a building, you should inquire what your responsibilities are in providing for it.

Entertainment Systems

Home entertainment systems are becoming more and more complex, and often fall into the realm of a highly technical specialty that lies outside the normal expertise of the power electrician. However, you may be called upon to leave fish wires or even run the antenna or speaker cables, so they can be concealed inside the walls at the same time as the rest of the building wiring.

Television and FM Radio Antennas

Cable and TV dishes now supply the signal for a good proportion of homes, and these are usually installed by the company that provides the service. But roof antennas are still an effective and economical source, widely used in many parts of the country, especially in rural areas. Normally, you can erect a single roof antenna to supply the signal for several TV sets. When you do this, the lead from the antenna goes first to a signal splitter, and then separate leads go from that splitter to the individual sets. The lines from the splitter to the various set locations are the ones that you'll be asked to conceal in the walls. You won't need a box at the set locations; a cover plate for the sake of appearance is enough.

Television signals are transmitted using the same principle as that used for FM radio. So, often a TV antenna will incorporate a section that covers the FM radio band as well. In that case, you don't need additional antennas for FM radio receivers.

In many cases, TV antennas are equipped with rotators so they can be directionally adjusted to get the best possible signal from each broadcast station. The rotator control is plugged into the 120-volt supply. It's also directly wired up to a small electric motor on the antenna mast. The power to that electric motor passes through a step-down transformer in the control box, and is only 24 volts. The rotator power cable is another line that's preferably concealed. You can pass it upward following the same route that the signal lead from the antenna follows coming down.

Inside the building, the antenna and antenna lead-ins shouldn't be run closer than 2 inches of other wiring systems, per *NEC* Section 810.18(B). Where it's available, cable systems have typically replaced traditional roof antennas.

Cable systems transmit television or data signal through coaxial cable, which holds fiber optics capable of carrying at least 1,000 times more information than the traditional copper twisted pair telephone wire.

Computers on Cable

The combination of computer software and hardware with the cable's distribution network has led to an increasing array of entertainment and high-speed information services for consumers and business. Cable can be connected to a modem which splits the signal from coaxial cable to a CAT 5, 6, or 7 cable, which then connects through a modem jack to a computer, or to a wired or wireless router which can deliver the signal to other computers in the building.

Wireless routers, while being the most popular and easiest to set up, may face signal blockage by large appliances or electromagnetic disturbances — such as electric motors, transformers, fluorescent lighting, power lines or even electrical wiring.

Wired signals are typically the most reliable method of carrying signals routed from cable to a computer. Most wired signals are transferred through CAT 5, 6, or 7 cable.

Some concerns in installation of CAT 5, 6, or 7 cable:

- Don't stretch or twist the cable during installation.
- Don't run cables over heat ducts or hot water pipes.

- Don't run cable where it could be subject to electromagnetic interference.
- Don't put any splices in horizontal runs.
- Remove only the minimum amount of jacket material on terminations — 2 inches for CAT 5; 1½ inches for CAT 5e and 6 cable.
- When running cable, allow 1 foot of slack in the work room and 10 feet of slack in the communication room. Follow manufacturer's recommendations for actual termination lengths.

Stereo and Remote Speaker Systems

To get a convincing stereo effect, you must place the speakers fairly far apart and generally far from the control center, too. With a bit of thought, you can easily and effectively conceal the connecting wires.

To run wiring around a room, take the baseboard off, put the wire behind it, and then replace the baseboard. Simpler yet, when the lead has to go to another side of the room, instead of following the walls, go under the carpet. And, rather than stapling wire around a door frame, take the trim off, put the wiring behind it, and then replace the trim.

For placing speakers remotely in other rooms or outside on a patio, any of the procedures for concealing wiring that we've discussed previously will do. As with other low-voltage wiring, keep speaker wires at least 2 inches away from any power wiring you encounter along the way.

In Conclusion

By now you're well aware that all electrical wiring of residential homes and apartments must be done in strict accordance with the provisions of the *National Electrical Code*, as well as with any local electrical codes that may apply in the specific area where you're working. Local electrical codes may differ considerably from place to place as a result of variations in climate conditions, such as temperature, humidity or severe weather. It's up to you to know the local codes and adjust your installation to fit the conditions and requirements of the community.

STUDY QUESTIONS

1. On what type of systems do signaling devices, such as door bells, chimes and buzzers, operate?

A) 120-volt circuits
B) Low-voltage systems, from 6 to 24 volts
C) They are all battery-operated systems
D) 120-volt circuits with battery back-up

2. Which of the following devices utilize an *internal* step-down transformer?

A) Door bells and telephones
B) Chimes and buzzers
C) Smoke detectors and burglar alarms
D) Intercoms and telephones

3. Where would you *not* install a smoke detector?

A) On a wall with its top more than 12 inches from the ceiling
B) On the bottom edge of a joist in an unfinished attic
C) In the center of a hallway ceiling near the stairs
D) On the ceiling of a room within 4 inches of the wall

4. How is a burglar alarm shut off once it is tripped?

A) Once tripped, it must be shut off manually
B) It shuts off once there are no longer any sensors in contact
C) Resetting the tripped sensor will turn off the alarm
D) Only the security company can reset the alarm, which they can do remotely

5. What is the electrician's primary role in wiring telephone systems?

A) Supplying the wiring
B) Providing routes for concealed wiring
C) Providing fittings and parts
D) Connecting to the outside telephone system

6. Which is one means of powering a home intercom system?

A) An AM-FM radio transmitter
B) Include it on the doorbell system transformer circuit
C) A stepped-down transformer connected to the 120-volt line and delivering 30 volts
D) The existing telephone system

7. Intercom systems in homes can be used for all *except* which of the following?

A) To provide communication between different rooms of the house
B) To contact emergency services
C) To broadcast music throughout the house
D) As baby monitors

8. Which statement *does not* apply to television antennas?

A) A single roof antenna can provide signals to several TV sets
B) A TV antenna can be coupled with an FM radio antenna
C) Reception can be improved by rotating the antenna
D) A 4-inch separation is required between the antenna and the building wiring

9. Which is a benefit of a wireless router for a computer system?

A) It can deliver the signal to a computer no matter where the computer is located
B) It is less affected by appliances or other plugged-in electronics than a wired router
C) It can deliver the signal to other computers in the building without special wiring
D) It provides the most secure signal

10. When installing CAT 5, 6 or 7 cable, what is the maximum number of splices allowed in any horizontal run?

A) Splices may be no closer than 6 feet, and there may not be more than three in any run
B) Two
C) One only
D) Splices are not allowed in horizontal runs

ANSWERS TO CHAPTER QUESTIONS

Following each answer is the page (or pages) in the book where the subject of that question is discussed. It's sometimes necessary to read through more than one page when the question asks for a concept, rather than a specific point.

Chapter 1	*See page*	Chapter 2	*See page*	Chapter 3	*See page*
1. C	10	1. C	28	1. C	48
2. D	12	2. B	29	2. A	50
3. B	13	3. D	23	3. B	53
4. A	14	4. A	33	4. D	54-55
5. A	16	5. C	34	5. A	55
6. B	16-17	6. D	35	6. D	55
7. C	18	7. C	36	7. A	55
8. A	21	8. B	37	8. C	57
9. C	21	9. D	38	9. B	58
10. D	23	10. A	38	10. C	59-60

Chapter 4	*See page*	Chapter 5	*See page*	Chapter 6	*See page*
1. D	63	1. D	80	1. B	97
2. A	67	2. B	81	2. D	98
3. B	68	3. C	82	3. C	100
4. C	69	4. D	84	4. B	102
5. D	69	5. A	85	5. C	102-103
6. B	70	6. D	87	6. A	104
7. C	70	7. C	89	7. A	107
8. A	72	8. A	89	8. D	108
9. B	73	9. B	90	9. C	108
10. D	74	10. D	90	10. D	108

ANSWERS continued

Chapter 7	See page
1. D	113-114
2. B	115
3. D	117
4. C	120
5. A	122
6. D	124
7. A	127
8. B	130
9. C	130-131
10. C	130-131

Chapter 8	See page
1. C	137
2. A	137
3. D	134
4. C	144
5. A	154
6. B	156
7. D	162
8. B	163
9. B	176
10. D	178

Chapter 9	See page
1. D	185-186
2. C	186
3. B	187
4. D	188
5. A	189
6. B	190
7. A	191
8. C	191-192
9. B	192
10. D	197

Chapter 10	See page
1. B	204
2. D	204
3. C	206
4. D	207
5. A	208
6. B	209
7. C	210
8. A	211
9. B	212
10. D	212

Chapter 11	See page
1. B	224
2. B	224
3. A	226
4. C	226
5. D	228
6. A	229
7. D	229
8. A	229
9. D	230
10. C	234

Chapter 12	See page
1. C	239-240
2. B	239
3. D	239
4. D	244
5. B	244-245
6. A	245
7. C	247
8. D	247-248
9. B	248
10. A	248-249

Chapter 13	See page
1. C	255-256
2. B	257
3. B	256-257
4. D	257
5. A	257
6. B	259
7. A	263
8. D	263
9. C	266
10. D	272-273

Chapter 14	See page
1. B	277
2. C	278
3. D	281-282
4. A	282
5. B	283
6. D	283
7. B	283-284
8. D	285
9. C	285
10. D	286

INDEX

A

B

G

H

I

J

K

L

M

N

S

T

U

V

W

X-Y-Z

Practical References for Builders

National Repair & Remodeling Estimator

The complete pricing guide for dwelling reconstruction costs. Reliable, specific data you can apply on every repair and remodeling job. Up-to-date material costs and labor figures based on thousands of jobs across the country. Provides recommended crew sizes; average production rates; exact material, equipment, and labor costs; a total unit cost and a total price including overhead and profit. Separate listings for high- and low-volume builders, so prices shown are specific for any size business. Estimating tips specific to repair and remodeling work to make your bids complete, realistic, and profitable. Includes a CD-ROM with an electronic version of the book with *National Estimator*, a stand-alone *Windows*™ estimating program, plus an interactive multimedia video that shows how to use the disk to compile construction cost estimates.
496 pages, 8½ x 11, $58.50. Revised annually

Electrician's Exam Study Guide

Here you'll find 1,500 exam-style multiple-choice and true/false questions and answers to help you pass the electrician's exam on the first try. Includes references to the *NEC* with plenty of illustrations to help you gain insight on the many mysteries of the Code. Filled with extensive tables and examples, this career-boosting guide presents a wealth of information on general definitions and requirements for installations, wiring methods, equipment, product safety standards, administration, enforcement, and much more. **370 pages, 8½ x 11, $39.95**

Contractor's Survival Manual Revised

The "real skinny" on the down-and-dirty survival skills that no one likes to talk about – unique, unconventional ways to get through a debt crisis: what to do when the bills can't be paid, finding money and buying time, conserving income, transferring debt, setting payment priorities, cash float techniques, dealing with judgments and liens, and laying the foundation for recovery. Here you'll find out how to survive a downturn and the key things you can do to pave the road to success. Have this book as your insurance policy; when hard times come to your business it will be your guide. **336 pages, 8½ x 11, $38.00**

National Electrical Estimator

This year's prices for installation of all common electrical work: conduit, wire, boxes, fixtures, switches, outlets, load-centers, panelboards, raceway, duct, signal systems, and more. Provides material costs, manhours per unit, and total installed cost. Explains what you should know to estimate each part of an electrical system. Includes a CD-ROM with an electronic version of the book with *National Estimator*, a stand-alone *Windows*™ estimating program, plus an interactive multimedia video that shows how to use the disk to compile construction cost estimates.
552 pages, 8½ x 11, $57.75. Revised annually

Code Check Electrical

In this handy flip chart, you'll find the answers to 600 common electrical code questions. You can see at a glance the differences between the 2002 and 2005 *National Electrical Code*. Quickly find answers to your questions on Ohm's law, grounding, ground faults and arcing faults, service equipment, temporary power, service and panel boards, calculations, multiwire circuits, conductor ampacity, conduit, tubing, cables, boxes, appliances, lighting, swimming pools, photovoltaics and generators.
30 pages, 8½ x 11, $17.95

Electrical Blueprint Reading Revised

Shows how to read and interpret electrical drawings, wiring diagrams, and specifications for constructing electrical systems. Shows how a typical lighting and power layout would appear on a plan, and explains what to do to execute the plan. Describes how to use a panelboard or heating schedule, and includes typical electrical specifications.
208 pages, 8½ x 11, $29.50

Commercial Electrical Wiring

Make the transition from residential to commercial electrical work. Here are wiring methods, spec reading tips, load calculations and everything you need for making the transition to commercial work: commercial construction documents, load calculations, electric services, transformers, overcurrent protection, wiring methods, raceways, boxes and fittings, wiring devices, conductors, electric motors, relays and motor controllers, special occupancies, and safety requirements. This book is written to help any electrician break into the lucrative field of commercial electrical work. Updated to the 1999 *NEC*. **320 pages, 8½ x 11, $36.50**

2006 *International Residential Code*

Replacing the CABO One- and Two-Family Dwelling Code, this book has the latest technological advances in building design and construction. Among the changes are provisions for steel framing and energy savings. Also contains mechanical, fuel gas and plumbing provisions that coordinate with the *International Mechanical Code* and the *International Plumbing Code*. **604 pages, 8½ x 11, $81.50**

Contractor's Guide to *QuickBooks Pro* 2008

This user-friendly manual walks you through *QuickBooks Pro*'s detailed setup procedure and explains step-by-step how to create a first-rate accounting system. You'll learn in days, rather than weeks, how to use *QuickBooks Pro* to get your contracting business organized, with simple, fast accounting procedures. On the CD included with the book you'll find a *QuickBooks Pro* file for a construction company (open it, enter your own company's data, and add info on your suppliers and subs). You also get a complete estimating program, including a database, and a job costing program that lets you export your estimates to *QuickBooks Pro*. It even includes many useful construction forms to use in your business.
344 pages, 8½ x 11, $54.75
Other QBP versions available. See order form.

CD Estimator

If your computer has *Windows*™ and a CD-ROM drive, CD Estimator puts at your fingertips over 150,000 construction costs for new construction, remodeling, renovation & insurance repair, home improvement, framing & finish carpentry, electrical, concrete & masonry, painting, earthwork & heavy equipment and plumbing & HVAC. Monthly cost updates are available at no charge on the Internet. You'll also have the *National Estimator* program — a stand-alone estimating program for *Windows*™ that *Remodeling* magazine called a "computer wiz," and *Job Cost Wizard*, a program that lets you export your estimates to *QuickBooks Pro* for actual job costing. A 60-minute interactive video teaches you how to use this CD-ROM to estimate construction costs. And to top it off, to help you create professional-looking estimates, the disk includes over 40 construction estimating and bidding forms in a format that's perfect for nearly any *Windows*™ word processing or spreadsheet program. **CD Estimator is $78.50**

2008 *National Electrical Code*

This new electrical code incorporates sweeping improvements to make the code more functional and user-friendly. Here you'll find the essential foundation for electrical code requirements for the 21st century. With hundreds of significant and widespread changes, this 2008 *NEC* contains all the latest electrical technologies, recently developed techniques, and enhanced safety standards for electrical work. This is the standard all electricians are required to know, even if it hasn't yet been adopted by their local or state jurisdictions. **784 pages, 8½ x 11, $75.00**

2008 Ugly's Electrical Reference

The most popular pocket-sized electrical book in America used by electricians, engineers, designers and maintenance workers. This unique book explains everything from bending conduit to complex electrical formulas. This 2008 edition contains all the electrical material that has made this reference famous, but also reflects 2008 *NEC* changes and new color-coded wiring diagrams. Also includes a Basic Math Review and a General First Aid Section. **162 pages, 5 x 7, $15.95**

Contractor's Guide to the Building Code

Explains in plain, simple English just what the 2006 *International Building Code* and *International Residential Code* require. Building codes are elaborate laws, designed for enforcement; they're not written to be helpful how-to instructions for builders. Here you'll find down-to-earth, easy-to-understand descriptions, helpful illustrations, and code tables that you can use to design and build residential and light commercial buildings that pass inspection the first time. Written by a former building inspector, it tells what works with the inspector to allow cost-saving methods, and warns what common building shortcuts are likely to get cited. Filled with the tables and illustrations from the *IBC* and *IRC* you're most likely to need, fully explained, with examples to guide you. Includes a CD-ROM with the entire book in PDF format, with an easy search feature. **408 pages, 8½ x 11, $66.75**

Construction Forms & Contracts

125 forms you can copy and use — or load into your computer (from the FREE disk enclosed). Then you can customize the forms to fit your company, fill them out, and print. Loads into *Word for Windows*, *Lotus 1-2-3*, *WordPerfect*, *Works*, or *Excel* programs. You'll find forms covering accounting, estimating, fieldwork, contracts, and general office. Each form comes with complete instructions on when to use it and how to fill it out. These forms were designed, tested and used by contractors, and will help keep your business organized, profitable and out of legal, accounting and collection troubles. Includes a CD-ROM for *Windows*™ and *Mac*™. **432 pages, 8½ x 11, $41.75**

Estimating Electrical Construction

Like taking a class in how to estimate materials and labor for residential and commercial electrical construction. Written by an A.S.P.E. National Estimator of the Year, it teaches you how to use labor units, the plan take-off, and the bid summary to make an accurate estimate, how to deal with suppliers, use pricing sheets, and modify labor units. Provides extensive labor unit tables and blank forms for your next electrical job. **272 pages, 8½ x 11, $35.00**

Illustrated Guide to the 1999 *National Electrical Code*

This fully-illustrated guide offers a quick and easy visual reference for installing electrical systems. Whether you're installing a new system or repairing an old one, you'll appreciate the simple explanations written by a code expert, and the detailed, intricately-drawn and labeled diagrams. A real time-saver when it comes to deciphering the 1999 *NEC*.
360 pages, 8½ x 11, $38.75

National Construction Estimator

Current building costs for residential, commercial, and industrial construction. Estimated prices for every common building material. Provides manhours, recommended crew, and gives the labor cost for installation. Includes a CD-ROM with an electronic version of the book with *National Estimator*, a stand-alone *Windows*™ estimating program, plus an interactive multimedia video that shows how to use the disk to compile construction cost estimates.
672 pages, 8½ x 11, $57.50. Revised annually

Form Builder: Contracts

This single CD-ROM has more than 70 of the most frequently-used builder forms and contracts — provides all the contracts you need to smoothly run your business. Customize them to fit your specific needs. Includes change orders, contracts with owners and suppliers, inspection reports, specification forms, and more. **CD-ROM, $54.95**

The Benchmark Report

If you do home inspections you have to file a report. The Benchmark's one-time reports are as comprehensive as you're likely to find. They come as checklists with complete illustrations of the area being inspected, on 4-part NCR paper, in a three-ring binder. Every area is covered: grounds, exterior, foundation, roof, plumbing, heating, cooling, electrical, interior, garage, kitchen, bathroom, well, septic, and more. Each aspect of an inspection is here to save you time and prevent mistakes. **60 pages, 8½ x 11, $21.00**

Fast Tabs for the 2008 *NEC*

Find what you're looking for fast in the *NEC* with these 48 self-adhesive large Fast Tabs. These inexpensive tabs, preprinted with revised article names and numbers and ready to stick onto the pages of your 2008 *National Electrical Code*, will let you quickly and easily get to the section you need. **$9.95**

ElectriCalc Pro Calculator

This unique calculator, based on the 2005 *National Electrical Code* and updateable to future *NEC* codes, solves electrical problems in seconds: Calculates wire sizes, gives you integrated voltage drop solutions, conduit sizing for 12 types of conduit, and finds motor full-load amps per the current *NEC*. Also offers one-button parallel and de-rated wire sizing, computes fuse and breaker sizes, sizes overload protection, calculates service and equipment grounding conductor sizes, finds NEMA starter sizes, works in volts, volt-amps, watts, kVA, kW, PF%, and DC resistance, and even operates as a math calculator. **3½ x 7, $99.95**

Peerless Institute Electrician's Exam to the 2005 *NEC* on CD-ROM

This interactive study reference has been carefully developed to maximize your retention and comprehension of the material you need to study to pass the electrician's exam. It includes thousands of multiple-choice questions and answers on principles of electricity, instrument reading, service entrances, electrical circuits, transformers, electrical material, electrical tools, and electrical code — every subject your electrician's exam is most likely to include. Contains eight different interactive timed exams that are instantly graded so you'll know the areas you need to prepare for in order to pass the actual electrician's exam. **CD-ROM, $195.00**

California Journeyman Electrician's Preparation & Study Guide

This book has just been published to meet the demands of graduating apprentices and journeymen electricians in the State of California who must now meet requirements of the new California Electrical Licensing Law that requires journeymen electricians pass a test. It's designed with sample questions and answers, definitions, illustrations, and study tips to help you pass the exam on the first try. Although written for the California exam, it can be used as a study guide for any state electrician's exam that's based on the *NEC*.
96 pages, 8½ x 11, $19.95

Electrician's Exam Preparation Guide

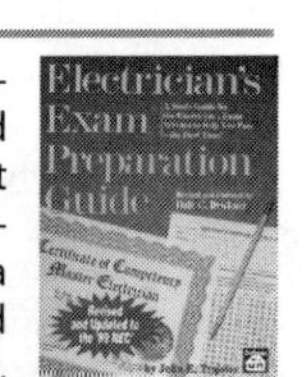

Need help in passing the apprentice, journeyman, or master electrician's exam? This is a book of questions and answers based on actual electrician's exams over the last few years. Almost a thousand multiple-choice questions – exactly the type you'll find on the exam – cover every area of electrical installation: electrical drawings, services and systems, transformers, capacitors, distribution equipment, branch circuits, feeders, calculations, measuring and testing, and more. It gives you the correct answer, an explanation, and where to find it in the latest *NEC*. Also tells how to apply for the test, how best to study, and what to expect on examination day. Includes a FREE CD-ROM with all the questions in the book in interactive test-yourself software that makes studying for the exam almost fun! Updated to the 2008 *NEC*.
352 pages, 8½ x 11, $49.50

Also available: **Electrician's Exam Preparation Guide 2005, $39.50**
Electrician's Exam Preparation Guide 2002, $37.50

2008 Dr. Watts Shirt Pocket Electrical Guide

This heavily constructed spiral-bound reference is filled with essential *NEC* tables, and the most-used basic electrical, voltage and transformer formulas, examples and illustrations that you can use on the job when you need it most. You'll find *NEC* tables that include allowable ampacities, derating, burial depths, grounding, and wire-fill, as well as basic electrical, power, voltage and transformer formulas, examples depicting bending, transformer installations, branch circuits and GFCIs. All this information fits in a shirt-pocket reference that is there, on the job, when you need it most. **46 pages, 3 x 5, $12.95**

Dr. Watts Pocket Datacom Guide

This quick reference guide is designed for those involved in the construction, cabling and testing of datacom systems. Here you'll find helpful tips, symbol explanations, clear illustrations, code clarifications, tables and diagrams, and an extensive glossary, all in color-coded, tabbed format. This compact shirt-pocket guide gives you detailed information on the common do's, and the don'ts to watch out for, blueprint symbols, specifics on building a telecom room, guidelines for distance, conduit fill, raceway cables, bending radiuses, and more. You'll find *National Electrical Code* articles and how they affect datacom projects, including a cable dictionary, cable substitutions, firestopping principles, and information on grounding. Provides standards in procedures for cabling, such as UPT cable categories, modular jack outlet connectors, patch and rollover cables, parallel and serial ports, and color coding instructions. You'll find guidelines and testing specifications, including twisted pair copper media. Provides an amazing amount of information that fits in your pocket for those on-the-job moments when uncertainty strikes. **46 pages, 3 x 5, $12.95**

Learn to Find it Fast in the 2002 *NEC*

Need help in passing the journeyman electrician's exam? The key to earning a good score on an open book test is often predicated on the ability to find the information you need quickly. This unique CD-ROM is designed to help you quickly navigate the 2002 *NEC* by using the Table of Contents rather than becoming mired searching in the Index. We're told that electrical apprentices using this training program scored an average grade higher than 84% on the actual exam. If your state bases its electrician's exam on the 2002 *NEC*, you should have this CD-ROM to help you pass.
CD-ROM $41.95

Peerless Institute Electrician's Exam to the 2005 *NEC*

This huge 712 page reference has to be the most comprehensive guide available for preparing to pass the electrician's exam. It includes a large glossary and thousands of multiple-choice questions and answers on principles of electricity, instrument reading, service entrances, electrical circuits, transformers, electrical material, electrical tools, and electrical code — every subject your electrician's exam is most likely to include. There are even five different sample tests in the back of this book so you'll be fully prepared to pass the actual electrician's exam. **Published by BNI/Peerless. 712 pages, 8½ x 11, $99.95**

Electrical Inspection Notes

In this pocket-sized flip chart, you'll find code compliance information to help you make sure that every part of your electrical work is up to code. Here you'll find checklists, calculations, diagrams, plain-English code explanations, tables and charts, and who is responsible for what task during each step of the project. It lists everything to check for in the design stage, what to check for in interior electrical work, conductors, grounding, wiring methods, conduits, outlets, circuit panels, lighting, testing methods, exterior lighting, electrical service, heating, low voltage and more.
234 Pages, 3 x 6, $24.95

Electrical Field Log Book

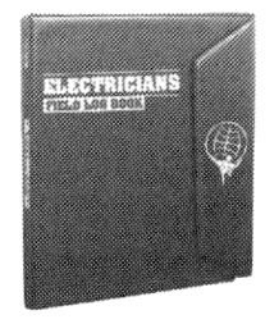

This is a complete package that carries everything you need when making a field call. It includes a pen holder, calculator, contact sheet, essential *NEC* tables, formulas, and more — all in one spiral-bound, Velcro-closing package. Inside, you'll find a daily log that can keep track of your activities, estimating forms and appointment scheduling forms in easy-tabbed format. In addition, the quick-view contact sheet has all your important phone numbers in one portable location. If you're doing electrical work in the field, this organizer can make your job a lot easier. **272 pages, 8½ x 11, $49.95**

2006 International Building Code

Updated means of egress and interior finish requirements, comprehensive roof provisions, seismic engineering provisions, innovative construction technology, revamped structural provisions, reorganized occupancy classifications and the latest industry standards in material design.
674 pages, 8½ x 11, $100.00

Troubleshooting Guide to Residential Construction

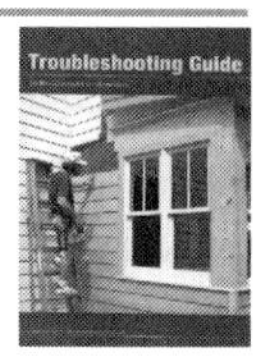

How to solve practically every construction problem – before it happens to you! With this book you'll learn from the mistakes other builders made as they faced 63 typical residential construction problems. Filled with clear photos and drawings that explain how to enhance your reputation as well as your bottom line by avoiding problems that plague most builders. Shows how to avoid, or fix, problems ranging from defective slabs, walls and ceilings, through roofing, plumbing & HVAC, to paint. **304 pages, 8½ x 11, $32.50**

Professional Kitchen Design

Remodeling kitchens requires a "special" touch - one that blends artistic flair with function to create a kitchen with charm and personality as well as one that is easy to work in. Here you'll find how to make the best use of the space available in any kitchen design job, as well as tips and lessons on how to design one-wall, two-wall, L-shaped, U-shaped, peninsula and island kitchens. Also includes what you need to know to run a profitable kitchen design business.
176 pages, 8½ x 11, $24.50

Residential Property Inspection Reports CD-ROM

This CD-ROM contains 50 pages of property inspection forms in both Rich Text and PDF formats. You can easily customize each form with your logo and address, and use them for your home inspections. Use the CD-ROM to write your inspections with your word processor, print them, and save copies for your records. Includes inspection forms for grounds and exterior, foundations, garages and carports, roofs and attics, pools and spas, electrical, plumbing and HVAC, living rooms, family rooms, dens, studies, kitchens, breakfast rooms, dining rooms, hallways, stairways, entries, and laundry rooms. **CD-ROM, $79.95**

Home Inspection Handbook

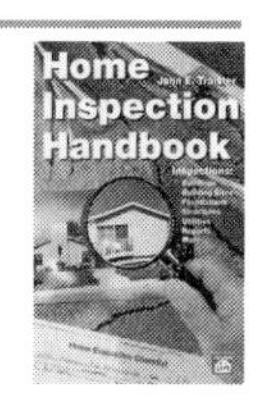

Every area you need to check in a home inspection - especially in older homes. Twenty complete inspection checklists: building site, foundation and basement, structural, bathrooms, chimneys and flues, ceilings, interior & exterior finishes, electrical, plumbing, HVAC, insects, vermin and decay, and more. Also includes information on starting and running your own home inspection business.
324 pages, 5½ x 8½, $24.95

The Complete Book of Home Inspections

This comprehensive manual covers every aspect of home inspection, from the tools required through the inspection of roofs, walls, interior rooms, windows and doors; garages, attics, basements and crawl spaces; paved areas around the structure, landscaping, insect damage and rot; electrical systems, HVAC systems, plumbing systems, and swimming pools. Covers energy considerations and environmental concerns such as radon and mold. Includes hundreds of photos and illustrations to help you understand, check, and identify potential problems.
290 pages, 8 x 10, $19.95

Home Inspection Checklists

Here are 111 illustrated checklists and worksheets to help you conduct home inspections for roofs, roof mounted structures, paved areas around the structure; walls, windows and doors; landscaping, garages, wood-destroying insects, and rot; attics, interior rooms, crawl spaces, basements, electrical systems, heating systems, plumbing systems, air conditioning systems, swimming pools, and even environmental concerns.
138 pages, 8 x 10, $12.95

National Renovation & Insurance Repair Estimator

Current prices in dollars and cents for hard-to-find items needed on most insurance, repair, remodeling, and renovation jobs. All price items include labor, material, and equipment breakouts, plus special charts that tell you exactly how these costs are calculated. Includes a CD-ROM with an electronic version of the book with *National Estimator*, a stand-alone *Windows*™ estimating program, plus an interactive multimedia video that shows how to use the disk to compile construction cost estimates.
576 pages, 8½ x 11, $59.50. Revised annually

Builder's Guide to Accounting Revised

Step-by-step, easy-to-follow guidelines for setting up and maintaining records for your building business. This practical guide to all accounting methods shows how to meet state and federal accounting requirements, explains the new depreciation rules, and describes how the Tax Reform Act can affect the way you keep records. Full of charts, diagrams, simple directions and examples to help you keep track of where your money is going. Recommended reading for many state contractor's exams. Each chapter ends with a set of test questions, and a CD-ROM included FREE has all the questions in interactive self-test software. Use the Study Mode to make studying for the exam much easier, and Exam Mode to practice your skills.
360 pages, 8½ x 11, $35.50

Green from the Ground Up

Green construction is the building trend of the decade – and being demanded by more and more home buyers in search of sustainable, healthy and energy-efficient homes. Here you'll learn how to apply the most forward-thinking and proven methods of green construction to the homes you build or remodel. You'll find details for planning, material selection, energy efficiency and indoor air quality – detailing every step in design and construction, from framing to finishes. This is a must-have reference for contractors who want to remain competitive and offer their customers the latest ideas for energy efficiency. Filled with hundreds of clear full-color photos that illustrate every aspect of building green – including, foundations, framing, roofs and attics, doors and windows, plumbing, HVAC, electrical, insulation, siding and decking, ventilation, solar energy, interior finishes, landscaping, and more. Even includes tips on reducing air leakage and sound transmission.
330 pages, 6 x 8½, $24.95

Home Builders' Jobsite Codes

A spiral-bound, quick reference to the 2006 *International Residential Code* that is filled with easy to read and understand code requirements for every aspect of residential construction. This user-friendly guide through the morass of the code is packed with illustrations, tables, and figures, to illuminate your path to inspection and approval.
281 pages, 5½ x 8½, $26.95

Drafting House Plans

Here you'll find step-by-step instructions for drawing a complete set of home plans for a one-story house, an addition to an existing house, or a remodeling project. This book shows how to visualize spatial relationships, use architectural scales and symbols, sketch preliminary drawings, develop detailed floor plans and exterior elevations, and prepare a final plot plan. It even includes code-approved joist and rafter spans and how to make sure that drawings meet code requirements. **192 pages, 8½ x 11, $34.95**

Finish Carpenter's Manual

Everything you need to know to be a finish carpenter: assessing a job before you begin, and tricks of the trade from a master finish carpenter. Easy-to-follow instructions for installing doors and windows, ceiling treatments (including fancy beams, corbels, cornices and moldings), wall treatments (including wainscoting and sheet paneling), and the finishing touches of chair, picture, and plate rails. Specialized interior work includes cabinetry and built-ins, stair finish work, and closets. Also covers exterior trims and porches. Includes manhour tables for finish work, and hundreds of illustrations and photos.
208 pages, 8½ x 11, $22.50

Builder's Guide to Room Additions

How to tackle problems that are unique to additions, such as requirements for basement conversions, reinforcing ceiling joists for second-story conversions, handling problems in attic conversions, what's required for footings, foundations, and slabs, how to design the best bathroom for the space, and much more. Besides actual construction, you'll even find help in designing, planning, and estimating your room addition jobs.
352 pages, 8½ x 11, $34.95

Steel-Frame House Construction

Framing with steel has obvious advantages over wood, yet building with steel requires new skills that can present challenges to the wood builder. This book explains the secrets of steel framing techniques for building homes, whether pre-engineered or built stick by stick. It shows you the techniques, the tools, the materials, and how you can make it happen. Includes hundreds of photos and illustrations, plus a FREE download with steel framing details and a database of steel materials and manhours, with an estimating program.
320 pages, 8½ x 11, $39.75

National Home Improvement Estimator

Current labor and material prices for home improvement projects. Provides manhours for each job, recommended crew size, and the labor cost for removal and installation work. Material prices are current, with location adjustment factors and free monthly updates on the Web. Gives step-by-step instructions for the work, with helpful diagrams, and home improvement shortcuts and tips from experts. Includes a CD-ROM with an electronic version of the book, and *National Estimator*, a stand-alone *Windows*™ estimating program, plus an interactive multimedia tutorial that shows how to use the disk to compile home improvement cost estimates.
520 pages, 8½ x 11, $58.75. Revised annually

Low Voltage Wiring

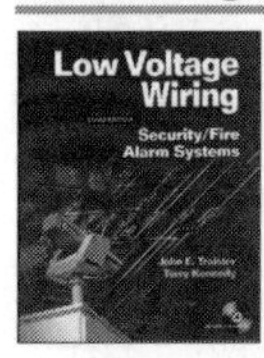

Take advantage of the boom in security and low voltage systems. Here you'll find answers to designing, installing, maintaining, and troubleshooting security and fire alarm systems in residential and commercial buildings. Explains how to understand blueprints for low voltage systems, what the code requirements are, and how to maximize your profit as a low voltage contractor. Includes a CD-ROM with the checklists and forms to help ensure your success as a low voltage contractor. **408 pages, 8 x 10, $39.95**

JLC Field Guide to Residential Construction, Vol. 1

Volume 1: The ultimate visual quick-reference guide for construction professionals. Over 400 precisely-detailed drawings with clear concise notes and explanations that show you everything from estimating and selecting lumber to foundations, roofing, siding and exteriors. Explains code requirements for all U.S. building codes. Spiral bound.
386 pages, 8½ x 11, $69.95

JLC Field Guide to Residential Construction, Vol. 2

Volume 2: Over 300 detailed technical drawings, illustrations and tables giving key dimensions and critical details from accessible kitchens to zoned heating systems. This new version of JLC Field Guide has the field-proven principles, methods and materials on all facets of interior construction and remodeling. Spiral bound. **350 pages, 8½ x 11, $69.95**

Craftsman's Construction Installation Encyclopedia

Step-by-step installation instructions for just about any residential construction, remodeling or repair task, arranged alphabetically, from *Acoustic tile* to *Wood flooring*. Includes hundreds of illustrations that show how to build, install, or remodel each part of the job, as well as manhour tables for each work item so you can estimate and bid with confidence. Also includes a CD-ROM with all the material in the book, handy look-up features, and the ability to capture and print out for your crew the instructions and diagrams for any job. **792 pages, 8½ x 11, $65.00**

Blueprint Reading for the Building Trades

How to read and understand construction documents, blueprints, and schedules. Includes layouts of structural, mechanical, HVAC and electrical drawings. Shows how to interpret sectional views, follow diagrams and schematics, and covers common problems with construction specifications. **192 pages, 5½ x 8½, $16.75**

Builder's Guide to Stucco, Lath & Plaster

Stucco is a durable, low-maintenance exterior surface widely used on homes throughout the world. This practical manual covers every aspect of stucco design, construction and repair. You'll find techniques for attaching stucco to wood frames, steel frames, sheathed material, and masonry. Includes step-by-step instructions on how to install flashing, corner beads, casing beads, control joints, weather-resistant barriers, exterior insulation systems, and one-coat stucco systems. Shows the recommended method for mixing and applying stucco, both manually and by machine, and illustrates techniques for producing different textures and for creating decorative plaster work such as implants and molds. Explains how to estimate labor and material costs for stucco, as well as equipment rental fees. Also includes practical instructions for stucco maintenance and repair. COMPANION CD-ROM – The book includes a CD-ROM with the complete text in Adobe PDF format, for jobsite reference and quick word search. (There is no print feature on the CD-ROM.)
Published by Builder's Book Inc. 284 pages, 8½ x 11, $49.95

Rough Framing Carpentry

If you'd like to make good money working outdoors as a framer, this is the book for you. Here you'll find shortcuts to laying out studs; speed cutting blocks, trimmers and plates by eye; quickly building and blocking rake walls; installing ceiling backing, ceiling joists, and truss joists; cutting and assembling hip trusses and California fills; arches and drop ceilings — all with production line procedures that save you time and help you make more money. Over 100 on-the-job photos of how to do it right and what can go wrong. **304 pages, 8½ x 11, $26.50**

National Concrete & Masonry Estimator

Since you don't get every concrete or masonry job you bid, why generate a detailed list of materials for each one? The data in this book will allow you to get a quick and accurate bid, and allow you to do a detailed material takeoff, only for the jobs on which you're the successful bidder. Includes assembly prices for bricks, and labor and material prices for brick bonds, brick specialties, concrete blocks, CMU, concrete footings and foundations, concrete on grade, concrete specialties, concrete beams and columns, beams for elevated slabs, elevated slab costs, and more. Includes a CD-ROM with an electronic version of the book with *National Estimator*, a stand-alone *Windows*™ estimating program, plus an interactive multimedia video that shows how to use the disk to compile construction cost estimates.
672 pages, 8½ x 11, $59.00. Revised annually

Contractor's Guide to Change Orders

This book gives you the ammunition you need to keep contract disputes from robbing you of your profit. You'll learn how to identify trouble spots in your contract, plans, specifications and site; negotiate and resolve change order disputes, and collect facts for evidence to support your claims. You'll also find detailed checklists to organize your procedures, field-tested sample forms and worksheets ready for duplication, and various professional letters for almost any situation.
382 pages, 8½ x 11, $79.00

Finish Carpentry: Efficient Techniques for Custom Interiors

Professional finish carpentry demands expert skills, precise tools, and a solid understanding of how to do the work. This book explains how to install moldings, paneled walls and ceilings, and just about every aspect of interior trim – including doors and windows. Covers built-in bookshelves, coffered ceilings, and skylight wells and soffits, including paneled ceilings with decorative beams. **276 pages, 8½ x 11, $34.95**

Daily Field Guide

Organization is the key to success. With this handy manual you can manage the details to keep each job running smoothly and on schedule. Use it to keep accurate records of jobsite activities, document events during the course of construction, and to schedule trade contractors and suppliers. Includes a computer disk to help you modify and personalize each form and checklist. **128 pages, 8½ x 11, $20.00**

Business Letters for the Construction Industry

Here you find over 100 professionally-written model letters for virtually every situation. Put your business in the best possible light with a well-written letter — especially when you can use that letter to improve a bad situation. Use these letters over and over again to resolve disputes, win new clients, clarify proposals, coordinate with architects, subcontractors, owners, and insurers, schedule meetings and inspections, and to respond to complaints or difficult situations. Here you find letters responding to threats of legal action, of commendation to workers, of job performance, apology for defective or delayed work; letters for justification of change orders and price increases; letters explaining your insurance liability, drug testing, injury at work, overtime, equipment use, and more. Practically every letter you'll have to write is in this book, already written, and available on *MS-Word* on the CD-ROM enclosed. Just load the letter you need, change a few phrases, print out and send, or e-mail, and you're free to spend your time on more productive endeavors.
376 pages, 8½ x 11, $59.95

Residential Construction Performance Guidelines

Created and reviewed by more than 300 builders and remodelers, this guide gives cut-and-dried construction standards that should apply to new construction and remodeling. It defines corrective action necessary to bring all construction up to standards. Standards are listed for sitework, foundations, interior concrete slabs, basement and crawl spaces for block walls and poured walls, wood-floor framing, beams, columns and posts, plywood and joists, walls, wall insulation, windows, doors, exterior finishes and trim, roofs, roof sheathing, roof installation and leaks, plumbing, sanitary and sewer systems, electrical, interior climate control, HVAC systems, cabinets and countertops, floor finishes and more.
Published by NAHB Remodelers Council. 120 pages, 6½ x 8½, $39.95

Contractor's Plain-English Legal Guide

For today's contractors, legal problems are like snakes in the swamp – you might not see them, but you know they're there. This book tells you where the snakes are hiding and directs you to the safe path. With the directions in this easy-to-read handbook you're less likely to need a $200-an-hour lawyer. Includes simple directions for starting your business, writing contracts that cover just about any eventuality, collecting what's owed you, filing liens, protecting yourself from unethical subcontractors, and more. For about the price of 15 minutes in a lawyer's office, you'll have a guide that will make many of those visits unnecessary. Includes a CD-ROM with blank copies of all the forms and contracts in the book. **272 pages, 8½ x 11, $49.50**

Smart Business for Contractors

In this book, a construction attorney explains how you should charge for your work, how to figure your overhead expenses, and how to calculate a realistic hourly rate to apply on each estimate. Includes how to bill and collect on your invoices, what you should always include in your contracts, and creative new ways of dealing with contract disputes. Shows how to keep customers happy so they'll hand you referrals, how best to handle subcontractors, and how to find a good accountant. You'll learn the pros and cons of incorporating, how to handle tax issues such as what you can and can't deduct, and what you're allowed to depreciate, and how to plan the future of your company. Reading this book is like getting good advice from a construction lawyer – at a fraction of the cost.
Published by: Taunton Press. 204 pages, 8½ x 11, $19.95

Working Alone

This unique book shows you how to become a dynamic one-man team as you handle nearly every aspect of house construction, including foundation layout, setting up scaffolding, framing floors, building and erecting walls, squaring up walls, installing sheathing, laying out rafters, raising the ridge, getting the roof square, installing rafters, subfascia, sheathing, finishing eaves, installing windows, hanging drywall, measuring trim, installing cabinets, and building decks.
152 pages, 5½ x 8½, $17.95

Craftsman Book Company
6058 Corte del Cedro
P.O. Box 6500
Carlsbad, CA 92018

☎ 24 hour order line
1-800-829-8123
Fax (760) 438-0398

In A Hurry?

We accept phone orders charged to your
❍ Visa, ❍ MasterCard, ❍ Discover or ❍ American Express

Card# ____________________

Exp. date__________Initials __________

Tax Deductible: Treasury regulations make these references tax deductible when used in your work. Save the canceled check or charge card statement as your receipt.

____
Name

____
e-mail address (for order tracking and special offers)

____
Company

____
Address

____
City/State/Zip ❍ This is a residence

Total enclosed__________(In California add 7.25% tax)
We pay shipping when your check covers your order in full.

Order online http://www.craftsman-book.com
Free on the Internet! Download any of Craftsman's estimating database for a 30-day free trial! www.craftsman-book.com/downloads

10-Day Money Back Guarantee

- ❍ 21.00 The Benchmark Report
- ❍ 16.75 Blueprint Reading for the Building Trades
- ❍ 35.50 Builder's Guide to Accounting Revised
- ❍ 34.95 Builder's Guide to Room Additions
- ❍ 49.95 Builder's Guide to Stucco, Lath & Plaster
- ❍ 59.95 Business Letters for the Construction Industry
- ❍ 19.95 California Journeyman Electrician's Preparation & Study Guide
- ❍ 78.50 CD Estimator
- ❍ 17.95 Code Check Electrical
- ❍ 36.50 Commercial Electrical Wiring
- ❍ 19.95 The Complete Book of Home Inspections
- ❍ 41.75 Construction Forms & Contracts
- ❍ 79.00 Contractor's Guide to Change Orders
- ❍ 54.75 Contractor's Guide to QuickBooks Pro 2008
- ❍ 53.00 Contractor's Guide to QuickBooks Pro 2007
- ❍ 49.75 Contractor's Guide to QuickBooks Pro 2005
- ❍ 48.50 Contractor's Guide to QuickBooks Pro 2004
- ❍ 47.75 Contractor's Guide to QuickBooks Pro 2003
- ❍ 45.25 Contractor's Guide to QuickBooks Pro 2001
- ❍ 66.75 Contractor's Guide to the Building Code
- ❍ 49.50 Contractor's Plain-English Legal Guide
- ❍ 38.00 Contractor's Survival Manual Revised
- ❍ 65.00 Craftsman's Construction Installation Encyclopedia
- ❍ 20.00 Daily Field Guide
- ❍ 12.95 Dr. Watts Pocket Datacom Guide
- ❍ 12.95 2008 Dr. Watts Shirt Pocket Electrical Guide
- ❍ 34.95 Drafting House Plans
- ❍ 29.50 Electrical Blueprint Reading Revised
- ❍ 49.95 Electrical Field Log Book
- ❍ 24.95 Electrical Inspection Notes
- ❍ 99.95 ElectriCalc Pro Calculator
- ❍ 49.50 Electrician's Exam Preparation Guide 2008
- ❍ 39.50 Electrician's Exam Preparation Guide 2005
- ❍ 37.50 Electrician's Exam Preparation Guide 2002
- ❍ 39.95 Electrician's Exam Study Guide
- ❍ 35.00 Estimating Electrical Construction
- ❍ 9.95 Fast Tabs for the 2008 *NEC*
- ❍ 22.50 Finish Carpenter's Manual
- ❍ 34.95 Finish Carpentry: Efficient Techniques for Custom Interiors
- ❍ 54.95 Form Builder: Contracts
- ❍ 24.95 Green from the Ground Up
- ❍ 26.95 Home Builders' Jobsite Codes
- ❍ 12.95 Home Inspection Checklists
- ❍ 24.95 Home Inspection Handbook
- ❍ 38.75 Illustrated Guide to the 1999 *National Electrical Code*
- ❍ 100.00 2006 *International Building Code*
- ❍ 81.50 2006 *International Residential Code*
- ❍ 69.95 JLC Field Guide to Residential Construction, Vol. 1
- ❍ 69.95 JLC Field Guide to Residential Construction, Vol. 2
- ❍ 41.95 Learn to Find it Fast in the 2002 *NEC*
- ❍ 39.95 Low Voltage Wiring
- ❍ 59.00 2008 National Concrete & Masonry Estimator w/FREE CD
- ❍ 57.50 2008 National Construction Estimator w/FREE CD
- ❍ 75.00 2008 *National Electrical Code*
- ❍ 57.75 2008 National Electrical Estimator w/FREE CD
- ❍ 58.75 2008 National Home Improvement Estimator w/FREE CD
- ❍ 59.50 2008 National Renovation & Insurance Repair Estimator
- ❍ 58.50 2008 National Repair & Remodeling Estimator w/FREE CD
- ❍ 99.95 Peerless Institute Electrician's Exam to the 2005 *NEC*
- ❍ 195.00 Peerless Institute Electrician's Exam to the 2005 *NEC* on CD
- ❍ 24.50 Professional Kitchen Design
- ❍ 39.95 Residential Construction Performance Guidelines
- ❍ 79.95 Residential Property Inspection Reports CD-ROM
- ❍ 26.50 Rough Framing Carpentry
- ❍ 19.95 Smart Business for Contractors
- ❍ 39.75 Steel-Frame House Construction
- ❍ 32.50 Troubleshooting Guide to Residential Construction
- ❍ 15.95 2008 Ugly's Electrical Reference
- ❍ 17.95 Working Alone
- ❍ 42.00 Residential Wiring to the 2008 *NEC*
- ❍ FREE Full Color Catalog

Prices subject to change without notice

Craftsman Book Company
6058 Corte del Cedro
P.O. Box 6500
Carlsbad, CA 92018

☎ 24 hour order line
1-800-829-8123
Fax (760) 438-0398

Name ______________________

Company ______________________

Address ______________________

City/State/Zip ______________________ ❍ This is a residence

In A Hurry?

We accept phone orders charged to your
❍ Visa, ❍ MasterCard, ❍ Discover or ❍ American Express

Card# ______________________

Exp. date ____________ Initials ____________

Total enclosed ______________________ (In California add 7.25% tax)

We pay shipping when your check covers your order in full.

Tax Deductible: Treasury regulations make these references tax deductible when used in your work. Save the canceled check or charge card statement as your receipt.

Order online http://www.craftsman-book.com **10-Day Money Back Guarantee** **Prices subject to change without notice**

❍ 21.00 The Benchmark Report
❍ 16.75 Blueprint Reading for the Building Trades
❍ 35.50 Builder's Guide to Accounting Revised
❍ 34.95 Builder's Guide to Room Additions
❍ 49.95 Builder's Guide to Stucco, Lath & Plaster
❍ 59.95 Business Letters for the Constr Industry
❍ 19.95 Calif Journeyman Elec Prep & Study Guide
❍ 78.50 CD Estimator
❍ 17.95 Code Check Electrical
❍ 36.50 Commercial Electrical Wiring
❍ 19.95 The Complete Book of Home Inspections
❍ 41.75 Construction Forms & Contracts
❍ 79.00 Contractor's Guide to Change Orders
❍ 54.75 Contr Guide to QuickBooks Pro 2008
❍ 53.00 Contr Guide to QuickBooks Pro 2007
❍ 49.75 Contr Guide to QuickBooks Pro 2005
❍ 48.50 Contr Guide to QuickBooks Pro 2004
❍ 47.75 Contr Guide to QuickBooks Pro 2003
❍ 45.25 Contr Guide to QuickBooks Pro 2001
❍ 66.75 Contractor's Guide to the Building Code
❍ 49.50 Contractor's Plain-English Legal Guide
❍ 38.00 Contractor's Survival Manual Revised
❍ 65.00 Craftsman's Constr Inst Encyclopedia
❍ 20.00 Daily Field Guide
❍ 12.95 Dr. Watts Pocket Datacom Guide
❍ 12.95 2008 Dr. Watts Shirt Pocket Elec Guide
❍ 34.95 Drafting House Plans
❍ 29.50 Electrical Blueprint Reading Revised
❍ 49.95 Electrical Field Log Book
❍ 24.95 Electrical Inspection Notes
❍ 99.95 ElectriCalc Pro Calculator
❍ 49.50 Electrician's Exam Prep Guide 2008
❍ 39.50 Electrician's Exam Prep Guide 2005
❍ 37.50 Electrician's Exam Prep Guide 2002
❍ 39.95 Electrician's Exam Study Guide
❍ 35.00 Estimating Electrical Construction
❍ 9.95 Fast Tabs for the 2008 *NEC*
❍ 22.50 Finish Carpenter's Manual
❍ 34.95 Finish Carpentry: Efficient Techniques for Custom Interiors
❍ 54.95 Form Builder: Contracts
❍ 24.95 Green from the Ground Up
❍ 26.95 Home Builders' Jobsite Codes
❍ 12.95 Home Inspection Checklists
❍ 24.95 Home Inspection Handbook
❍ 38.75 Illust Guide to the 1999 *Natl Elec Code*
❍ 100.00 2006 *International Building Code*
❍ 81.50 2006 *International Residential Code*
❍ 69.95 JLC Field Guide to Res Construction, Vol. 1
❍ 69.95 JLC Field Guide to Res Construction, Vol. 2
❍ 41.95 Learn to Find it Fast in the 2002 *NEC*
❍ 39.95 Low Voltage Wiring
❍ 59.00 2008 Natl Conc & Masonry Est w/FREE CD
❍ 57.50 2008 Natl Construction Est w/FREE CD
❍ 75.00 2008 *National Electrical Code*
❍ 57.75 2008 Natl Electrical Est w/FREE CD
❍ 58.75 2008 Natl Home Impr Est w/FREE CD
❍ 59.50 2008 Natl Renovation & Insurance Repair Est w/FREE CD
❍ 58.50 2008 Natl Repair & Rem Est w/FREE CD
❍ 99.95 Peerless Inst Elect Exam to the 2005 *NEC*
❍ 195.00 Peerless Inst Elect Exam to the 2005 *NEC* on CD
❍ 24.50 Professional Kitchen Design
❍ 39.95 Residential Const Perf Guidelines
❍ 79.95 Res Property Inspection Reports CD-ROM
❍ 26.50 Rough Framing Carpentry
❍ 19.95 Smart Business for Contractors
❍ 39.75 Steel-Frame House Construction
❍ 32.50 Troubleshooting Guide to Res Constr
❍ 15.95 2008 Ugly's Electrical Reference
❍ 17.95 Working Alone
❍ 42.00 Residential Wiring to the 2008 *NEC*
❍ FREE Full Color Catalog

Craftsman Book Company
6058 Corte del Cedro
P.O. Box 6500
Carlsbad, CA 92018

☎ 24 hour order line
1-800-829-8123
Fax (760) 438-0398

Name ______________________

Company ______________________

Address ______________________

City/State/Zip ______________________ ❍ This is a residence

In A Hurry?

We accept phone orders charged to your
❍ Visa, ❍ MasterCard, ❍ Discover or ❍ American Express

Card# ______________________

Exp. date ____________ Initials ____________

Total enclosed ______________________ (In California add 7.25% tax)

We pay shipping when your check covers your order in full.

Tax Deductible: Treasury regulations make these references tax deductible when used in your work. Save the canceled check or charge card statement as your receipt.

Order online http://www.craftsman-book.com **10-Day Money Back Guarantee** **Prices subject to change without notice**

❍ 21.00 The Benchmark Report
❍ 16.75 Blueprint Reading for the Building Trades
❍ 35.50 Builder's Guide to Accounting Revised
❍ 34.95 Builder's Guide to Room Additions
❍ 49.95 Builder's Guide to Stucco, Lath & Plaster
❍ 59.95 Business Letters for the Constr Industry
❍ 19.95 Calif Journeyman Elec Prep & Study Guide
❍ 78.50 CD Estimator
❍ 17.95 Code Check Electrical
❍ 36.50 Commercial Electrical Wiring
❍ 19.95 The Complete Book of Home Inspections
❍ 41.75 Construction Forms & Contracts
❍ 79.00 Contractor's Guide to Change Orders
❍ 54.75 Contr Guide to QuickBooks Pro 2008
❍ 53.00 Contr Guide to QuickBooks Pro 2007
❍ 49.75 Contr Guide to QuickBooks Pro 2005
❍ 48.50 Contr Guide to QuickBooks Pro 2004
❍ 47.75 Contr Guide to QuickBooks Pro 2003
❍ 45.25 Contr Guide to QuickBooks Pro 2001
❍ 66.75 Contractor's Guide to the Building Code
❍ 49.50 Contractor's Plain-English Legal Guide
❍ 38.00 Contractor's Survival Manual Revised
❍ 65.00 Craftsman's Constr Inst Encyclopedia
❍ 20.00 Daily Field Guide
❍ 12.95 Dr. Watts Pocket Datacom Guide
❍ 12.95 2008 Dr. Watts Shirt Pocket Elec Guide
❍ 34.95 Drafting House Plans
❍ 29.50 Electrical Blueprint Reading Revised
❍ 49.95 Electrical Field Log Book
❍ 24.95 Electrical Inspection Notes
❍ 99.95 ElectriCalc Pro Calculator
❍ 49.50 Electrician's Exam Prep Guide 2008
❍ 39.50 Electrician's Exam Prep Guide 2005
❍ 37.50 Electrician's Exam Prep Guide 2002
❍ 39.95 Electrician's Exam Study Guide
❍ 35.00 Estimating Electrical Construction
❍ 9.95 Fast Tabs for the 2008 *NEC*
❍ 22.50 Finish Carpenter's Manual
❍ 34.95 Finish Carpentry: Efficient Techniques for Custom Interiors
❍ 54.95 Form Builder: Contracts
❍ 24.95 Green from the Ground Up
❍ 26.95 Home Builders' Jobsite Codes
❍ 12.95 Home Inspection Checklists
❍ 24.95 Home Inspection Handbook
❍ 38.75 Illust Guide to the 1999 *Natl Elec Code*
❍ 100.00 2006 *International Building Code*
❍ 81.50 2006 *International Residential Code*
❍ 69.95 JLC Field Guide to Res Construction, Vol. 1
❍ 69.95 JLC Field Guide to Res Construction, Vol. 2
❍ 41.95 Learn to Find it Fast in the 2002 *NEC*
❍ 39.95 Low Voltage Wiring
❍ 59.00 2008 Natl Conc & Masonry Est w/FREE CD
❍ 57.50 2008 Natl Construction Est w/FREE CD
❍ 75.00 2008 *National Electrical Code*
❍ 57.75 2008 Natl Electrical Est w/FREE CD
❍ 58.75 2008 Natl Home Impr Est w/FREE CD
❍ 59.50 2008 Natl Renovation & Insurance Repair Est w/FREE CD
❍ 58.50 2008 Natl Repair & Rem Est w/FREE CD
❍ 99.95 Peerless Inst Elect Exam to the 2005 *NEC*
❍ 195.00 Peerless Inst Elect Exam to the 2005 *NEC* on CD
❍ 24.50 Professional Kitchen Design
❍ 39.95 Residential Const Perf Guidelines
❍ 79.95 Res Property Inspection Reports CD-ROM
❍ 26.50 Rough Framing Carpentry
❍ 19.95 Smart Business for Contractors
❍ 39.75 Steel-Frame House Construction
❍ 32.50 Troubleshooting Guide to Res Constr
❍ 15.95 2008 Ugly's Electrical Reference
❍ 17.95 Working Alone
❍ 42.00 Residential Wiring to the 2008 *NEC*
❍ FREE Full Color Catalog

Mail This Card Today For a Free Full Color Catalog

Over 100 books, annual cost guides and estimating software packages at your fingertips with information that can save you time and money. Here you'll find information on carpentry, contracting, estimating, remodeling electrical work, and plumbing.

All items come with an unconditional 10-day money-back guarantee. If they don't save you money, mail them back for a full refund.

Name ______________________

Company ______________________

Address ______________________

City/State/Zip ______________________

Craftsman Book Company / 6058 Corte del Cedro / P.O. Box 6500 / Carlsbad, CA 92018

NO POSTAGE NECESSARY IF MAILED IN THE UNITED STATES

BUSINESS REPLY MAIL

FIRST CLASS MAIL PERMIT NO. 271 CARLSBAD, CA

POSTAGE WILL BE PAID BY ADDRESSEE

Craftsman Book Company
6058 Corte del Cedro
P.O. Box 6500
Carlsbad, CA 92018-9974

Download all of Craftsman's most popular costbooks
for one low price with the Craftsman Site License.
http://www.craftsmansitelicense.com

NO POSTAGE NECESSARY IF MAILED IN THE UNITED STATES

BUSINESS REPLY MAIL

FIRST CLASS MAIL PERMIT NO. 271 CARLSBAD, CA

POSTAGE WILL BE PAID BY ADDRESSEE

Craftsman Book Company
6058 Corte del Cedro
P.O. Box 6500
Carlsbad, CA 92018-9974

Download all of Craftsman's most popular costbooks
for one low price with the Craftsman Site License.
http://www.craftsmansitelicense.com

NO POSTAGE NECESSARY IF MAILED IN THE UNITED STATES

BUSINESS REPLY MAIL

FIRST CLASS MAIL PERMIT NO. 271 CARLSBAD, CA

POSTAGE WILL BE PAID BY ADDRESSEE

Craftsman Book Company
6058 Corte del Cedro
P.O. Box 6500
Carlsbad, CA 92018-9974